COMPUTATIONAL INTELLIGENCE IN HEALTHCARE APPLICATIONS

COMPUTATIONAL INTELLIGENCE IN HEALTHCARE APPLICATIONS

Edited by

RAJEEV AGRAWAL

Llyod Institute of Engineering and Technology, Greater Noida, Uttar Pradesh, India

M. A. ANSARI

School of Engineering, Gautam Buddha University, Greater Noida, Uttar Pradesh, India

R. S. ANAND

Department of Electrical Engineering, Indian Institute of Technology–Roorkee, Roorkee, Uttarakhand, India

SWETA SNEHA

Department of Information Systems and Security, Kennesaw State University, Kennesaw, GA, United States

RAJAT MEHROTRA

Department of Electrical & Electronics Engineering, GL Bajaj Institute of Technology and Management, Greater Noida, Uttar Pradesh, India

Academic Press is an imprint of Elsevier
125 London Wall, London EC2Y 5AS, United Kingdom
525 B Street, Suite 1650, San Diego, CA 92101, United States
50 Hampshire Street, 5th Floor, Cambridge, MA 02139, United States
The Boulevard, Langford Lane, Kidlington, Oxford OX5 1GB, United Kingdom

ISBN 978-0-323-99031-8

For information on all Academic Press publications
visit our website at https://www.elsevier.com/books-and-journals

Publisher: Stacy Masucci
Acquisitions Editor: Rafael E Teixeira
Editorial Project Manager: Samantha Allard
Production Project Manager: Sreejith Viswanathan
Cover Designer: Greg Harris

Typeset by STRAIVE, India

Contents

6. A review of diabetes management tools and applications

Hossain Shahriar, Sweta Sneha, Yesake Abaye, Talha Hashmi, Shakaria Wilson, and Usen Usen

7. Recent advancements of pelvic inflammatory disease: A review on evidence-based medicine

Arshiya Sultana, Sumbul Mehdi, Khaleequr Rahman, M.J.A. Fazmiya, Md Belal Bin Heyat, Faijan Akhtar, and Atif Amin Baig

8. A review of amenorrhea toward **Unani** to modern system with emerging technology: Current advancements, research gap, and future direction

Sumbul Mehdi, Arshiya Sultana, Md Belal Bin Heyat, Channabasava Chola, Faijan Akhtar, Hirpesa Kebede Gutema, Dawood M.R. Al-qadasi, and Atif Amin Baig

9. Wearable EEG technology for the brain-computer interface

Meenakshi Bisla and R.S. Anand

10. Automatic epileptic seizure detection based on the discrete wavelet transform approach using an artificial neural network classifier on the scalp electroencephalogram signal

Pragati Tripathi, M.A. Ansari, Faijan Akhtar, Md Belal Bin Heyat, Rajat Mehrotra, Akhter Hussain Yatoo, Bibi Nushrina Teelhawod, Ashamo Betelihem Asfaw, and Atif Amin Baig

11. Event identification by fusing EEG and EMG signals

Kashif Sherwani and Munna Khan

12. Hand gesture recognition for the prediction of Alzheimer's disease

R. Sivakani and Gufran Ansari

13. A frequency analysis-based apnea detection algorithm using photoplethysmography

G. Gaurav and R.S. Anand

14. Noninvasive health monitoring using bioelectrical impedance analysis

Mahmood Aldobali, Kirti Pal, and Harvinder Chhabra

15. Detection of cancer from histopathology medical image data using ML with CNN ResNet-50 architecture

Shadan Alam Shadab, M.A. Ansari, Nidhi Singh, Aditi Verma, Pragati Tripathi, and Rajat Mehrotra

16. Performance analysis of augmented data for enhanced brain tumor image classification using transfer learning

Preet Sanghavi, Shrey Dedhia, Siddharth Salvi, Pankaj Sonawane, and Sonali Jadhav

17. Brain tumor detection through MRI using image thresholding, k-means, and watershed segmentation

Aditi Verma, M.A. Ansari, Pragati Tripathi, Rajat Mehrotra, and Shadan Alam Shadab

18. An intelligent diagnostic technique using deep convolutional neural network

Shrabana Saha, Rajarshi Bhadra, and Subhajit Kar

19. Design of a biosensor for the detection of glucose concentration in urine using 2D photonic crystals

Sanjeev Sharma, Arvind Kumar, and Rajeev Agrawal

20. Classification of pneumonic infections through chest radiography using textural features analysis and the pattern recognition system

Rajat Mehrotra, M.A. Ansari, and Rajeev Agrawal

21. Convolutional bi-directional long-short-term-memory based model to forecast COVID-19 in Algeria

Sourabh Shastri, Kuljeet Singh, Astha Sharma, Mohamed Lounis, Sachin Kumar, and Vibhakar Mansotra

Contributors

Yesake Abaye Department of Information Systems and Security, Kennesaw State University, Kennesaw, GA, United States

Rajeev Agrawal Department of Electronics and Communications, Llyod Institute of Engineering and Technology, Greater Noida, UP, India

Faijan Akhtar School of Computer Science and Engineering, University of Electronic Science and Technology of China, Chengdu, Sichuan, China

Mahmood Aldobali Department of Electrical Engineering, Gautam Buddha University, Greater Noida, Uttar Pradesh, India

Dawood M.R. Al-qadasi School of Material Science, Tongji University, Shanghai, China

R.S. Anand Department of Electrical Engineering, Indian Institute of Technology–Roorkee, Roorkee, Uttarakhand, India

Gufran Ansari Department of Computer Applications, B.S. Abdur Rahman Crescent Institute of Science & Technology, Chennai, India

M.A. Ansari Department of Electrical Engineering, School of Engineering, Gautam Buddha University, Greater Noida, UP, India

Ashamo Betelihem Asfaw School of Information and Software Engineering, University of Electronic Science and Technology of China, Chengdu, Sichuan, China

Atif Amin Baig Faculty of Medicine, Universiti Sultan Zainal Abidin, Kuala, Terengganu, Malaysia

Kahkashan Baig Department of Ilmul Qabalat wa Amraze Niswan (Gynecology and Obstetrics), National Institute of Unani Medicine, Ministry of AYUSH & Rajiv Gandhi University of Health Sciences, Bengaluru, Karnataka, India

Rajarshi Bhadra Electrical Engineering, Future Institute of Engineering and Management, Kolkata, West Bengal, India

Meenakshi Bisla Department of Electrical Engineering, Indian Institute of Technology–Roorkee, Roorkee, Uttarakhand, India

Naveen Chauhan KIET Group of Institutions, Delhi-NCR, Ghaziabad, India

Harvinder Chhabra Indian Spinal Injuries Centre, New Delhi, India

Channabasava Chola Department of Computer Science and Engineering, Indian Institute of Information Technology, Kottayam, Kerala, India

Shrey Dedhia Dwarkadas J. Sanghvi College of Engineering, Computer Engineering, Mumbai, India

M.J.A. Fazmiya Department of Ilmul Qabalat wa Amraze Niswan (Gynecology and Obstetrics), National Institute of Unani Medicine, Ministry of AYUSH & Rajiv Gandhi University of Health Sciences, Bengaluru, Karnataka, India

Arindam Ganguly Department of Microbiology, Bankura Sammilani College, Bankura, West Bengal, India

Kanika Garg SRM Institute of Science and Technology, Delhi NCR Campus, Ghaziabad, India

G. Gaurav Faculty of Engineering and Technology, Datta Meghe Institute of Medical Sciences University, Wardha, Maharashtra, India

Hirpesa Kebede Gutema School of Information and Software Engineering, University of Electronic Science and Technology of China, Chengdu, China

Talha Hashmi Department of Information Systems and Security, Kennesaw State University, Kennesaw, GA, United States

Md Belal Bin Heyat IoT Research Center, College of Computer Science and Software Engineering, Shenzhen University, Shenzhen, Guangdong, China; Department of Science and Engineering, Novel Global Community Education Foundation, Hebersham, NSW, Australia; International Institute of Information Technology, Hyderabad, Telangana, India

Sonali Jadhav Computer Department, Thadomal Shahani Engineering College, Mumbai, India

K.B. Jayanthi School of Electrical Sciences, K.S. Rangasamy College of Technology, Tiruchengode, Tamil Nadu, India

Subhajit Kar Electrical Engineering, Future Institute of Engineering and Management, Kolkata, West Bengal, India

Munna Khan Department of Electrical Engineering, Jamia Millia Islamia, New Delhi, India

Arvind Kumar Department of Physics, GL Bajaj Institute of Technology and Management, Greater Noida, UP, India

Sachin Kumar Department of Computer Science & IT, University of Jammu, Jammu, Jammu & Kashmir, India

Mohamed Lounis Department of Agro-veterinary Science, Faculty of Natural and Life Sciences, University of Ziane Achour, Djelfa, Algeria

Vibhakar Mansotra Department of Computer Science & IT, University of Jammu, Jammu, Jammu & Kashmir, India

Sumbul Mehdi Department of Ilmul Qabalat wa Amraze Niswan (Gynecology and Obstetrics), National Institute of Unani Medicine, Ministry of AYUSH & Rajiv Gandhi University of Health Sciences, Bengaluru, Karnataka, India

Rajat Mehrotra Department of Electrical & Electronics Engineering, GL Bajaj Institute of Technology & Management; Department of Electrical Engineering, School of Engineering, Gautam Buddha University, Greater Noida, UP, India

Pradeep Mohapatra Department of Microbiology, Raiganj University, Raiganj, Uttar Dinajpur, West Bengal, India

Debanjan Mitra Department of Microbiology, Raiganj University, Raiganj, Uttar Dinajpur, West Bengal, India

Benjir Nachhmin Department of Microbiology, Raiganj University, Raiganj, Uttar Dinajpur, West Bengal, India

Kirti Pal Department of Electrical Engineering, School of Engineering, Gautam Buddha University, Greater Noida, UP, India

Khaleequr Rahman Department of Ilmul Saidla, National Institute of Unani Medicine, Ministry of AYUSH & Rajiv Gandhi University of Health Sciences, Bengaluru, Karnataka, India

Shrabana Saha Electrical Engineering, Future Institute of Engineering and Management, Kolkata, West Bengal, India

Siddharth Salvi Dwarkadas J. Sanghvi College of Engineering, Computer Engineering, Mumbai, India

Preet Sanghavi Dwarkadas J. Sanghvi College of Engineering, Computer Engineering, Mumbai, India

Shadan Alam Shadab Department of Electrical Engineering, Gautam Buddha University, Greater Noida, UP, India

Hossain Shahriar Department of Information Technology, Kennesaw State University, Kennesaw, GA, United States

Astha Sharma Department of Electronics and Communication Engineering, G.L. Bajaj Institute of Technology and Management, Greater Noida, Uttar Pradesh, India

Sanjeev Sharma Department of Physics, GL Bajaj Institute of Technology and Management, Greater Noida, UP, India

Sourabh Shastri Department of Computer Science & IT, University of Jammu, Jammu, Jammu & Kashmir, India

Kashif Sherwani Department of Electrical Engineering, Jamia Millia Islamia, New Delhi, India

Kuljeet Singh Department of Computer Science & IT, University of Jammu, Jammu, Jammu & Kashmir, India

Nidhi Singh Department of Electrical Engineering, Gautam Buddha University, Greater Noida, UP, India

Punit Kumar Singh Department of BioEngineering, Integral University Lucknow, Lucknow, Uttar Pradesh, India

Sudhakar Singh Department of Biomedical Engineering, Lovely Professional University, Phagwara, Punjab, India

R. Sivakani Computer Science and Engineering, B.S. Abdur Rahman Crescent Institute of Science and Technology, Chennai, India

Sweta Sneha Department of Information Systems and Security, Kennesaw State University, Kennesaw, GA, United States

Pankaj Sonawane Dwarkadas J. Sanghvi College of Engineering, Computer Engineering, Mumbai, India

Sudha Subramaniam Department of Biomedical Engineering, Velalar College of Engineering and Technology, Erode, Tamil Nadu, India

Arshiya Sultana Department of Ilmul Qabalat wa Amraze Niswan (Gynecology and Obstetrics), National Institute of Unani Medicine, Ministry of AYUSH & Rajiv Gandhi University of Health Sciences, Bengaluru, Karnataka, India

Bibi Nushrina Teelhawod School of Information and Software Engineering, University of Electronic Science and Technology of China, Chengdu, Sichuan, China

Pragati Tripathi Department of Electrical Engineering, Gautam Buddha University, Greater Noida, UP, India

Usen Usen Department of Information Systems and Security, Kennesaw State University, Kennesaw, GA, United States

Aditi Verma Department of Electrical Engineering, Gautam Buddha University, Greater Noida, UP, India

Shakaria Wilson Department of Information Systems and Security, Kennesaw State University, Kennesaw, GA, United States

Akhter Hussain Yatoo School of Mathematical Sciences, School of Electronics Science and Engineering, University of Electronic Science and Technology of China, Chengdu, Sichuan, China

About the editors

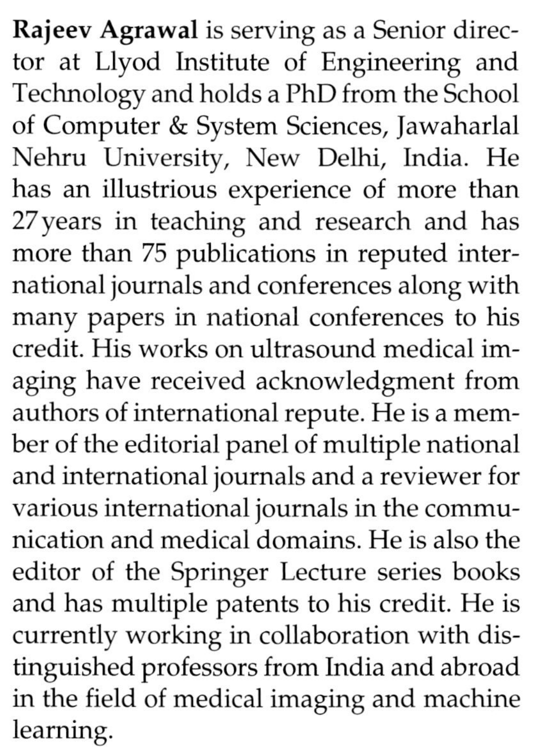

Rajeev Agrawal is serving as a Senior director at Llyod Institute of Engineering and Technology and holds a PhD from the School of Computer & System Sciences, Jawaharlal Nehru University, New Delhi, India. He has an illustrious experience of more than 27 years in teaching and research and has more than 75 publications in reputed international journals and conferences along with many papers in national conferences to his credit. His works on ultrasound medical imaging have received acknowledgment from authors of international repute. He is a member of the editorial panel of multiple national and international journals and a reviewer for various international journals in the communication and medical domains. He is also the editor of the Springer Lecture series books and has multiple patents to his credit. He is currently working in collaboration with distinguished professors from India and abroad in the field of medical imaging and machine learning.

M.A. Ansari earned his BTech in Electrical Engineering from Aligarh Muslim University, Aligarh, India, and MTech and PhD from Indian Institute of Technology Roorkee, India, in the area of medical image processing. He has taught in several national and international universities. He has more than 20 years of teaching experience in engineering and technology. His research interests include digital image processing, medical imaging, biomedical instrumentation, artificial intelligence, machine learning, soft computing, and wavelet applications to engineering problems. He has published more than 200 research papers in reputed national and international journals/conferences and holds several patents. He is currently associated with the School of Engineering, Gautam Buddha University, Greater Noida, NCR Delhi, India. He is a senior member of IEEE, ISIAM, and chair of HAC/SIGHT, UP Section, India.

R.S. Anand earned his BE (Electrical), ME (M&I), and PhD (Digital Signal and Image Processing) from University of Roorkee, currently Indian Institute of Technology Roorkee, India. He is currently working as a professor in the Department of Electrical Engineering, Indian Institute of Technology Roorkee, India. He has wide research and teaching experience and has research interests in medical imaging, ultrasonic imaging in nondestructive evaluation (NDE) and medical diagnosis, biomedical instrumentation, biomedical signal and image processing. He has published more than 300 papers in reputed national and international journals and conferences and has written several books. He is a life member of Ultrasonic Society of India.

Sweta Sneha is the director of Healthcare Management and Informatics and an associate professor of Information Systems and Security at Kennesaw State University, United States. She received her doctorate in Computer Information Systems from Georgia State University and a bachelor's degree in Computer Science from the University of Maryland, College Park, MD, United States. She is a leader and a visionary who spearheaded the development of a premier interdisciplinary graduate program in Healthcare Management and Informatics with the objective to meet the workforce needs and economic development imperative in Health IT. Her experience in healthcare and information technology gleaned over a decade center around a wide array of technical and behavioral challenges. She has published over 50 research papers in premier journals and conferences both nationally and internationally. She has also authored a book titled *Revolutionizing Health Monitoring*.

Rajat Mehrotra is working as an assistant professor in the Department of Electrical & Electronics Engineering at GL Bajaj Institute of Technology& Management, Greater Noida, India. He has more than 12 years of experience in the field of academics and has a keen interest in the fields of deep learning, machine learning, and biomedical image processing. He has received his BTech in Electrical and Electronics Engineering from Dr. A.P.J. Abdul Kalam Technical University (AKTU), Lucknow, India, and MTech in Telecommunication Engineering from AKTU Lucknow. Currently, he is associated with research work in the School of Engineering at Gautam Buddha University, Greater Noida, India. He is a member of IEEE, UP Section India.

Preface

Computational Intelligence in Healthcare Applications is a distinctive effort to describe a variety of techniques designed to represent, enhance, and empower multidisciplinary and interdomain research based on *computational intelligence in healthcare*. The primary objective of this book is to provide neoteric and latest developments in the area of computational intelligence and their practical applications in healthcare. The book also focuses on identifying challenges and solutions through a multidisciplinary approach, shaping the path for new research dimensions. The post-COVID-19 era will bring to light different issues affecting national and international progress with healthcare being the center of attention for research and development.

The convergence of artificial intelligence, machine learning, big data, data analytics, and high-performance computing provides the capability to solve complex problems; leveraging these platforms to create pervasive healthcare systems is the major task of this book. We expect this book to serve as a reference for the pervasive healthcare domain, which takes into consideration the future and beyond the new convergent computing and other applications. This book will provide topics for in-depth research and discussions for researchers, application designers, academicians, students, doctors, and health practitioners in the area of *m-health*.

For the convenience of the reader, this book is divided into three sections. The first section of the book reviews the use of computational intelligence in biomedical imaging and signal processing techniques such as MRI, CT, X-ray, EEG, EMG, ECG, in analyzing and diagnosing various medical conditions, particularly diseases related to the heart and the brain. In the second section, the application of deep learning and machine learning tools in medical signal and image processing is presented for the analysis of biomedical signals and images that greatly benefit the healthcare sector by improving patient outcomes through early and reliable detection. The third section of the book presents the application of various machine learning and deep learning tools in the field of biomedical imaging using different imaging modalities to promote a better understanding and analysis of biomedical image and signal processing for the detection and identification of a specific disease.

In brief, this book will:

- present neoteric and latest developments in the area of *computational intelligence and its practical applications in healthcare*;
- help identify and focus the challenges and solutions through an integrated approach, shaping the path for new research dimensions;
- describe the implementation of deep learning techniques for the detection and classification of diseases;
- describe an advanced procedure to address and enhance available diagnostic methods;
- facilitate a deeper understanding of current trends in artificial intelligence and machine learning within the healthcare domain;

demonstrate the prospect of computational intelligence in healthcare through best practices;
describe the evolution of intelligent healthcare systems based on knowledge/data engineering.

Today, technology is saving lives in every part of the globe. We are looking forward to a time when medical treatment can be provided virtually. This sector is limited today, but shortly it will be the new face of medical facilities. Thus, the primary aim of this book is to put forward some new application-oriented techniques for healthcare using artificial intelligence and machine learning to better understand and analyze the biomedical image and signal processing techniques for the in-time detection and identification of chronic and deadly diseases.

Editors

Acknowledgments

We extend our heartfelt thanks to all those whose work, research, and support have helped us to edit this book on *Computational Intelligence in Healthcare Applications*. Editing such a book was harder than we thought and definitely more rewarding than we could have ever imagined. To start with, we express our gratitude to Elsevier for having faith in us and providing the opportunity and platform to edit a book on this topic, which is the most pivotal and talked about topic in today's scenario in the area of healthcare with computational intelligence.

We thank the authors, contributors, and researchers for sharing with us their valuable knowledge in the form of their research work and supporting us to follow the timelines set by the publisher. We are equally grateful to the reviewers who have invested their valuable time in reviewing the contributed chapters and helped us to bring out the best with their valuable suggestions in time to improve the quality of the chapters and the book.

We are thankful to the premier affiliating institutes of the editors, namely Indian Institute of Technology Roorkee, India; GL Bajaj Institute of Technology & Management, Greater Noida, India; Gautam Buddha University, Greater Noida, India; and Kennesaw State University, United States, for their support in the successful completion of this book.

We are also thankful to the hospitals/medical institutes, frontline workers, and doctors who have helped us by providing their constant support and motivation for the completion of this book.

And, last but not least, we are eternally grateful to our parents, teachers, mentors, colleagues, research scholars, and friends for their constant reinforcement that has helped us to complete this book successfully.

Thanks to everyone in the team for their support, cooperation, and motivation!

Editors

CHAPTER

1

Clinical decision support systems: Benefits, potential challenges, and applications in pneumothorax segmentation

Sudha Subramaniam[a] *and K.B. Jayanthi*[b]

[a]Department of Biomedical Engineering, Velalar College of Engineering and Technology, Erode, Tamil Nadu, India [b]School of Electrical Sciences, K.S. Rangasamy College of Technology, Tiruchengode, Tamil Nadu, India

Introduction

The most significant digital transformative technology of the 21st century is artificial intelligence (AI). The penetration of digital technology started generating an ocean of data points that accelerated the power of AI. The transformation is fueled by techniques such as machine learning (ML), deep convolution networks, recurrent neural network, reinforcement learning, generative adversarial network, and natural language processing. The adoption of AI is not limited to the traditional business and technology sector, but it significantly impacts its economic growth.

The impact of AI has its roots in the healthcare sector, helping patients and clinicians in different ways. Deep learning (DL) (Haque & Neubert, 2020) automatically identifies the data's complex patterns and helps radiologists derive intelligent decisions. The performance of a CDSS has shown to be equivalent to that of experienced radiologists in the early detection and diagnosis of disease. The availability of a tremendous volume of medical data due to advancements in image acquisition devices paved the way for opportunities and challenges in healthcare research.

Image information plays a crucial role in disease diagnosis and treatment. X-ray is the first image modality that starts appearing in 1895. Various other imaging modalities such as

https://doi.org/10.1016/B978-0-323-99031-8.00009-0

computed tomography, ultrasound, magnetic resonance imaging (MRI), positron emission tomography (PET) are invented soon after. Multimodality images dramatically increase the volume of the data and the workload of the radiologists and physicians. The development of computer-aided diagnosis assists radiologists in the interpretation of an image. Computer-assisted image analysis and CDSSs have been a significant research area in medical imaging in the past few decades. AI-based CDSS helps to analyze biomedical image data and derive medical outcomes from data. It can improve the diagnosis, prognosis, and treatment of an individual.

Benefits of AI clinical decision support system

CDSS effectively access the clinical data, improves clinical decision-making, and delivers quality care to patients with reduced healthcare costs. AI creates an intelligent system better than human experts to solve complex problems. Embedding AI in CDSS utilizes a large quantity of data to provide quality decisions and optimal treatment for each patient. CDSS (Shaikh et al., 2021) empower radiologist and medical practitioners to make a sound decision in health care.

Enhancing diagnostic accuracy

One of the critical issues that account for a legal claim against a physician and healthcare organization is the risk of delayed or misdiagnosis. In the healthcare sector, diagnostic error may lead to life-threatening consequences beyond medical expenses. It increases the mortality in patients and prolonged hospitalization. Early diagnosis of disease through AI tools and techniques brings a paradigm shift to health care. AI techniques can unseal the relevant clinical information hidden in the massive volume of data and can assist clinicians in making sound decisions. The AI-based diagnostic system helps to reduce the therapeutic and diagnostic error that is unavoidable in human practice. AI helps to identify the pathology using medical images.

Making informed decisions

AI-assisted clinical decision systems aid medical practitioners to make informed decisions in a short period. It also recommends medications for the postsurgical patient in discharge and ensures optimal care by recommending periodic follow-up for patients. Clinical decisions help patients to choose an alternative treatment of their choice. AI aids in effective decision-making by reducing medical errors and assist physicians to double-check their decisions and hence provide recommendations.

Assist physicians

The new barrier that hampers patient care is a shortage of physicians. The Association of American Medical College (AAMC) predicted the shortage of physicians in the United States to be between 40,800 and 104,900 by the year 2030. World Health Organization (WHO) recommends one government doctor for every 1000 people. India reported the ratio to be 1:10,189, implying a shortage of 600,000 doctors and 2 million nurses.

In recent years there is a rising prevalence of stress-related burnout among clinicians due to long working hours, night work, time pressures, chaotic environment, and low control of the pace. AI-based CDSS leverages information from unstructured medical data and assists physicians in the early diagnosis of disease. This aids physicians to understand the needs of the patient better.

Potential challenges associated with AI system

Getting users on board

Implementing AI-based CDSS in the hospitals and diagnostic centers can lead to changing procedures and protocols of the hospital, which is a challenging task. Physicians, nurses, and other concerned staff members of the hospital have to be trained. Entailing the medical practitioners to design and deliver AI-based CDSS (Walczak, 2020) brings considerable benefits to AI engineers and healthcare teams. Including medical professionals in the design of AI systems brings better usage of data, better design of AI systems and better performance of AI algorithms.

Performance attainment

The biggest concern of AI in health care is the unexpected and negative consequences of the AI system. Medical practitioners usually have high expectations in terms of the performance of AI systems associated with the sensitivity and specificity of the model. The error that physicians are willing to tolerate depends on the complexity of the medical problem and the risk involved in clinical decisions. The error tolerance depends on medical applications. It is not easy to define the system performance in case of treatment decisions.

Quality algorithm with improved data

Promising AI-based CDSS systems (Sutton et al., 2020) can be designed to provide better patient outcomes. Accurate data collection and information leverage the AI industry in sound decision-making, and better results/data collection is an ongoing process of gathering information from various sources and analyzing and interpreting them for better decisions. It gives the complete picture of the problem and is the foundation of decision-making.

The data collected should be of better quality and quantitative in number. Data collection aids to create a holistic view of patient, treatment and to enhance health outcomes. Digital devices enhance the amount of data delivered, thereby influence the AI system. Petabytes of data are being collected by digital devices and contribute to the growing database. The collected data help the doctor in the early diagnosis of disease and better treatment through intelligent AI algorithms.

U-Net architecture for the decision support system and clinical diagnosis

Convolution neural networks (CNNs) have been proven to be the state-of-art architecture for image analysis tasks such as image classification, segmentation and generation. Deep learning networks have taken digital technology by storm in the past years. It has an immensely positive impact in the area of medical image analysis. It entails tasks like disease classification, quantifying the severity of anomalies in MRI, and segmenting human organs from the body. Medical images are separated from traditional natural images. Natural images are generally RGB images. Medical images are of varying forms, such as two-dimensional grayscale images, 2D images with three-channel and four-channel, 3D and 4D volume images. The pixel value of raw images typically does not convey information, whereas, in medical images, pixel value conveys information about the tissue present. Medical images, in particular, need a special type of analysis to extract information from unstructured data.

Apart from distinguishing the diseases in an individual through image, it instead requires localization of region of interest of abnormality. Various traditional methods such as thresholding, edge detection, region-based segmentation, clustering method, and morphological approaches are used for segmentation. Deep learning networks are data hunger networks and require a large dataset for training the model. Affording the amount of data needed by deep networks is quite expensive since the data need to be collected more systematically. It requires a lot of domain expertise for the classification of images. CNN focuses on classification, which needs image as an input and label as an output. Biomedical analysis requires localization of the area of interest of abnormalities rather than classifying the disease.

U-Net architecture (Siddique, Paheding, Elkin, & Devabhaktuni, 2021) evolved from traditional CNN architecture to provide better performance for biomedical images with limited dataset images. The framework of CNN architecture is shown in Fig. 1.1. CNN takes medical images as an input and passes them through intermediate hidden layers to produce a segmented map as an output. The segmented map visualization helps to understand features that CNN detects to discriminate the uniqueness of the image.

U-Net architecture is asymmetric architecture consisting of three parts as shown in Fig. 1.2.

Contracting part
Bottleneck part
Expanding part

The contracting path present on the left side of the architecture is an encoder architecture. A typical CNN architecture consists of a convolution block with a filter for extracting the feature information followed by rectified linear unit (RELU) activation function and a downsampling layer.

FIG. 1.1 State-of-art CNN architecture. *No Permission Required.*

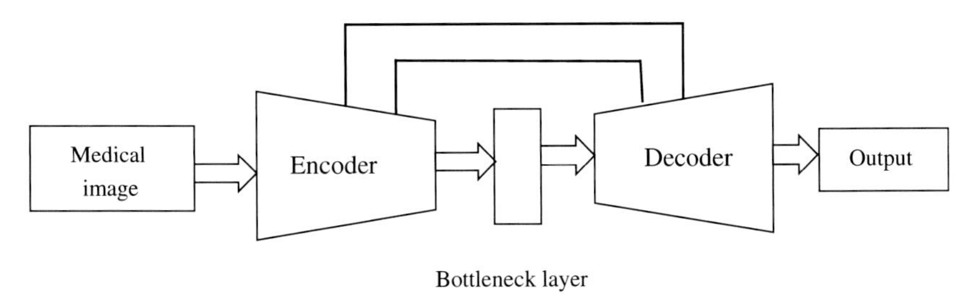

FIG. 1.2 U-Net architecture. *No Permission Required.*

The contraction path reduces the spatial size of the input while increasing the feature map to extract useful information. The bottleneck layer consisting of two FC layers predicts the segmented map output. The expanding path present on the right side of the architecture is a decoder architecture. It consists of an upsampling layer with a deconvolution layer to restore the spatial dimension of the image. The skip connections are added to concatenate the feature map of the encoder with the corresponding decoder to extract the spatial information present in the image.

The base U-Net is 2D-UNet with a 2D convolution layer and 2D max-pooling layer for prediction of segmentation map to leverage the context of the image across its height and width. 3D volumetric images can be segmented with 3D-UNet, and it resides the structure of the base U-Net. 3D U-Net has extensive applications in CT and MRI images. The architecture is obtained by replacing the 2D convolution layer, pooling layer, and upsampling layer with 3D convolution, pooling layer, and upsampling layer, as shown in Fig. 1.3.

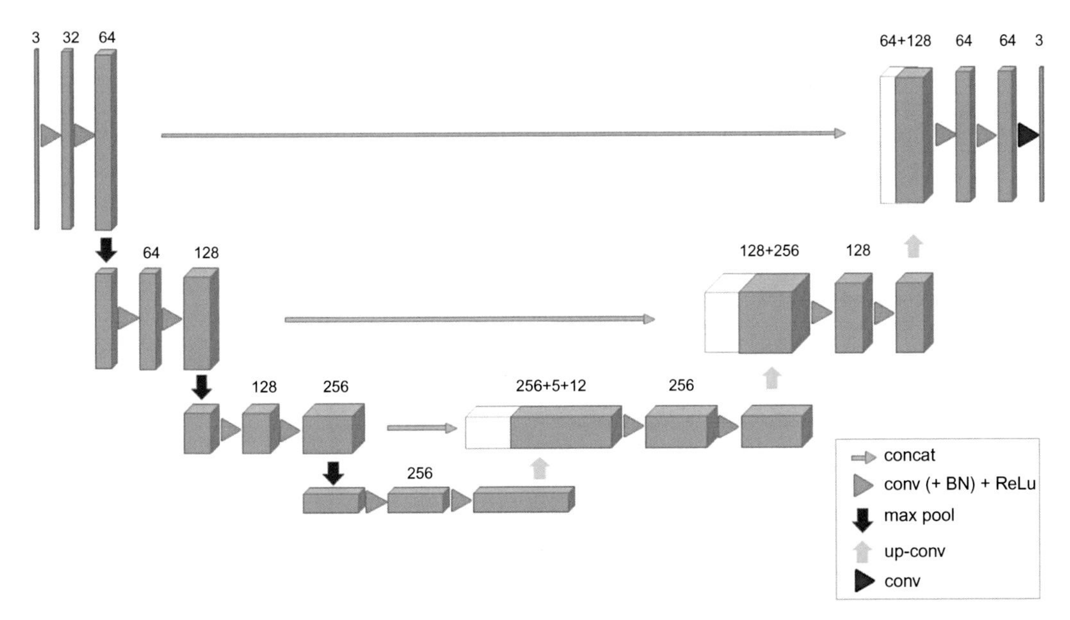

FIG. 1.3 3D U-Net architecture. *No Permission Required.*

Attention is a strategy to highlight salient features by suppressing irrelevant information for a specific task. This significantly increases the quality of the segmented output by focusing on the region of interest instead of nonuseful background information. This further result in the effective use of computational resources and provides better generalization power. Attention is obtained by assigning higher weightage to high relevance contextual information and lower weightage to areas of lower relevance contextual information.

The spatial information recreated in the expansion path is imprecise and is counteracted by the usage of skip connection between compression and expansion path of U-Net. The low-level features extracted through the initial layer of the contraction path are of inadequate representation. Concatenation of spatial information across each layer of the encoder-decoder part results in the redundant extraction of low-level feature information. Attention network aids in the suppression of redundant information captured by U-Net architecture.

Attention U-Net architecture is designed by the addition of attention gates across the skip connection. Significant performance improvement is obtained through repeated use of attention gates across each layer of skip connection. The weights are assigned to the features based on the context vector generated by the attention gate. Thus highlights the features containing salient information and suppressing the irrelevant information on counterpart. Fig. 1.4 shows the schematic attention gate.

Case study on segmentation of pneumothorax from chest X-ray

Pneumothorax, also called collapsed lungs, is a life-threatening disease caused by the accumulation of air in the pleural space. The area between the lungs and chest wall is called pleural space. The accumulated air pressures the lungs, thereby collapsing the lungs—chest X-rays more commonly used to screen and diagnose pneumothorax. Chest radiography, computed tomography, or ultrasonography are also used for the diagnosis of the disease. It commonly results in respiratory failure or cardiac arrest.

Diagnosis of pneumothorax from chest X-ray images is challenging as they are low contrast images with highly variable shapes and are overlapped with ribcage and clavicles. Automatic detection of pneumothorax and segmentation assists the radiologist in the timely

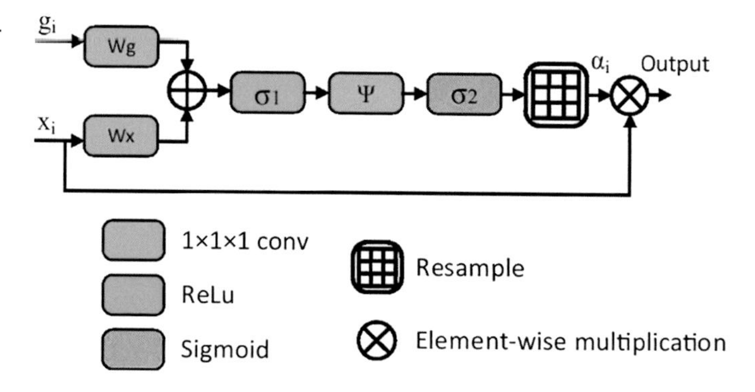

FIG. 1.4 Schematic of attention gate. *No Permission Required.*

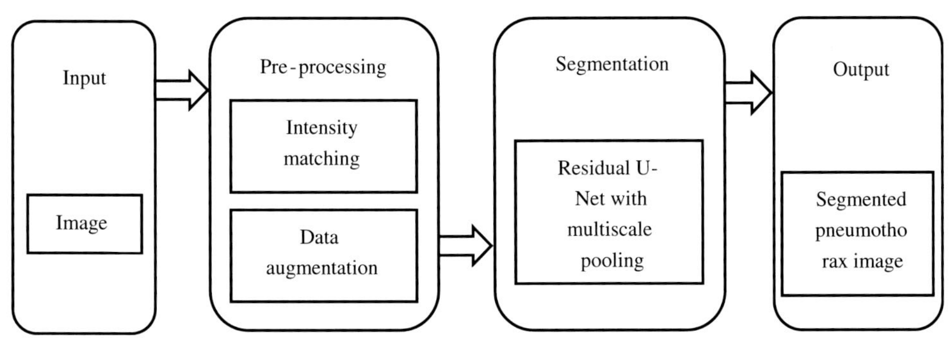

FIG. 1.5 A framework of the proposed model. *No Permission Required.*

diagnosis and treatment of disease. The accuracy of diagnosis is highly dependent on the experience of the radiologist. The lack of radiologists in diagnosing disease often results in the delay of treatment, thereby creating a harmful impact on the patient's health, including death.

Deep learning architectures are a popular choice for pneumothorax segmentation (Jakhar, Kaur, & Gupta, 2019). The framework of the proposed work is shown in Fig. 1.5. The chest X-ray image is used as an input image. The X-ray images obtained are of varying intensity. Intensity matching is carried out to match the intensity of the image to a standard template. The image obtained is later normalized and filtered out for the removal of boundary noise. Data augmentation is carried out to increase the volume of training data. Segmentation of pneumothorax is carried out through U-Net architecture with residual connection in the encoder and multiscale pooling to obtain the constant output. Before the proposed U-Net architecture, various methods were adapted, resulting in inaccurate and inefficient results.

U-Net architecture works well for medical images. Segmenting the region affected by the disease comes under the problem of object detection. The pneumothorax segmentation comes under the umbrella of semantic segmentation (Tolkachev, Sirazitdinov, Kholiavchenko, Mustafaev, & Ibragimov, 2021). It is carried out through the classification of each pixel in the image. The pixels are classified into pneumothorax pixels and nonpneumothorax pixels. The output image is a segmented image of black and white, in which the white region indicates the pneumothorax region. The encoder part of U-Net downsample the image to learn the feature representation and the decoder part extracts the spatial information, which potentially causes pneumothorax.

The availability of a massive volume of medical data is critical to attaining the goals of a deep learning network. Most of the time, labeled data are not available for training. In this context, labeled data refer to segmented pneumothorax images. Experts in the medical field are required for labelling the data, which is time-consuming and expensive. The problem of overfitting occurs when complex deep neural networks are trained with limited data. This problem is solved by using data augmentation that considerably increases the volume of training data. New images are created by transforming the images that exist in the training dataset. Transformation techniques such as rotation, shifting, scaling, and flipping are used to increase training volume. The original and transformed data are fed to the neural network for better performance attainment.

Comparison of fully connected network (FCN) and U-Net for pneumothorax segmentation

The objective of semantic segmentation is to identify the specific object in the body. An object is identified by labelling each pixel in the image with corresponding class values. Segmentation architecture is a modified form of classification architecture. Classification is obtained by predicting the entire image to one of the classes, whereas segmentation is obtained by predicting each pixel into distinct classes. CNN architecture used for carrying semantic segmentation has significant drawbacks. The dense layer present in CNN has an enormous parameter that needs to be learned while training the model and is computationally expensive. It also has constraints over the input dimension.

Fully convolution network (FCN) counteracts this drawback by replacing a dense layer with a stack of convolution layers. FCN preserves the dimension of the output image to be the same as that of the input image. The architecture of FCN is shown in Fig. 1.6.

A slight modification of FCN obtains U-Net architecture. FCN has only one layer in the decoder and upsamples the features once. U-Net architecture has multiple upsampling layers with a symmetrical pattern across encoder and decoder. In FCN, the spatial information from the encoder is summed with a decoder, whereas in U-Net, the downsampling feature maps are concatenated with upsampling features.

The U-Net and FCN architecture are trained using chest X-ray images. The dataset consisting of 285 chest X-ray images augmented to 2565 images are used for training, while 53 X-ray images are used for validation. The training is carried out using an Adam optimizer with a learning rate of 0.0001. The resultant value of accuracy and loss plotted are shown in Fig. 1.7. The validation accuracy of U-Net and FCN for the chest X-ray dataset is calculated to be 95.30% and 89.23%, respectively.

Similarly, the loss value is found to be 0.18 and 0.02 for FCN and U-Net, respectively. Sample segmentation images obtained by the U-Net model and FCN model are shown in Fig. 1.8. FCN architecture fails to extract the spatial information more precisely, whereas U-Net architecture gives good segmentation output.

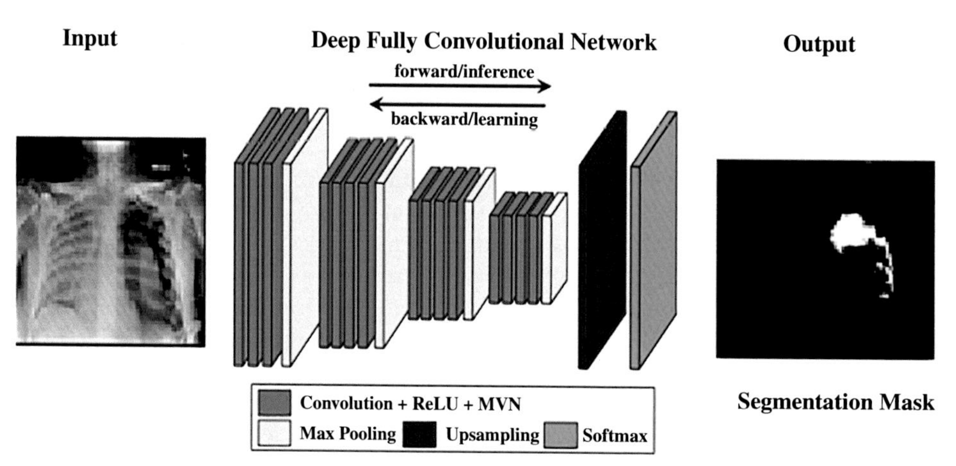

FIG. 1.6 Fully convolution network. *No Permission Required.*

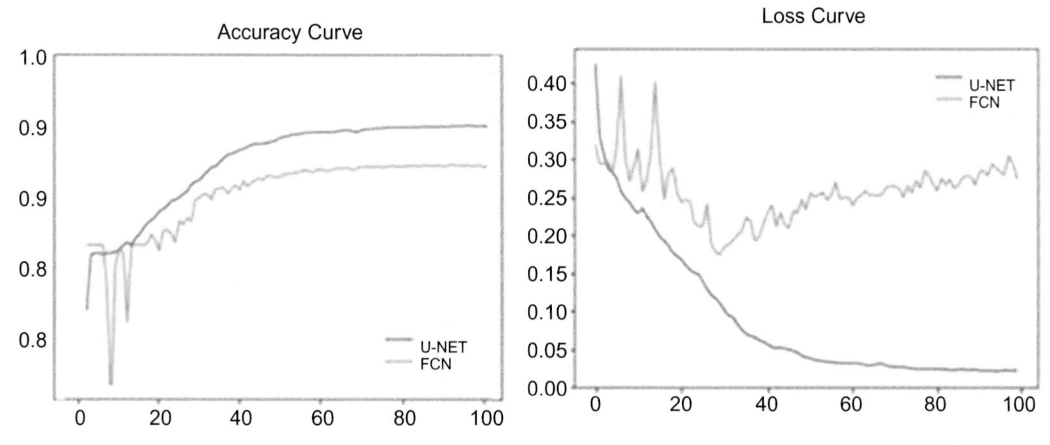

FIG. 1.7 Accuracy and loss curve. *No Permission Required.*

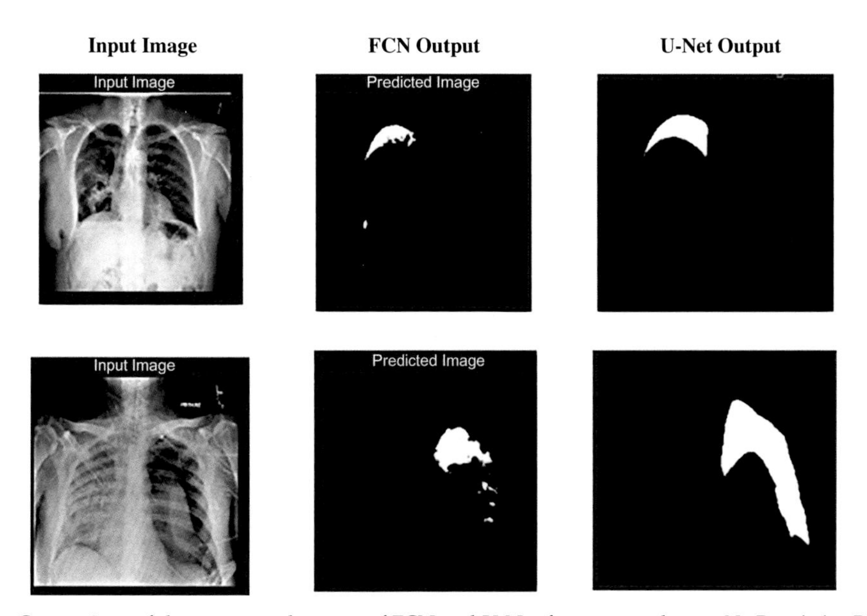

FIG. 1.8 Comparison of the segmented output of FCN and U-Net for pneumothorax. *No Permission Required.*

Performance metric

The effectiveness of the segmentation system is evaluated using performance metrics, which aids in the comparison of the proposed system with the existing technique. It projects the progress of the system in terms of numbers. Commonly used metrics for evaluation of segmentation system are accuracy, intersection over union, dice similarity coefficient, and loss.

Accuracy

Accuracy is perhaps the most straightforward evaluation metric for classification and segmentation. It represents the percentage of image pixels that are correctly classified among the total image pixel. It has the limitation of providing biased results in case of a class imbalance problem. The accuracy measure of the dominant class will overshadow the accuracy associated with other classes. Hence average per-class accuracy is calculated as a performance metric for images with class imbalance problems. Pneumothorax segmentation falls under binary class problems consisting of two classes. Pixel containing pneumothorax is represented as pneumothorax pixel and belongs to class 1, and another pixel where there is no pneumothorax is represented as nonpneumothorax pixel and belongs to class 0.

$$Accuracy_{Pneumothorax} = \frac{Correctly\ classified\ pneumothorax\ pixel}{Total\ number\ of\ pneumothorax\ pixel} \tag{1.1}$$

$$Accuracy_{non-Pneumothorax} = \frac{Correctly\ classified\ non-pneumothorax\ pixel}{Total\ number\ of\ nonpneumothorax\ pixel} \tag{1.2}$$

$$Average\ per\ class\ accuracy = \frac{1}{2}\left(Accuracy_{Pneumothorax} + Accuracy_{non-Pneumothorax}\right) \tag{1.3}$$

Loss

Loss perhaps represents a deviation in the predicted output from the actual output. The cross-entropy loss function is a commonly evaluated loss function for segmentation problems. This function evaluates the loss for each pixel vector in the image and averages all pixels. It has limitations in class imbalance problem; hence weighted cross-entropy loss function is used for the segmentation task.

Intersection over union (IoU)

IoU quantifies the overlap between the predicted mask and the ground truth mask. It measures the number of pixels common between the ground truth mask and predicted mask among the total number of pixels present across the two masks.

$$IoU = \frac{Ground\ Truth \cap Prediction}{Ground\ Truth \cup Prediction} \tag{1.4}$$

Dice similarity coefficient (DSC)

DSC is the most commonly used metric in evaluating segmentation tasks. It calculates the similarity between the ground truth mask and the predicted mask.

$$DSC = \frac{2|Ground\ Truth \cap Prediction|}{|Ground\ Truth| + |Prediction|} \tag{1.5}$$

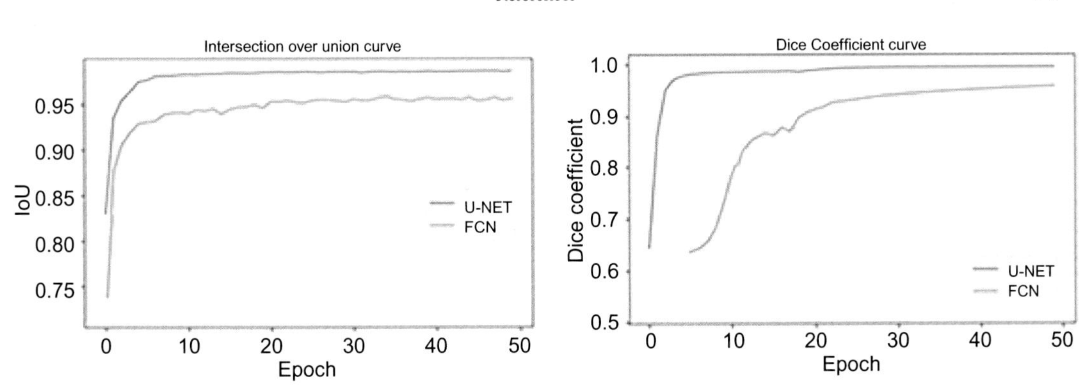

FIG. 1.9 Performance measure of IoU and DSC for pneumothorax segmentation. *No Permission Required.*

The IoU and DSC are measured for two models. Fig. 1.9 shows the performance of the chest X-ray image. The IoU of the FCN model is found to be 93.4%, and U-Net is found to be 98.76%. The results show that the U-Net model segments the pneumothorax region better than the FCN model. The performance of DSC for the FCN model is found to be 92.3%, and the U-Net model is found to be 98.23%.

Conclusion

The emergence of AI and CNN has made enormous progress in medical image procession and ineffective decision-making. A CDSS helps medical practitioners to enhance health care and its outcomes. The case study on pneumothorax segmentation clearly states the use of AI in CDSS. Integration of AI with CDSS helps healthcare providers to enable integrated workflow and to provide assistance at the time of care. U-Net architecture provides accurate results when the volume of image data is significant. Integration of AI in the CDS system helps health care reach rural areas where radiologists' availability is sparse.

References

Haque, I. R. I., & Neubert, J. (2020). Deep learning approaches to biomedical image segmentation. *Informatics in Medicine Unlocked, 18*, 2352–9148.

Jakhar, K., Kaur, A., & Gupta, M. (2019). Pneumothorax segmentation: Deep learning image segmentation to predict pneumothorax. *arXiv*. https://arxiv.org.

Shaikh, F., Dehmeshki, J., Bisdas, S., Roettger-Dupont, D., Kubassova, O., Aziz, M., et al. (2021). Artificial intelligence-based clinical decision support systems using advanced medical imaging and radiomics. *Current Problems in Diagnostic Radiology, 50*(2), 262–267. https://doi.org/10.1067/j.cpradiol.2020.05.006.

Siddique, N., Paheding, S., Elkin, C. P., & Devabhaktuni, V. (2021). U-Net and its variants for medical image segmentation: A review of theory and applications. *IEEE Access, 9*, 82031–82057. https://doi.org/10.1109/access.2021.3086020.

Sutton, R. T., Pincock, D., Baumgart, D. C., Sadowski, D. C., Fedorak, R. N., & Kroeker, K. I. (2020). An overview of clinical decision support systems: Benefits, risks, and strategies for success. *Npj Digital Medicine, 3*(1). https://doi.org/10.1038/s41746-020-0221-y.

Tolkachev, A., Sirazitdinov, I., Kholiavchenko, M., Mustafaev, T., & Ibragimov, B. (2021). Deep learning for diagnosis and segmentation of pneumothorax: The results on the Kaggle competition and validation against radiologists. *IEEE Journal of Biomedical and Health Informatics, 25*(5), 1660–1672. https://doi.org/10.1109/JBHI.2020.3023476.

Walczak, S. (2020). *The role of artificial intelligence in clinical decision support systems and a classification framework* (pp. 390–409). IGI Global. https://doi.org/10.4018/978-1-7998-1204-3.ch021.

CHAPTER

2

Opportunities and challenges for smart healthcare system in fog computing

Naveen Chauhan[a], *Rajeev Agrawal*[b], *and Kanika Garg*[c]

[a]KIET Group of Institutions, Delhi-NCR, Ghaziabad, India [b]Department of Electronics and Communications, Llyod Institute of Engineering and Technology, Greater Noida, UP, India [c]SRM Institute of Science and Technology, Delhi NCR Campus, Ghaziabad, India

Introduction

For enhancing the quality of service (QoS), fog computing has become the central infrastructure on the Internet. The proliferation in the number of end devices (e.g., intelligent sensors, smartphones, smartwatch, control systems) in a communicating network creates the Internet of Things (IoT) expansion in multiple domains. IoT offers a smooth platform that integrates hardware, intelligent computing devices, software's to connect people, and objects over a network that enables them for interaction, communication, and enriching their lives. According to Gartner (2021), 500 billion IoT devices are expected to be connected to the Internet by 2030, the massive growth of IoT devices is probably in all dimensions of industry 4.0 expansions (2019). A vast amount of data is generated by these devices; IoT devices' growth and market share are shown in Fig. 2.1; thus, the extreme growth in data generation can be seen in the next couple of years. The massive data generation emerges the need for highly computation effective machines and uninterrupted storage space that can be accumulated in real-time. Cloud systems and IoT, jointly called Cloud of Things (CoTs), are the key standard for next-generation intelligent communication and computation (Yi, Huang, & Cai, 2021). CoTs included many key benefits under industry reformation in eHealth, transportation, manufacturing, communications sectors. However, industry reformation has started facing challenges in several domains; for example, high mobility in transportation may also produce the challenge for the availability of data and processing at the traditional data centers/cloud systems. The revolution of Industry 4.0 changes, from mechanical engineering to electrical engineering, to telecommunication and information communication, to intelligent

https://doi.org/10.1016/B978-0-323-99031-8.00014-4

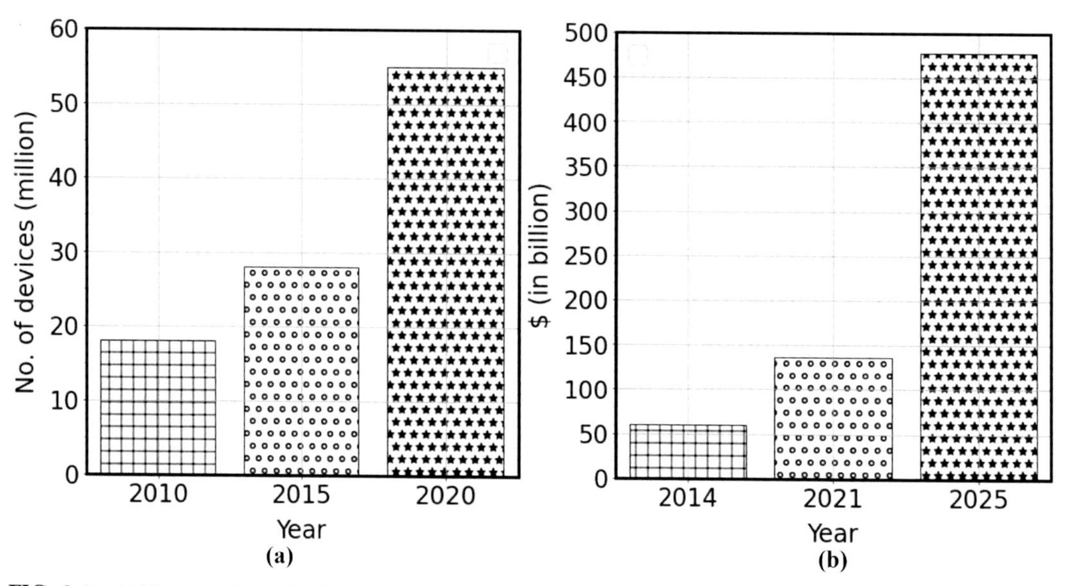

FIG. 2.1 (A) Expected growth of IoT devices. (B) Expected growth of IoT in Healthcare Market (USD Billion). *No permission required.*

devices deployment. In Kumari, Tanwar, Tyagi, and Kumar (2018), the involvement of IoT devices in the healthcare industry is proliferating, and the revenue generated in 2016 was more than 15%. Healthcare 4.0 is facing many challenges in volume, variety, and velocity of data. Moreover, a vast amount of data is triggered by healthcare devices at regular intervals; thus, simultaneous services cannot be performed operations such as data preprocessing, data filtering, and normalizing, data storage, data security on a large volume of data. Due to the limitation in capability and capacity of devices, the initial processing of data at the device layer has been become a challenging task for the researcher (Farahani, Firouzi, Chang, & Badaroglu, 2018).

The devices have some intrinsic deficiencies, such as limited power source, low CPU speed, and limited storage space (Chauhan, Banka, & Agrawal, 2021b). Additionally, Cloud computing (CC) emerges as an unlimited resource, with a pay and use model. The domain of CC comes up with many IoT services such as computation of resources, data processing, and data analytics. However, several issues can be seen in CC concerning the high waiting time of processing of data, which has an unfavorable effect on the IoT services requiring a real-time response (Baranwal & Vidyarthi, 2021). The precarious deviation in the subsequent communication path from the device layer to the cloud layer becomes a hazardous situation for time-sensitive applications (Chauhan, Banka, & Agrawal, 2021a). eHealth or intelligent healthcare system is a key sensitive area that cannot afford unpredictable delay in response and colossal latency. Therefore, it is impractical to depend on traditional centralized computing for storing, analyzing, and processing data at the far end server. The overdue time taken by the healthcare sensor enables devices to transfer the patient's data to a cloud network for acquiring the response back to the applications that can adversely affect the output of emergency and e-health services (Shakarami, Ghobaei-Arani, Masdari, & Hosseinzadeh, 2020). The difficulty in transmission of processed data over the cloud and dependency on the final outcome back to users is a serious matter of concern in a real-time environment. In emergency situations (healthcare network) lack of availability of an application or a little delay in data processing can put a high risk on life. Chaing and Zhang (2016) stated that several industrial control systems, manufacturing, smart grids, and IoT applications such as driverless cars, financial trading, and drone flight system might also require very low latency in milliseconds. To please the applications which demand a modest latency rate with real-time data processing, such as healthcare IoT systems, the concept of fog computing was developed (Bilal, Khalid, Erbad, & Khan, 2018; Seferagić, Famaey, De Poorter, & Hoebeke, 2020). eHealth is an emerging field in IoT that requires a low latency rate and high response time; however, to make it balanced, computation near to device, fog computing is considered a suitable match for these applications (Li, Yuan, Palaniswami, & Moessner, 2015; Mouradian, Naboulsi, Morrow, & Pola, 2018; Rahmani, Gia, Negash, & Anzanpour, 2018).

International enterprises and startups throughout the world are using seamless computing, storage, and greater network capacity. Their workstations have been migrated to controllable cloud servers with a limitless resource bundle. Cloud-based IoT solutions, on the other hand, fall short of solving a number of challenges when applications and services are time-sensitive. In preparation for further expansion in this domain, some of the issues are covered here:

1. In cloud data centers, the application receives a higher response time compare to mobile edge computing.
2. The reliance on the public network may have negative consequences for bandwidth, the privacy of sensitive data, and the security of IoT devices.

Fog node can solve the above-stated challenges, but due to limited resources, the study must be carried out in management and optimal allocation of resources.

Industry 4.0 is an emergent architecture of numerous sensor devices; that sends a huge volume of data regularly. In some cases, a critical situation can be seen as an obstacle in data loading at the far end. In real-time data collections and processing, it is cumbersome to rely on traditional cloud architecture, which is unacceptable in some time-sensitive healthcare applications. In order to reform the above-stated issues of traditional CC architecture, the Fog computing system can collaborate with a cloud system, i.e., a three-level architecture model Fig. 2.2. Wu, Sun, and Wolter (2018) proposed 3 level architecture that includes end devices, fog computing middleware, and cloud center. The performance of 3-tier model outperforms in terms of energy and latency (Li, Santos, Delicato, & Primex, 2017). The result showed energy reduction when the tasks were executed on the fog node (Jalali, Hinton, Ayre, & Alpcan, 2016). Liu and Fan (2018) and Lyu, Vinel, Maharjan, and Gjessing (2018), the study highlighted the offloading decisions to minimize resource consumption requirements. The main pitfalls of CC, like low latency rate and response time, can be overcome with fog computing, which empowers the delivery of services on time with consistency.

This article is organized as follows; the feasibility of fog computing in the healthcare system is discussed in "Preliminaries" section. A fog-based healthcare application and case studies on ECG (Electrocardiogram) monitoring and feature extraction are presented in "Related work" and "Case study" sections. Distinct discussions and open issues are highlighted in "Discussion and open issues" section subsequently.

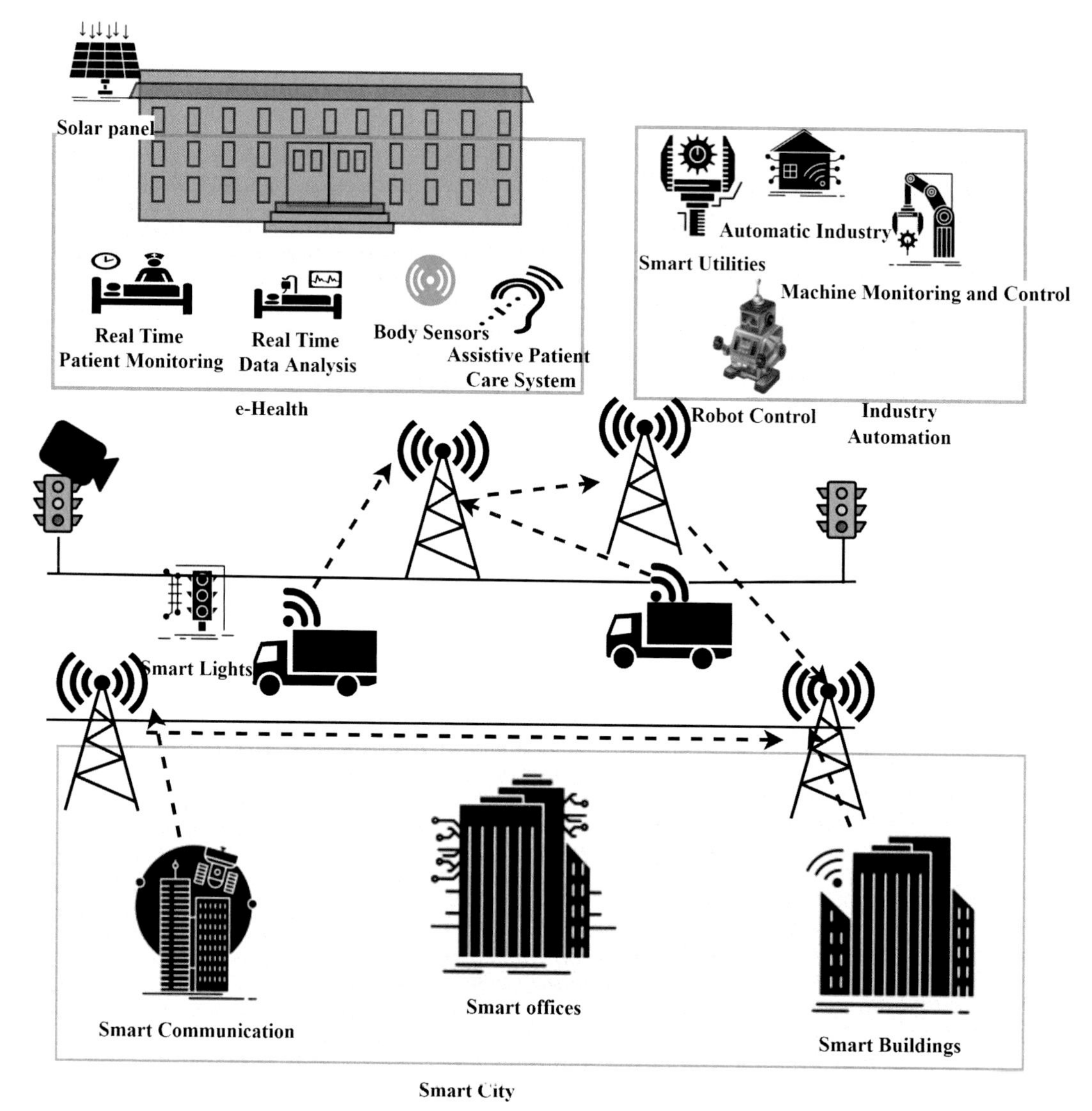

FIG. 2.2 IoT and its applications, infrastructure, and communication. *No permission required.*

Preliminaries

This section of the chapter looks at some of the background material that readers will need to understand the proposed concept and model, such as the evolution of smart healthcare and the crucial role of fog computing in healthcare.

Evolution of smart healthcare

This section highlights the works that focus on implementing healthcare applications to make it an intelligent healthcare system. Internet technology is acquiring the revolution of industry 3.0 to 4.0 standards; meanwhile, that includes numerous sensors, highly computable machines, extensive storage, and scalable network bandwidth. Many studies have earlier discussed the reformation of the healthcare industry in past years. This reformation is being feasible because of the emerging technology, i.e., fog computing. Some authors have considered it as edge computing or micro data centers with limited resources. It was incorporated through the use of a fog-based patient monitoring system (Vora, Tanwar, Tyagi, Kumar, & Rodrigues, 2017) to enable ambient assisted living. In order to scale the network bandwidth usage, a group of fog nodes is connected by a cloudlet for further processing of data.

For lower waiting time and increase in data rates, a fog-based model was proposed and simulated through discrete event system specification (Etemad, Aazam, & St-Hilaire, 2017), in which distributed fog nodes were jointly connected to a broker to achieve the performance. A new computing model, firework, is discussed in Chaing and Zhang (2016), which processed vast amounts of data by involving a collaborative edge environment. Data integrity and data seclusion are maintained by applying a firework manager node in the middle of the model. Kubernetes model is discussed in Dupont, Giaffreda, and Capra (2017); the model is organized in three layers (Fig. 2.3): Edge as the middle layer, IoT gateways as the manager layer, and Cloud. Singh, Tripathi, Alberti, and Jara (2017) a semantic edge model was presented to cope with latency-sensitive services while transmitting the information over the network. As discussed in Chauhan et al. (2021a), intelligent gateways are devised to implement the delay-sensitive services; moreover, the model is equipped with notification service, dispersed storage, and data processing at the network edge. A fog-based low power model is presented to process the clinical speech data (Monteiro, Dubey, Mahler, & Tang, 2016); moreover, a low-cost local computation model was suggested (Gia, Jiang, Sarker, Rahmani, & Westerlund, 2017). This computational model enabled remote monitoring of patients in real-time; additionally, the structure of the model included majorly recording sensors for respiration rate, ECG, and body temperature. A separate fog-based model was also described in many studies, which does not use the collaborative structure of mesh fog nodes or smart gateways. As presented in Neware and Shrawankar (2020), a mobile device-centric alert-based service is devised that executes the requested task according to the criticality of the situation. The benefits of including fog and cloud layers in the model are highlighted (Chen, Zhang, & Shi, 2017).

eWALL, a monitoring system, included distributed computation similar to the fog layer (Fratu, Pena, & Halunga, 2015). The information was collected from healthcare sensors and subsequently loaded to local storage to perform further processing; besides, it helps achieve data privacy and minimizes the communication overload. A three-layer architecture is discussed and implemented in Akrivopoulos, Chatzigiannakis, Tselios, and Antoniou (2017),

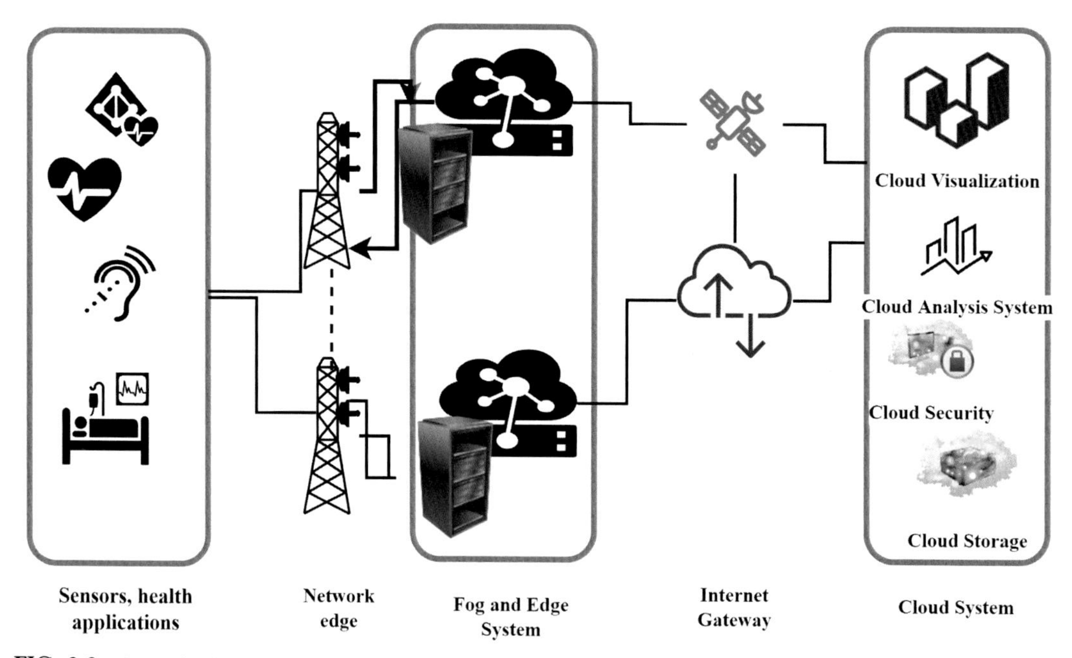

FIG. 2.3 Smart healthcare computation architecture using cloud and fog-based system. *No permission required.*

Masouros, Bakolas, Tsoutsouras, Siozios, and Soudris (2017), and Yi, Hao, Qin, and Li (2015) by incorporating the end-user device layer, fog as edge layer, and remote cloud layer separately. The proposed model was enabled with the processing capability of healthcare data and real-time decision-making. A medical cloud equipped with wearable healthcare sensors and personalized healthcare applications is jointly devised that transfer sensor data to the cloudlet. Cloudlet servers are assisted to process data and transmit compressed information to the mobile cloud (Tasic, Gusev, & Ristov, 2016). An automatic healthcare management architecture system is evolved with the help of the fog layer that mainly focuses on fall detection and model built over services for reporting and taking emergency actions; furthermore, an architectural model is proposed (Nikoloudakis et al., 2016; Ozdemir, Tunc, & Hariri, 2017). A fog layer is implemented that orchestrates the fog system's resources and services to ensure the resiliency and robustness of the system.

Importance of fog computing in healthcare

The importance and feasibility of employing fog computing in health care are discussed in this section. A fog-based low-cost remotely monitored system is presented in Khattak et al. (2019). Moreover, ECG signals, respiration system, body temperature are collected and monitored by sensors and sent to gateways as the middle layer for further processing. As a fog computing system, the computation-enabled middle layer is introduced in Ahmad, Amin, Hussain, Kang, and Cheong (2016). An essential aspect of healthcare application, latency sensitivity, is discussed in Chakraborty, Bhowmick, Talaga, and Agrawal (2016). A distributed programming model was included in the architecture to manage the geographically distributed nodes; furthermore, data consistency and data accuracy can be achieved, and service time can be improved. As the middle layer is condensed with limited resource powers, it is not easy to operate all the tasks at the fog node. Importantly, the identification of task sensitive and based on the output these were forwarded to cloud layer (Dubey et al., 2015). A middle layer is fog computing that is processed, analyzed, and stored locally. This local computation and storage are responsible for providing services to geographically distributed IoT devices (Negash et al., 2017; Rahmani et al., 2018). A fog-based monitoring and diagnosing system is presented in Sood and Mahajan (2017), which considered the patient data infected with Chikungunya virus. The monitoring system incorporates the three-layer architecture, which is similar to the objective of Jalali et al. (2016), Li et al., 2017, Lyu et al. (2018), and Wu et al. (2018). An early alert-based healthcare system is presented in Zohara, Khan, Bhuiyan, and Das (2017); brain strokes and heart attacks, acute sensitive diseases are examined to minimize the response time of alert generation. Moreover, the system outperforms several essential points, such as execution time, network usage, and energy consumption gain positive results. The limitation of fog computing is that it is implemented at the middle layer to offer services to limited devices or some limited tasks.

A distributed computational model is implemented (Vilela, Rodrigues, Righi, Kozlov, & Rodrigues, 2020) to reduce the communication load and include patient data privacy. Healthcare-centric applications face diverse issues such as scalability, energy consumption, high mobility, and reliability. These issues are addressed in Atalm, Walters, and Wills (2018), Dubey et al. (2015), Elmisery, Rho, and Aborizka (2019), Sahni, Cao, Zhang, and Yang (2017), and Zohara et al. (2017) and to overcome above issues, the distributed intelligent gateways are deployed as mesh fog servers near the network edges. Data processing over a patient's health data is executed at a fog node in the proximity of IoT devices. A fog selection model is presented that picks the nearest fog whenever it is in proximity; in another situation, it uses the shortest path among the available path options. In Rajagopalan, Jagga, Kumari, and Ali (2017), smart e-health gateways are presented; those are accountable for processing the data locally; moreover, further processing is performed over the cloud after compressing the data overhead, which helps minimize the data service delivery time. Resource optimization and sharing are discussed in Sun, Liu, Yue, and Wang (2020) and Sabireen and Neelanarayanan (2021) to maximize the corresponding utility. A new scalable, high computation model is presented in Wu et al. (2017) for performing prognosis and diagnosis. Additionally, it facilitates remote real-time monitoring. In Aazam and Huh (2015) presented a solution for the services such as emergency, health care, and time-sensitive, that needs a vital real-time response. Earlier the CC was used as a computation and storage layer. Microdata centers are included in the form of smart gateways to rectify the above issues. The proposed model performed better than the two-layer architecture.

Related work

This section focused on the numerous services that fog computing can provide. Many services were covered in the preceding section, but this section focuses on complex and time-sensitive applications including health monitoring systems and fall detection services. In Li et al. (2015), end-to-end response time is considered a critical challenge to achieve in the distributed environment of smart devices. Fog computing is formulated as an analysis and data processing layer to improve living that addresses the deficiencies of CC. A real-time, heart attack mobile-based detection system is first introduced in Ali and Ghazal (2017) that facilitated the patient to provide emergency aid from heart attack situations; moreover, SDN controllers are implemented in the architecture to achieve the minimum response time fog assisted health care system. In Aazam and Huh (2015), fog-based emergency alert service architecture is proposed to ensure efficient response time. Similar work on ECG monitoring service is presented in Gia, Rahmani, Westerlund, and Liljeberg (2015) to gain the optimality in cost and data storage. Nastic et al. (2017) and Yi et al. (2015) addressed the issues for cloud systems such as high response time and inappropriate latency; however, those issues are rectified through the new platform as fog-assisted systems. Resource management and data analytics are included and executed in the middle layer. Table 2.1 shows the work in fog driven healthcare systems around delay sensitivity as an important concern:

TABLE 2.1 Fog computing in healthcare: latency, response time, security, resource management, and service cost

	Problem	Techniques
Faaris et al.	As IoT devices are having limited storage and CPU capability, these devices cannot perform analysis at device layer; moreover, analysis at cloud layer is expensive task in terms of network bandwidth, and service response time	A prototype called MIFaaS is proposed that supports delay sensitive application to support fifth generation network
Farahani et al.	eHealth applications such as myocardial infarction (MI) required latency in seconds. Therefore, it is impractical to lean on cloud layer to support analysis of vital signs, and biosignals across the globe	In the proposed eHealth platform, analysis for highly time sensitive data, and decision-making for offloading are performed at near to edges. In addition to, overcome the load at middle layer, data storage, and other computation are performed at cloud layer
Chakraborty et al.	Service delivery time, delay sensitivity is an important concern in healthcare system	A dynamic, distributed fog-based healthcare architecture is presented for time sensitive applications. This model was tested and trained on heart rate datasets
Gia et al. (2015)	ECG signals analyses are required low latency	Analysis and feature extraction of ECG signals including heart rate model is presented. The proposed model achieved more than 90% efficiency of network bandwidth and usage. The results claimed that the model outperformed with very low latency real-time response
Rahmani et al. (2018)	Real-time evaluation of streaming based data in eHealth must require low response time	In addition, fog-based system is addressed to design health care applications by taking care of latency sensitivity issue, low response time, and vast volume of data
Chauhan, Banka, and Agrawal (2021b)	Offloading to the traditional centralized cloud system may not be a perfect solution for latency sensitive mobile network specially emerging IoT paradigm	The idea of fog computing emerges as a solution to provide computational services especially to the IoT applications with major essentiality of latency and high resilience
Barbarossa, Sardellitti, and Lorenzo (2014)	In MCC, the study of offloading is an interesting challenging topic which includes decision-making of execution of mobile requests and allocation of computational resources	A series of offloading mechanism was reviewed by the author to frame mathematical formulation for optimizing network and resource usage with strict constraint to service delivery time

TABLE 2.1 Fog computing in healthcare: latency, response time, security, resource management, and service cost—cont'd

	Problem	Techniques
Wang, Wang, Wang, and Chen (2017)	Jointly addressed the issues such as delay performance, and service cost; while offloading to cloud servers	Opens the door to adopt edge/gateway as a middle-layer computation node to sort out the issues related to energy consumption, cost, and service time
Liu and Fan (2018)	Energy consumption, delay performance, and service cost are highlighted in respect of mobile fog computing system	Optimizing the offloading probability and transmission power for mobile devices is improved to minimize the energy consumption ratio, delay performance, and total service cost
Nan et al. (2017)	Due to limited capacity of data, and computation servers at fog node, a need to balance the data processing and offloading between fog and cloud layer is recommended	Lyapunov optimization is utilized to make control decision on application offloading by adjusting the two way trade-off between avg. response time and avg. service cost
Etemad et al. (2017)	Limitation of cloud services such as latency sensitivity, and other time sensitivity issues	Discrete event system specification (DEVS) is used to perform the simulation
Moosavi (2016)	Security for ubiquitous healthcare devices	End-to-end security schemes are addressed
Sood and Mahajan (2017)	End users information should not be accessed through unauthorized agencies	Data privacy and protection schemes are analyzed at device layer, fog layer, and cloud layer including their integration
Anthi, Williams, Słowińska, Theodorakopoulos, and Burnap (2019)	Cloud service and application are supposed to under threat	eHealth services must have to take efficient step to tackle DoS attack, SQL injection, malicious, unrestricted file uploading vulnerabilities
Bhardwaj and Krishna (2021)	Device layer security threats	Device identity, authentication, authorization, and secure booting of IoT devices are addressed
Liu, Lee, and Zheng (2016)	Multiresource allocation problem in the middle layer	Resource intensive and latency sensitive applications are analyzed on multiresource allocation constraints in cloudlet environment
Jia, Cao, and Liang (2017)	The avg. transmission delay between the mobile devices and central cloud can be relatively long	Cloudlet/fog as middle layer is deployed to gain shortest response time
Sahni et al. (2017)	Existing paradigms do not utilize low-level devices for any decision-making process, even it utilized for communication interoperability	A computing structure is proposed, Edge mesh, which distributes the decision tasks among edge devices. Edge mesh provides many benefits such as distributed processing, fault tolerance, scalability
Guo, Liu, Chang, and Ristaniemi (2018)	Geographically distributed computational resources have caused high latency when connected with remote cloud	Cloudlet-based distributed computational system is designed to solve the latency issue that distributed geographically and connected with remote cloud system
Dastjerdi and Buyya (2016)	Mobile devices faced challenges to opt resources on remote cloud	Multicloudlet placement study is addressed to reduce power consumption and processing latency

Case study

The first section consists of real-time analysis of Magnetic Resonance Imaging (MRI) by developing a general linear model to highlight active brain areas. In contrast, the second part is included subsequent studies that highlighted the implementation and architectural work in healthcare applications. The result analysis of the model is taken and shown in Fig. 2.4. Similarly, network and processing time are analyzed on edge and cloud servers separately (Fig. 2.5). The Microsoft Azure standard A4m (4 vCPUs, 32GB memory), windows OS machine is used to design the cloud server; whereas, for deploying edge server Intel core i5 2.90GHz machine is utilized. The model is integrated with Raspberry pi 3 model B (1.4GHz 64-bit quad-core processor), and the architecture is given in Fig. 2.4).

The model is implemented in Java language and the general linear model is designed in python. The collected outputs from the model are presented in Fig. 2.6A and B. When the number of devices rises, the processing time at the fog server increases rapidly; additionally, network time is lower than cloud network time. The exhibited result demonstrates that fog servers can operate effectively in low-traffic environments. In Gia et al. (2015), the ECG is taken as a case study to analyses various cardiac diseases. A flexible template is implemented to extract the ECG features such as preprocessing of ECG signals, conversion from signals to digital form, feature extraction, and wavelet transformation. Various filters are accumulated for preprocessing of ECG signals to be easily converted into digital data. High computational methods and algorithms are programmed to complete the wavelet transformation step. Various features such as P-R interval, Q-T interval, S-T interval segment, QRS area, and QRS energy are extracted with the help of feature retrieval algorithms. Pandaboard (Ramirez et al., 2012) was selected for this study because of its low power consumption and high-performance chipboard. Pandaboard is equipped with an ARM Cortex-A9 MPCore dual-core processor that is suitable for symmetric multiprocessing, hardware acceleration, and power efficiency. MIT-BIT Arrhythmia database (Moody & Mark, 2001) collects ECG signals for

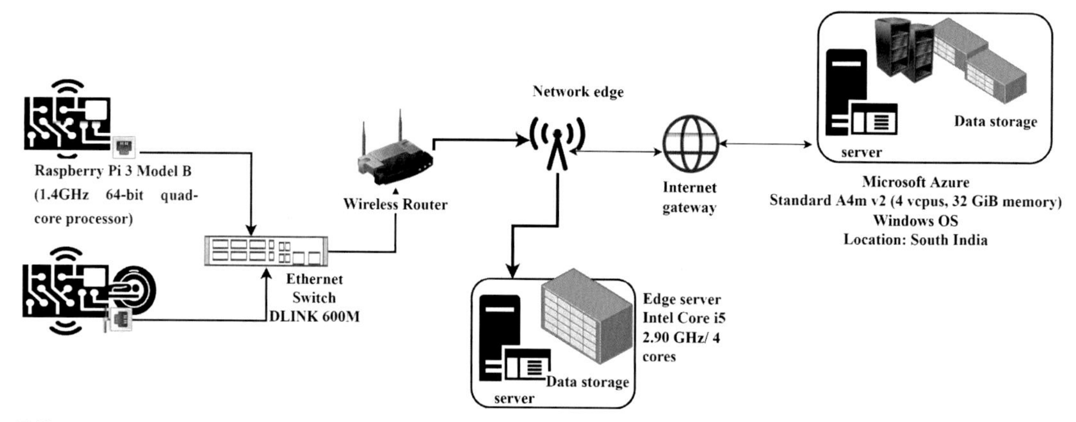

FIG. 2.4 Magnetic resonance imaging computation architecture using cloud and fog-based system. *No permission required.*

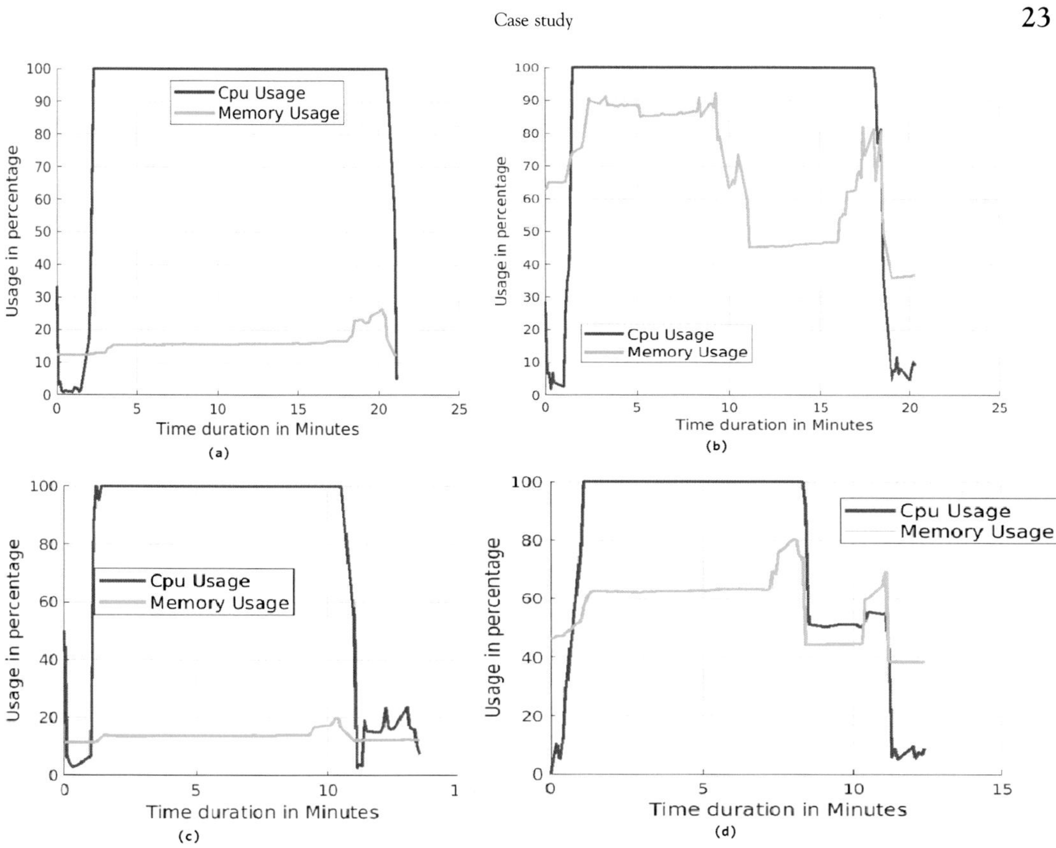

FIG. 2.5 MRI system performance CPU usage and memory usage (A) computation over cloud server with 100 devices (B) computation over the cloud and fog server with 100 devices (C) computation over cloud server with 50 devices (D) computation over cloud and fog server with 50 devices. *No permission required.*

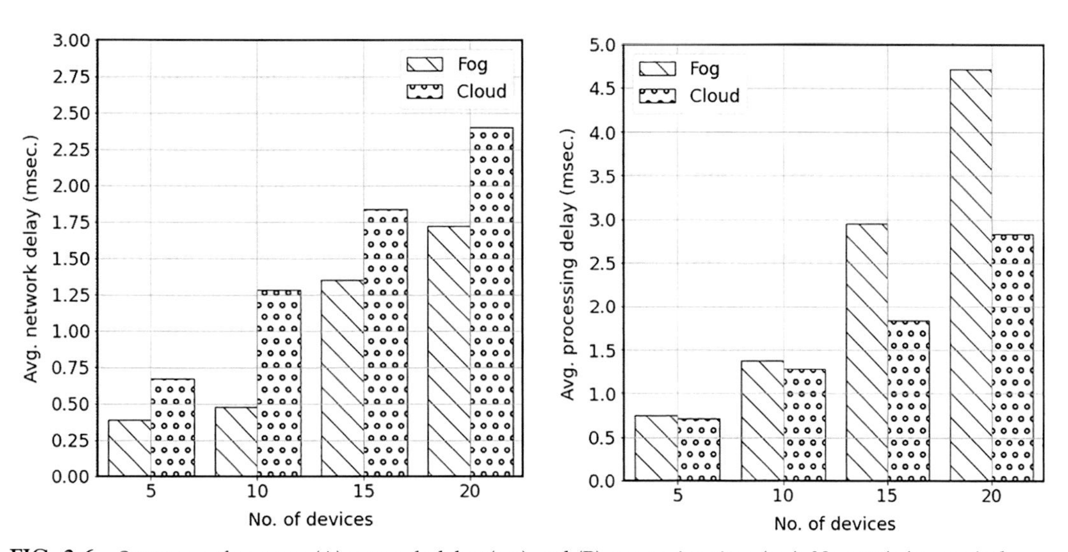

FIG. 2.6 System performance (A) network delay (ms) and (B) processing time (ms). *No permission required.*

digitalization, processing, and feature extraction. The gateway server is utilized to store the stream of ECG data, and afterward, it will be sent for further processing to fog servers. The proposed model has achieved 90% bandwidth efficiency, and results were indicated that the model outperformed for low latency real-time responsive applications. Another case study on ECG stated in Zao et al. (2014), an augmented brain-computer interface incorporated electroencephalogram (EEG) monitoring system is presented and demonstrated. Wireless EEG headsets, smartphones, and extensive devices are utilized to detect brain states in real-time scenarios. This cloud-based model is integrated with multiple fog servers that concurrently execute the classification algorithm in the available fog servers, while further evaluation of the classification model must be performed on the cloud server.

In order to demonstrate the model, the synchronous EEG signals and motion streams of data are tested on a multiplayer online BCI game EEG tractor beam has been played. MINDO-4S wireless EEG headsets and smartphones were used to connect with the local fog server. Raw EEG data streams were sent to the local fog server for real-time processing and prediction. For speech disorders (Monteiro et al., 2016), a fog-driven interface named FIT is proposed to test speech analysis. In this proposed model, smartwatches are used to collect patient clinical speech data with Parkinson's disease; furthermore, smartwatches are utilized to gather patient data when performing an exercise at home. Later on, the FIT layer received data from the smartwatches and perform processing and analysis. The FIT is equipped with an Intel Edison board that supports dual-core, dual threaded Intel Atom CPU at 500 MHZ with 1 GB random access memory. A trial is conducted for a month over six patients, and the acoustic features such as perceptual loudness, zero-crossing rate, spectral centroid, and short-time energy are extracted from the speech signal. FIT processes five speech files to compute the features; in addition, the demonstrated model outperforms the processing time of files. Case studies were carried out on telehealth services, speech disorders, and ECG monitoring. The patients are examined through the data collected from wearable medical sensors. For this study, ECG and activity monitoring sensors are taken. The fog system is implemented over an Intel Edison board with a 22 nm dual-core, dual threaded Atom CPU configuration. The performance is evaluated on ten speech files collected from Parkinson's disease patients (Dubey, 2015). The proposed model outperformed in processing time, and all files were executed below 15 s.

Discussion and open issues

There are many important research challenges beyond those discussed here that need to be addressed for IoT to see its full potential. Specifically, data management, storage, communication, resources management, standardization, operability, security, and data privacy are also important but not discussed here. In the neoteric years of the technology revolution, the use of fog computing systems is exponentially increased to overcome the limitations of traditional architecture. However, fog computing has attracted many researchers and policymakers to adopt the advantage of the system. In concern of the healthcare industry, several limitations

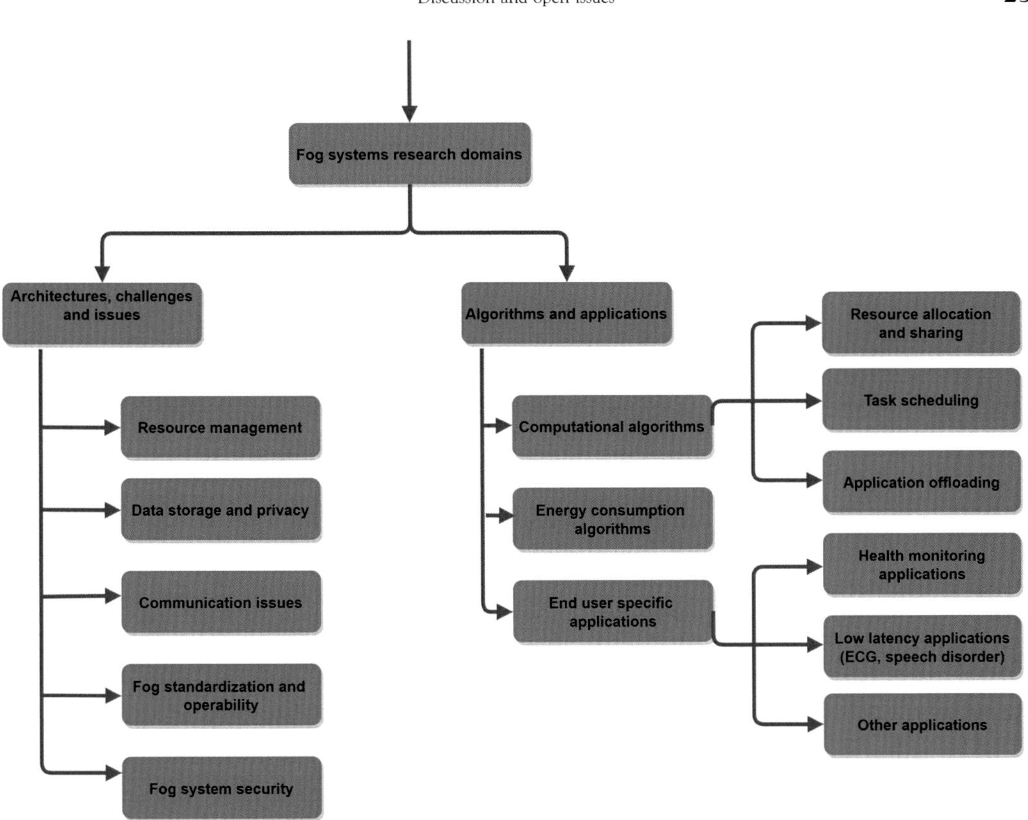

FIG. 2.7 Fog computing challenges, issues, applications, and research domains. *No permission required.*

and challenges are persisting in the fog computing system. The important characteristics of highlighted challenges (Fig. 2.7) are discussed in the subsequent part of this section.

Computation offloading and resource management

Mobile edge computing (MEC) has recently attracted the researcher to solve the computation challenges (Mao, Zhang, Song, & Letaief, 2017; Wang, Yu, Liang, Chen, & Tang, 2017; Zhao, Tian, Qin, & Nie, 2017). MEC has shifted the computation to near the network edge. Edge system has utilized task partitioning and offloading techniques to boost the system performance impressively. Collaborative work to manage offloading and resource management are discussed in Wang and Guo (2019). An energy-efficient offloading model is proposed to minimize total computation cost by selecting optimum offloading decisions and resource allocation. The proposed model is integrated with cloud servers to handle the delay-sensitive tasks. However, the categorization of tasks in offloading decisions can minimize the queuing delay at the fog server. Resource allocation in real-time for heterogeneous tasks is formulated to manage resource balancing, maximize the task completion ratio, and the system's overall throughput (Liu & Fan, 2018). A cost-effective resource management strategy was discussed; however, the problem was addressed to satisfy the QoS requirement, optimum task distribution to a group of edge servers, and virtual machine deployment in the fog systems. Some of the work has highlighted the task distribution and task scheduling issues in fog computing.

When the tasks are executed on a fog system, due to the limiting resources, it is a foremost function to efficiently schedule the tasks (Wang, Wei, Tang, & Fan, 2017). A heuristic multitask scheduling based on an ant colony optimization solution is applied to preserve the task profit, task deadline, task dependency, and load balance. Espouse the end-user experience, especially in the healthcare system; fog computing emerges as a key enabling technology. But mobile edge computing systems are coercion with limited resources that create the platform for efficient resource allocation schemes. Wang, Wei, Liang, and Fan (2018) presented a dynamic tasks assignment and scheduling algorithm based on the Weighted Bi-graph model. Another similar task is offloading, and the resource allocation problem is solved through the fruit fly optimization algorithm (Lin, Pankaj, & Wang, 2018). The proposed model is simulated to improve offloading issue to acquire the nominal energy consumption. The work is validated through the comparison with heuristic ant colony optimization and genetic algorithm.

Data storage and privacy

Data management and privacy of patient data will be a big challenge in the coming years. Cisco reports state that 300% proliferation of IoT devices is expected in the next 10 years. The data are collected through body sensors in the healthcare domain and are partially stored at a fog server; moreover, eHealth data is dynamic. It continuously changes with time and is dependent on the situation of a patient. Different data formats are to be processed distinctively, such as ECG data is collected in XML format, skin diseases are in high definition pictures collected in the form of pixels. To minimize the service delivery time, the required data sets must be stored at the device layer or near to device layer. Limitations of IoT devices are addressed in Wei, Fan, and Wang (2016), where the storing of datasets at the device layer is not a worthwhile option. The confrontation of health care data in volume, velocity, and varieties is closer to the fog system's potentiality to process, analyze, and offload the high precision and correctness in the data collecting from healthcare IoT devices. Data privacy is another dimension of the challenges facing data security. A third party should not access the collected patient data by any means; moreover, it puts the critical data at high risk. In addition to addressing, the prominent research is an open issue in data storage and management.

Operability, standardization, and security threats

Standards and operational guidelines for IoT devices are separately addressed for infrastructure, data protocols, and data communication (Salman & Jain, 2017) While there are no proper standards and regulations for healthcare devices; moreover, a wide range of drafting the regulations and protocols for communication, data sharing and storing, and operations are needed to come in.

Device security and threats

Unauthorized access to devices, attacks, and threats that causes breach the security and repudiation in patient's data are significant challenges for device safety and security. Related work in a similar domain is presented in this section. Distributed denial of service attacks can adversely affect the fog system because of their resource-constrained limitations. An unauthorized user can hit the system to take illegitimate access to system resources, or it might busy the resources of the fog system. These inadequacies of edge/fog computing have opened multidimensional research challenges in security slants. These must be revived to make healthcare devices bug-free for potential users (Zahra et al., 2017). Several IoT applications and services such as the ubiquitous healthcare system, intelligent grid system, and driverless auto vehicles are essentially operated over security and privacy measures. Healthcare body sensors are producing a continuous stream of data to a large extent. Similarly, in the context of location-aware data, a robust security mechanism must be operated to secure data's user identity and privacy. Therefore, the fog security layer should be upgraded to provide significant security and privacy functions to safeguard the data before offloading from device to computation layer (Aazam, Zeadally, & Harras, 2018).

Communication issues

Fog system involves various heterogeneous network services, from device layer to fog layer; similarly, if fog is unable to process the execution of the application at their end, data must be loaded to a cloud server. This phenomenon dramatically increases the bandwidth requirement that requires consistency and is high throughout the network. Optimization of network bandwidth is an open issue (Ge et al., 2010). Data processing services have faced several issues while executed at data centers, such as slow data rate, low network bandwidth, high operational cost, fault tolerance, security vulnerabilities. These challenges adversely impact real-time data analysis; however, it is cumbersome to process massive amounts of clinical data such as continuous stream of ECG signals and health monitoring applications. The emergence of the fifth-generation network structure helps fill the gap as high network bandwidth, low latency to meet the essentiality for the real-time control system, and automation applications (Chaudhary, Kumar, & Zeadally, 2017). Kerberos' reliable authentication system can authenticate healthcare devices to send a continuous stream of data sending from devices to fog (Peng, Zhang, Jiang, Wang, & Wang, 2015). The Network service chaining model is utilized in the fifth-generation network to achieve the best quality of service in the context of data offloading. In addition, data offloading is an effective solution when appropriate resources at fog servers are unavailable. The distribution of computation load to nearby servers is a prominent solution to avail the low latency attributes. Network offloading is considered where either replace the cellular networks with Wi-Fi networks, long-term evaluation networks, or delay-tolerant network.

Due to the limitations of the work, a few of the key research issues are discussed here, whereas many issues and challenges are still open for discussion in the fog computing system.

Conclusion

The integration of fog computing in healthcare 4.0 has started a big leap of computation and intelligent health services. Various health monitoring applications and real-time data analysis are required for a high computational system with low latency in service delivery time. This paper focuses on the research study that brings attention to fog computing in the healthcare system. The leverage of fog systems, such as low latency, localized data processing and analysis, location awareness, and low operating expenses, is exploited to benefit from it. This review article explicitly discusses the techniques and methods that steer the deployment of fog systems in healthcare applications; however, the maximum reward can be achieved. The challenges and issues are collected to make the efforts to deploy healthcare applications and services in the fog system. A cloud and fog-based architecture is discussed in various previous articles, and to support the feasibility and performance of the architecture, the case studies are presented in this article. ECG monitoring and feature extraction and speech disorder applications are considered in this work because of their low latency requirement. The presented work is highlighted on various key issues cognate as architectural, algorithm, and application challenges and issues, which were faced while merging the healthcare 4.0 revolutions into industrial automation. This study's limitations are discussed here, such as resource management and allocation issues in resource-constrained devices.

References

Aazam, M., & Huh, E.-N. (2015). *E-HAMC: Leveraging fog computing for emergency alert service* (pp. 518–523). IEEE.
Aazam, M., Zeadally, S., & Harras, K. A. (2018). Fog computing architecture, evaluation, and future research directions. *IEEE Communications Magazine, 56*(5), 46–52. https://doi.org/10.1109/MCOM.2018.1700707.
Ahmad, M., Amin, M. B., Hussain, S., Kang, B. H., & Cheong, T. (2016). Health fog: A novel framework for health and wellness applications. *The Journal of Supercomputing, 72*, 3677–3695. https://doi.org/10.1007/s11227-016-1634-x.
Akrivopoulos, O., Chatzigiannakis, I., Tselios, C., & Antoniou, A. (2017). On the deployment of healthcare applications over fog computing infrastructure. *COMPSAC* (pp. 288–293). IEEE.
Ali, S., & Ghazal, M. (2017). *Heart attack mobile detection service (RHAMDS): An IoT use case for software defined network* (pp. 1–6). IEEE.
Anthi, E., Williams, L., Słowińska, M., Theodorakopoulos, G., & Burnap, P. (2019). A supervised intrusion detection system for smart home IoT devices. *IEEE Internet of Things Journal, 6*, 9042–9053.
Atalm, H. F., Walters, R. J., & Wills, G. B. (2018). Fog computing and the internet of things: A review. *Big Data and Cognitive Computing*, 1–18.
Baranwal, G., & Vidyarthi, D. P. (2021). Computation offloading model for smart factory. *Journal of Ambient Intelligence and Humanized Computing, 12*(8), 8305–8318. https://doi.org/10.1007/s12652-020-02564-0.
Barbarossa, S., Sardellitti, S., & Lorenzo, P. D. (2014). Communicating while computing: Distributed mobile cloud computing over 5G heterogeneous networks. *IEEE Signal Processing Magazine, 31*(6), 45–55. https://doi.org/10.1109/MSP.2014.2334709.
Bhardwaj, A., & Krishna, C. R. (2021). Virtualization in cloud computing: Moving from hypervisor to containerization—A survey. *Arabian Journal for Science and Engineering, 46*, 8585–8601. https://doi.org/10.1007/s13369-021-05553-3.

Bilal, K., Khalid, O., Erbad, A., & Khan, S. U. (2018). Potentials, trends, and prospects in edge technologies: Fog, cloudlet, mobile edge, and micro data centers. *Computers*, 99–120.
Chaing, M., & Zhang, T. (2016). Fog and IoT: An overview of research opportunities. *IEEE Internet of Things Journal*, 854–864.
Chakraborty, S., Bhowmick, S., Talaga, P., & Agrawal, D. P. (2016). *Fog networks in healthcare application*. IEEE.
Chaudhary, R., Kumar, N., & Zeadally, S. (2017). Network service chaining in fog and cloud computing for the 5G environment: Data management and security challenges. *IEEE Communications Magazine*, 114–122.
Chauhan, N., Banka, H., & Agrawal, R. (2021a). Adaptive bandwidth adjustment for resource constrained services in fog queueing system. *Cluster Computing*, *24*(4), 3837–3850. https://doi.org/10.1007/s10586-021-03378-1.
Chauhan, N., Banka, H., & Agrawal, R. (2021b). Delay-aware application offloading in fog environment using multi-class Brownian model. *Wireless Networks*, *27*(7), 4479–4495. https://doi.org/10.1007/s11276-021-02724-w.
Chen, S., Zhang, T., & Shi, W. (2017). Fog computing. *IEEE Internet Computing*, 4–6.
Dastjerdi, A. V., & Buyya, R. (2016). Fog computing: Helping the internet of things realize its potential. *Computer*, *49*(8), 112–116. https://doi.org/10.1109/MC.2016.245.
Dubey, H. (2015). *EchoWear: Smartwatch technology for voice and speech treatments of patients with Parkinson's disease.*
Dubey, H., Yang, J., Constant, N., Amiri, A. M., Yang, Q., & Makodiya, K. (2015). *Fog data: Enhancing telehealth big data through fog computing* (pp. 14:1–14:6).
Dupont, C., Giaffreda, R., & Capra, L. (2017). *Edge computing in IoT context: Horizontal and vertical Linux container migration.*
Elmisery, A. M., Rho, S., & Aborizka, M. (2019). A new computing environment for collective privacy protection from constrained healthcare devices to IoT cloud services. *Cluster Computing*, 1611–1638.
Etemad, M., Aazam, M., & St-Hilaire, M. (2017). *Using DEVS for modeling and simulating a fog computing environment.*
Farahani, B., Firouzi, F., Chang, V., & Badaroglu, M. (2018). Towards fog-driven IoT eHealth: Promises and challenges of IoT in medicine and healthcare. *Future Generation Computer Systems*, 659–676.
Fratu, O., Pena, R., & Halunga, S. (2015). *Fog computing system for monitoring mild dementia and COPD patients—Romanian case study.*
Gartner. (2021). *Forecast: IT services for IoT, worldwide, 2019–2025.* CISCO. https://www.cisco.com/c/dam/en/us/products/collateral/se/internet-of-things/at-a-glance-c45-731471.pdf. https://www.gartner.com/en/documents/4004741/forecast-it-services-for-iot-worldwide-2019-2025.
Ge, F., Lin, H., Khajeh, A., Chiang, C. J., Ahmed, M. E., Charles, W. B., et al. (2010). *Cognitive radio rides on the cloud.*
Gia, T. N., Jiang, M., Sarker, V. K., Rahmani, A. M., & Westerlund, T. (2017). *Low-cost fog-assisted health-care IoT system with energy-efficient sensor nodes.*
Gia, T., Rahmani, A. M., Westerlund, T., & Liljeberg, P. (2015). *Fog computing in healthcare internet of things: A case study on ECG feature extraction.* IEEE.
Guo, X., Liu, L., Chang, Z., & Ristaniemi, T. (2018). Data offloading and task allocation for cloudlet-assisted ad hoc mobile clouds. *Wireless Networks*, *24*, 79–88. https://doi.org/10.1007/s11276-016-1322-z.
Jalali, F., Hinton, K., Ayrc, R., & Alpcan, T. (2016). Fog computing may help to save energy in cloud computing. *IEEE Journal on Selected Areas in Communications*, 1728–1739.
Jia, M., Cao, J., & Liang, W. (2017). Optimal cloudlet placement and user to cloudlet allocation in wireless metropolitan area networks. *IEEE Transactions on Cloud Computing*, *5*(4), 725–737. https://doi.org/10.1109/TCC.2015.2449834.
Khattak, H. A., Arshad, H., ul Islam, S., Ahmed, G., Jabbar, S., Sharif, A. M., et al. (2019). Utilization and load balancing in fog servers for health applications. *EURASIP Journal on Wireless Communications and Networking*, *2019*(1), 91. https://doi.org/10.1186/s13638-019-1395-3.
Kumari, A., Tanwar, S., Tyagi, S., & Kumar, N. (2018). Fog computing for healthcare 4.0 environment: Opportunities and challenges. *Computers and Electrical Engineering*, 1–13.
Li, W., Santos, I., Delicato, F. C., & Primex, L. (2017). System modelling and performance evaluation of a three-tier cloud of things. *Future Generation Computer Systems*, 104–125.
Li, J., Yuan, D., Palaniswami, M., & Moessner, K. (2015). EHOPES: Data-centered fog platform for smart living. In *Int. telecommunication networks and applications conf.*
Lin, K., Pankaj, S., & Wang, D. (2018). Task offloading and resource allocation for edge-of-things computing on smart healthcare systems. *Computers and Electrical Engineering*, 348–360.
Liu, L., & Fan, Q. (2018). Resource allocation optimization based on mixed integer programming in the multi-cloudlet environment. *IEEE Access*, 24533–24542.

Liu, Y., Lee, M. J., & Zheng, Y. (2016). Adaptive multi-resource allocation for cloudlet-based mobile cloud computing system. *IEEE Transactions on Mobile Computing, 15*(10), 2398–2410. https://doi.org/10.1109/TMC.2015.2504091.

Lyu, X., Vinel, A., Maharjan, S., & Gjessing, S. (2018). Selective offloading in mobile edge computing for the green internet of things. *IEEE Network*, 54–60.

Mao, Y., Zhang, J., Song, S. H., & Letaief, K. B. (2017). Stochastic joint radio and computational resource management for multi-user mobile-edge computing systems. *IEEE Transactions on Wireless Communications*, 5994–6009.

Masouros, D., Bakolas, I., Tsoutsouras, V., Siozios, K., & Soudris, D. (2017). *From edge to cloud: Design and implementation of a healthcare Internet of Things infrastructure.*

Monteiro, A., Dubey, H., Mahler, L., & Tang, Q. (2016). *Fit: A fog computing device for speech teletreatments.* IEEE.

Moody, G. B., & Mark, R. G. (2001). The impact of the mit-bih arrhythmia database. In *Engineering in medicine and biology magazine IEEE* (pp. 45–50).

Moosavi, S. R., Gia, T. N., Nigussie, E., Rahmani, A. M., Virtanen, S., Tenhunen, H., & Isoaho, J. (2016). End-to-end security scheme for mobility enabled healthcare Internet of Things. *Future Generation Computing Systems, 64*, 108–124. https://doi.org/10.1016/j.future.2016.02.020.

Mouradian, C., Naboulsi, D., Morrow, M. J., & Pola, P. A. (2018). A comprehensive survey on fog computing: State-of-the-art and research challenges. *IEEE Communication Surveys and Tutorials*, 416–464.

Nan, Y., Li, W., Bao, W., Delicato, F. C., Pires, P. F., Dou, Y., & Zomaya, A. (2017). Adaptive energy-aware computation offloading for cloud of things systems. *IEEE Access, 5*, 23947–23957. https://doi.org/10.1109/ACCESS.2017.2766165.

Nastic, S., Rausch, T., Scekic, O., Dustdar, S., Gusev, M., Koteska, B., et al. (2017). A serverless real-time data analytics platform for edge computing. *IEEE Internet Computing*, 64–71.

Negash, B., Gia, T. N., Anzanpour, A., Azimi, I., Jiang, M., Westerlund, T., et al. (2017). Leveraging fog computing for healthcare IoT. In A. Rahmani, P. Liljeberg, J. S. Preden, & A. Jantsch (Eds.), *Fog computing in the Internet of Things* (pp. 145–169). Springer.

Neware, R., & Shrawankar, U. (2020). Fog computing architecture, applications and security issues. *International Journal of Fog Computing, 3*(1), 31. https://doi.org/10.4018/IJFC.2020010105.

Nikoloudakis, Y., Panagiotakis, S., Markakis, E., Pallis, E., Mastorakis, G., Mavromoustakis, C. X., et al. (2016). A fog-based emergency system for smart enhanced living environments. *IEEE Cloud Computing*, 54–62.

Ozdemir, A. T., Tunc, C., & Hariri, S. (2017). *Autonomic fall detection system.* IEEE.

Peng, M., Zhang, K., Jiang, J., Wang, J., & Wang, W. (2015). Energy-efficient resource assignment and power allocation in heterogeneous cloud radio access networks. *IEEE Transactions on Vehicular Technology*, 5275–5287.

Rahmani, A. M., Gia, T., Negash, B., & Anzanpour, A. (2018). Exploiting smart e-Health gateways at the edge of healthcare internet-of-things: A fog computing approach. *Future Generation Computer Systems*, 641–658.

Rajagopalan, A., Jagga, M., Kumari, A., & Ali, S. T. (2017). *A DDoS prevention scheme for session resumption SEA architecture in healthcare IoT.* IEEE.

Ramirez, A. H., Schildcrout, J. S., Blakemore, D. L., Masys, D. R., Pulley, J. M., Basford, M. A., et al. (2012). Modulators of normal ECG intervals identified in a large electronic medical record. *Heart Rhythm*, 271–277.

Sabireen, H., & Neelanarayanan, V. (2021). A review on fog computing: Architecture, fog with IoT, algorithms and research challenges. *ICT Express, 7*(2), 162–176. https://doi.org/10.1016/j.icte.2021.05.004.

Sahni, Y., Cao, J., Zhang, S., & Yang, L. (2017). Edge mesh: A new paradigm to enable distributed intelligence in internet of things. *IEEE Access*, 16441–16458.

Salman, T., & Jain, R. (2017). A survey of protocols and standards for internet of things. *Advanced Computing and Communications*, 1–20.

Seferagić, A., Famaey, J., De Poorter, E., & Hoebeke, J. (2020). Survey on wireless technology trade-offs for the industrial internet of things. *Sensors, 20*(2). https://doi.org/10.3390/s20020488.

Shakarami, A., Ghobaei-Arani, M., Masdari, M., & Hosseinzadeh, M. (2020). A survey on the computation offloading approaches in mobile edge/cloud computing environment: A stochastic-based perspective. *Journal of Grid Computing, 18*(4), 639–671. https://doi.org/10.1007/s10723-020-09530-2.

Singh, D., Tripathi, G., Alberti, A., & Jara, A. (2017). *Semantic edge computing and IoT architecture for military health services in battlefield* (pp. 185–190). Institute of Electrical and Electronics Engineers Inc.

Sood, S. K., & Mahajan, I. (2017). Wearable IoT sensor based healthcare system for identifying and controlling chikungunya virus. *Computers in Industry*, 33–44.

Sun, W., Liu, J., Yue, Y., & Wang, P. (2020). Joint resource allocation and incentive design for blockchain-based mobile edge computing. *IEEE Transactions on Wireless Communications*, *19*(9), 6050–6064. https://doi.org/10.1109/TWC.2020.2999721.

Tasic, J., Gusev, M., & Ristov, S. (2016). *A medical cloud* (pp. 400–405). IEEE.

Vilela, P. H., Rodrigues, J. J. P. C., Righi, R.d. R., Kozlov, S., & Rodrigues, V. F. (2020). Looking at fog computing for E-Health through the lens of deployment challenges and applications. *Sensors*, *20*(9). https://doi.org/10.3390/s20092553.

Vora, J., Tanwar, S., Tyagi, S., Kumar, N., & Rodrigues, J. P. C. (2017). *FAAL: Fog computing-based patient monitoring system for ambient assisted living*. IEEE.

Wang, Q., & Guo, S. (2019). Energy-efficient computation offloading and resource allocation for delay-sensitive mobile edge computing. *Sustainable Computing: Informatics and Systems*, 154–164.

Wang, T., Wei, X., Liang, T., & Fan, J. (2018). Dynamic tasks scheduling based on weighted bi-graph in Mobile cloud computing. *Sustainable Computing: Informatics and Systems*, 214–222.

Wang, T., Wei, X., Tang, C., & Fan, J. (2017). Efficient multi-tasks scheduling algorithm in mobile cloud computing with time constraints. *Peer-to-Peer Networking and Applications*, 793–807.

Wang, C., Yu, F. R., Liang, C., Chen, Q., & Tang, L. (2017). Joint computation offloading and interference management in wireless cellular networks with mobile edge computing. *IEEE Transactions on Vehicular Technology*, 7432–7445.

Wang, X., Wang, J., Wang, X., & Chen, X. (2017). Energy and delay tradeoff for application offloading in mobile cloud computing. *IEEE Systems Journal*, *11*(2), 858–867. https://doi.org/10.1109/JSYST.2015.2466617.

Wei, X., Fan, J., & Wang, T. (2016). Efficient application scheduling in mobile cloud computing based on MAX-MIN ant system. *Soft Computing*, 2611–2625.

Wu, D., Liu, S., Zhang, L., Terpenny, J., Gao, R. X., Kurfess, T., et al. (2017). A fog computing-based framework for process monitoring and prognosis in cyber manufacturing. *Journal of Manufacturing Systems*, 25–34.

Wu, H., Sun, Y., & Wolter, K. (2018). Energy-efficient decision making for mobile cloud offloading. *IEEE Transactions on Cloud Computing*, 1–15.

Yi, C., Huang, S., & Cai, J. (2021). Joint resource allocation for device-to-device communication assisted fog computing. *IEEE Transactions on Mobile Computing*, *20*(3), 1076–1091. https://doi.org/10.1109/TMC.2019.2952354.

Yi, S., Hao, Z., Qin, Z., & Li, Q. (2015). *Fog computing: Platform and applications*.

Zahra, S., Alam, M., Javaid, Q., Wahid, A., Javaid, N., Malik, S. U. R., et al. (2017). Fog computing over IoT: A secure deployment and formal verification. *IEEE Access*, 27132–27144.

Zao, J. K., Gan, T. T., You, C. K., Méndez, S. J. R., Chung, C. E., Te Wang, Y., et al. (2014). *Augmented brain computer interaction based on fog computing and linked data*.

Zhao, P., Tian, H., Qin, C., & Nie, G. (2017). Energy-saving offloading by jointly allocating radio and computational resources for mobile edge computing. *IEEE Access*, 11255–11268.

Zohara, F. T., Khan, M. R. R., Bhuiyan, M. F. R., & Das, A. K. (2017). *Enhancing the capabilities of IoT based fog and cloud infrastructures for time sensitive events*.

CHAPTER

3

Contemporary overview of bacterial vaginosis in conventional and complementary and alternative medicine

Arshiya Sultana[a], *Kahkashan Baig*[a], *Khaleequr Rahman*[b], *Sumbul Mehdi*[a], *Md Belal Bin Heyat*[c,d,e], *Faijan Akhtar*[f], *and Atif Amin Baig*[g]

[a]Department of Ilmul Qabalat wa Amraze Niswan (Gynecology and Obstetrics), National Institute of Unani Medicine, Ministry of AYUSH & Rajiv Gandhi University of Health Sciences, Bengaluru, Karnataka, India [b]Department of Ilmul Saidla, National Institute of Unani Medicine, Ministry of AYUSH & Rajiv Gandhi University of Health Sciences, Bengaluru, Karnataka, India [c]IoT Research Center, College of Computer Science and Software Engineering, Shenzhen University, Shenzhen, Guangdong, China [d]International Institute of Information Technology, Hyderabad, Telangana, India [e]Department of Science and Engineering, Novel Global Community Education Foundation, Hebersham, NSW, Australia [f]School of Computer Science and Engineering, University of Electronic Science and Technology of China, Chengdu, Sichuan, China [g]Faculty of Medicine, Universiti Sultan Zainal Abidin, Kuala, Terengganu, Malaysia

Introduction

Vaginal infection is the commonest gynecological infection for which women pursue remedial care. Thirty per cent prevalence rate of abnormal vaginal discharge in reported in India. In developing countries, reproductive tract infections (RTIs) are a silent epidemic that distresses women's lives and epitomizes as the main health condition (Bhat & Begum, 2017). *"The term abnormal vaginal flora (AVF) is used to indicate women with diminished lactobacillary morphotypes and overgrowth of pathogenic microorganisms"* (Donders, 2010).

Computational Intelligence in Healthcare Applications
https://doi.org/10.1016/B978-0-323-99031-8.00024-7

Vaginitis is an inflammation of the vagina, commonest RTIs usually categorized by abnormal vaginal discharge containing numerous white blood cells, vulvar irritation, vulvar itching, malodor, dysuria dyspareunia, and vaginal congestion (Ranjit, Raghubanshi, Maskey, & Parajuli, 2018). Approximately two-thirds of female's experience vaginitis at least once during their lifetime (Najafi et al., 2019) and recurrence of symptoms are seen within a year in 50% of women (Bagnall & Rizzolo, 2017). It was reported that vaginal candidiasis, bacterial vaginosis (BV), and *Trichomonas vaginalis* lead to secondary infection in 90% of cases (Najafi et al., 2019). Trichomonal infection and bacterial vaginosis (BV) often exist simultaneously and BV is also diagnosed in 60% of patients with this disease (Abdali et al., 2015).

BV is the commonest vaginal infection in reproductive age that causes abnormal vaginal discharge (Abdali et al., 2015; Hay, 2017; Ranjit et al., 2018) seen in the Gynecological OPDs, if not treated, it leads to severe complications (Rao, Pindi, Rani, Sasikala, & Kawle, 2016). The term "vaginosis" is preferred not "vaginitis" as there is no vaginal inflammation (Ranjit et al., 2018). It is related to the change of vaginal flora leading to a noteworthy absence or reduction of *lactobacillus* species that produces normal hydrogen peroxide and an upsurge in anaerobic bacteria (Najafi et al., 2019). The aetiological organisms of BV are *Bacteroides ureolyticus*, *Clostridium* species, *Gardnerella vaginalis*, *Prevotella* species, *Porphyromonas* species, *Peptostreptococcus* species, *Mycoplasma hominis*, *Mobiluncus* species, and *Fusobacterium* species (Muthusamy & Elangovan, 2016). The BV clinical manifestations are vaginal discharge, mostly with a fishy odor, that is exacerbated usually after sexual intercourse (Najafi et al., 2019). Metronidazole is a *"first line"* for the management of bacterial vaginosis (Abdali et al., 2015). *"BV is a polymicrobial syndrome categorized by the replacement of normal vaginal lactobacilli by anaerobic bacteria,* (*Atopobium vaginae*, *Mobiluncus* spp., *G. vaginalis*, *Prevotella* spp., and *M. hominis*) (Muthusamy & Elangovan, 2016). It is also known as *Gardnerella* anaerobic vaginitis or nonspecific vaginitis or *Corynebacterium* vaginitis or *Haemophilus* vaginitis (Schorge et al., 2014).

A systematic literature exploration in PubMed, EBSCO, Scopus, and other electronic databases for bacterial vaginosis in bacterial vaginosis an overview, "risk and bacterial," "epidemiology and bacterial vaginosis," "Diagnosis of bacterial vaginosis," "evidence-based studies for management of BV" "herbs and BV," "complementary and alternative medicine in BV." All articles were methodically appraised without any language or time restriction. A data-charting in electronic form was mutually established by the first author to decide which variables to extract. The first author (AS) and other authors continuously updated the data-charting form. A total of 124 articles and textbooks were retrieved to collect the information regarding BV. The articles that were included in the manuscripts encompassed research articles, review articles, and textbooks. This review was planned under the following headings: Introduction, epidemiology, aetiopathogenesis, clinical characteristics, diagnosis, investigations, management and complementary and alternative medicine.

Epidemiology

Prevalence

The incidence of BV varies from 4% to 61% from asymptomatic to symptomatic women who visit sexually transmitted disease clinics (Baery et al., 2018) that leads to vaginal discharge and odor that affect overall 29% of all women (Bagnall & Rizzolo, 2017; Ellington & Saccomano, 2020). Depending on the geographical setting, the prevalence variation is observed (Ellington & Saccomano, 2020; Tidbury, Langhart, Weidlinger, & Stute, 2020). The global prevalence of BV was estimated to vary between 22% and 50%; 20%–49%, 11% and 15%–30% in Africa, the UK and the USA, respectively, as it is more commonly prevalent in developing countries (Najafi et al., 2019). The prevalence in Caucasian women is comparatively higher in African and American Blacks than in Western Europe (5%–15%) (Donders, Zodzika, & Rezeberga, 2014). BV affects >21 million women in USA (Ellington & Saccomano, 2020). *"National Health and Nutrition Survey in the United States"* stated a 29% incidence of BV in between 17- and 49-year-old women whereas its prevalence was 12% in an Australian study (Vodstrcil et al., 2013). The reviews show that though BV is observed worldwide, however, the prevalence is more common in developing countries.

A systematic review of 1692 articles described the worldwide epidemiology of BV and its disparities between ethnic groups within countries, and between countries, diagnosed by Nugent scoring and 86 articles were included. They observed that its prevalence was diversely significant among ethnic groups in South America, Europe, North America, Asia, and the Middle East. The prevalence was low in Africa whereas some areas of Asia and Europe showed high prevalence (Kenyon, Colebunders, & Crucitti, 2013). This shows that BV is more common in low socioeconomic status.

The prevalence of BV as per the study conducted in Iran was 16.2% among reproductive-aged women (Pazhohideh et al., 2018). Peebles et al. in their systemic review and meta-analysis found that globally the prevalence ranges from 23% to 29% in different regions such as 24% in East Asia, and Caribbean, 23% in Europe and Central Asia (Peebles, Velloza, Balkus, McClelland, & Barnabas, 2019) and it varies from 17.8% to 63.7% in India as per *National AIDS Control Programme III* (Rao et al., 2016).

Seth et al. in Karnataka (India) conducted a study on 250 reproductive-aged women, they found the prevalence rate of BV was 44.8% whereas 40.66% in Chhattisgarh and 33% in 1995 in Barbados (Seth, Chaitra, Vaishnavi, & Sharath Chandra, 2017). In India, the prevalence of BV in Goa (2001), Mysore (2005), and Chennai (2002) was 17.8%, 19.1%, and 24.6%, respectively (Kenyon et al., 2013). The prevalence was 48% in Hyderabad (Rao et al., 2016). However, it is tough to decide the actual prevalence of BV because asymptomatic episodes are frequent and happen during menstruation (Darmayanti, Murti, & Susilawati, 2017).

Cost of BV

A meta-analysis was conducted for the global burden of BV prevalence in the USA. *"They predicted direct medical costs for treatment of symptomatic BV, a causal relationship, was also estimated to find the potential costs of BV-associated preterm births and human immunodeficiency virus cases."* The US $4.8 billion annual global economic burdens for treating symptomatic BV was estimated and the same was thrice times approximately when costs of *"BV-associated preterm births* and *human immunodeficiency virus"* cases were included. The authors concluded that globally its high prevalence, with an associated high economic burden and noticeable racial differences in prevalence. Moreover, the remarkably high cost of BV-related sequelae places the interest to prioritize research to comprehend possible fundamental relationships between BV and its adverse health outcomes (Peebles et al., 2019).

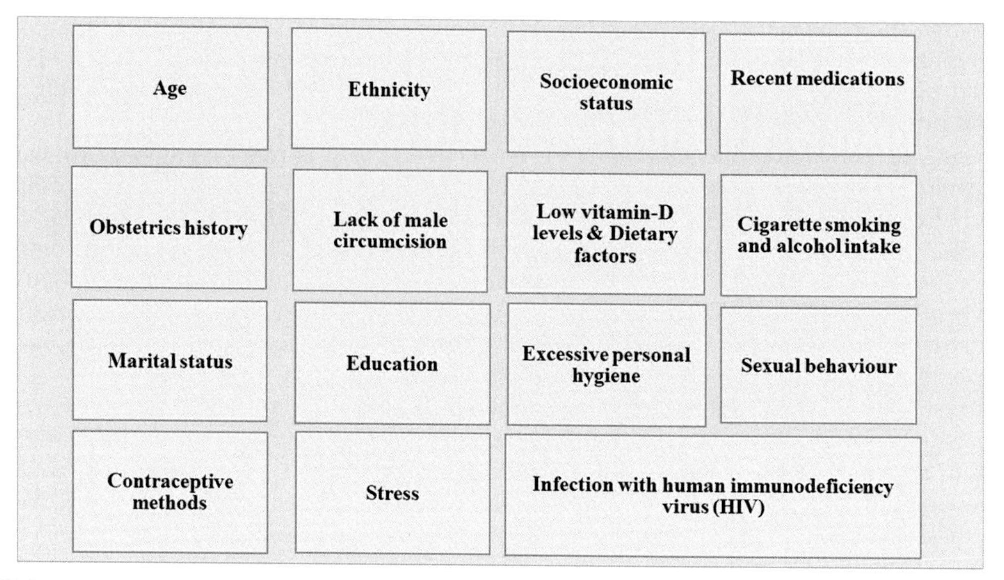

FIG. 3.1 Risk factors of bacterial vaginosis.

The risk factors are summarized in Fig. 3.1.

- *Age*: A study in Nigeria, found that the prevalence of BV is the most prevalent among the 26–30 years of age group (Garba, Zabaze, Tabitha, James, & Makshwar, 2014). Rao et al. that BV was most common in 24–29 years (Rao et al., 2016) because of high sexual acquaintance, and at this age is the utmost reproductively active age group. Hence, it is more common in the childbearing age group (Bhilwar, Lal, Sharma, Bhalla, & Kumar, 2015; Pazhohideh et al., 2018; Ranjit et al., 2018). However, a study showed that above 45 years of age BV was highest because of alkaline pH in women, the decline in estrogen level, as a result, an ideal state is created for the growth of anaerobic bacteria in place of *lactobacilli* (Bitew, Abebaw, Bekele, & Mihret, 2017; Mascarenhas et al., 2012) (Fig. 3.1).

Another study showed that 15 to 24 years' young women had a BV prevalence of 41.5% and between 47.8% and 60.0% in the 25 years and above. The authors concluded the adjusted odds ratio showed no significant relation between BV and age (Bitew et al., 2017).

Seth et al. in Karnataka (India) conducted a study on 250 reproductive-aged women, they found the prevalence rate of BV is highest among 20–29 years of women (61.6%) (Seth et al., 2017).

- *Marital status*: Unmarried women (100%) were more predisposed to BV when compared with married women (24.2%) (Ranjit et al., 2018). The tight clothing, improper perineal care, inadequate care toward menstrual hygiene, food habits, lifestyle change, and sedentary factors are probable causes for the gaining of BV in unmarried women (Ranjit et al., 2018). Another study observed that the unmarried study women (53.8%) had a high prevalence of BV compared to married women (44.8%) or divorced (50.0%) with no statistical significance (Bitew et al., 2017).

- *Ethnicity*: Black or Hispanic ethnicity showed a higher rate of BV (Kenyon et al., 2013; Tidbury et al., 2020). A systematic review and meta-analysis found that within North America, compared to another racial group (23% white and 11% Asian, $P < 0.01$) the prevalence rate was higher among Hispanic and black women 31% and 33%, respectively (Peebles et al., 2019). A study in India showed that a higher prevalence of BV was seen among another ethnicity (57.1%) followed by 50% in Dalit, 25% in Brahmin, 22.7% in Janajati, and 19.4% in Chhetri 19.4% amongall women (Ranjit et al., 2018).
- *Education*: A study showed that 35.9% of women with a college-level education had a lower prevalence whereas high school education had a 44.7%–55.3% prevalence of BV (Bitew et al., 2017). Another study showed that the highest number of total women (21.9%) were illiterate (Ranjit et al., 2018).
- *Socioeconomic status*: The employment status, occupations (Ranjit et al., 2018), and low socioeconomic status are also risk factors (Bagnall & Rizzolo, 2017; Kenyon et al., 2013; Tidbury et al., 2020). Ranjit et al., in their study, concluded that low socioeconomic status/ poverty is the risk factor. They observed that illiterate, women between the age group 30 and 40 years, and farmers were more predisposed to BV (Ranjit et al., 2018). However, another survey conducted in Nepal showed the highest rate of BV in Dalit (Manandhar, Sharma, Pokharel, Shrestha, & Pradhan, 2005). Other studies also highlighted that the rate of BV is higher in low socioeconomic status (Bhat & Begum, 2017; Bhilwar et al., 2015; Chaudhary et al., 2012). This indicates that poverty and ignorance, along with the availability of few health care facilities played an important role in women morbidity in India (Chaudhary et al., 2012). One study showed smoking status, partner age, occupation, and education were not linked to BV (Madhivanan et al., 2008).Another study showed that the BV was higher among farmers (38.9%) whereas lowers in the business class (14.7%) (Ranjit et al., 2018).
- *Excessive personal hygiene*: Excessive personal hygiene, e.g., vaginal douching regularly (Bagnall & Rizzolo, 2017; Tidbury et al., 2020). The use of marketable douching products modifies alters the vaginal flora and changes the vaginal pH and creates a more vulnerable environment for pathogen overgrowth thereby upsurging the threat of BV (Ellington & Saccomano, 2020). It was noted that 32.1% of cases who had a habit of daily vaginal douching showed BV infection, confirming that vaginal douching represents a risk factor (Ranjit et al., 2018).
- *Obstetrics history*: The prevalence of BV was higher in women with abortion history (53.8%) in comparison to women with no abortion history (46.8%) (Bitew et al., 2017). Seth al. found that prevalence (83.9%) of BV was higher in parity between 1 and 3 (Seth et al., 2017).
- *Sexual behavior*: Early age of sexual intercourse (Ranjit et al., 2018; Schorge et al., 2014), number and frequency of sexual intercourse (Ranjit et al., 2018; Tidbury et al., 2020), multiple sex partners, homosexuality in women were both are affected) (Bagnall & Rizzolo, 2017; Donders, 2010), and working as a sex worker. Vaginal douching and sexual behavior are modifiable risk factors among the most prominent risk factors. Several studies have shown that sexual activity is not a precondition for BV as the incidence of BV in sexually inactive females or virgins was also observed (Ranjit et al., 2018). A meta-analysis and systematic review found a significant link between multiple and new sexual female and male partners. Further, unprotected sexual encounters reduction may decrease incident and recurrent infection (Fethers, Fairley, Hocking, Gurrin, & Bradshaw, 2008).

A study reported that those who described more than one-lifetime male sex partner had a 43.4% rate of BV women whereas, 58% who stated ≥4-lifetime male sex partners had BV (Bitew et al., 2017).

- Recent medications (antibiotics and immunosuppressants) (Ranjit et al., 2018; Tidbury et al., 2020) decreased estrogen production in the host, spermicide use, and STDs (Ranjit et al., 2018).
- *Contraceptive methods*: The intrauterine device (IUCD) increases whereas barrier methods may reduce the risk of BV, though, accessible data are debatable. In Sweden, research showed that OC, as well as condom users, had a lower risk of BV and IUCD users showed no relationship with BV. Another study conducted in Italy on 1341 women, showed similar results (Chiaffarino, Parazzini, De Besi, & Lavezzari, 2004). Not using condoms (Bagnall & Rizzolo, 2017) but this finding is contradicted by one study that showed 42.9% who used condoms daily had a higher risk of BV than that 29.2% who used condoms occasionally (Ranjit et al., 2018). One possible reason for that condom may irritate that leads the bacteria to invade the vagina and increases the risk of BV. The lower prevalence of BV was noted in women who used a COC pill is supposed to foster the effect on the *lactobacilli* in the vagina (Bagnall & Rizzolo, 2017). A study noted an indirect risk factor for BV, and poor genital hygiene and irregular condoms use is adaptive to the development of this ailment (Mascarenhas et al., 2012). Seth et al. conducted a study that found 32.14% of IUCD users had BV (Seth et al., 2017). Vodstrcil et al. in their meta-analysis reported the link between hormonal contraception (HC) and BV. They found that hormonal contraceptive use had reduced the risk of BV (Vodstrcil et al., 2013). The maximum prevalence of BV was observed reported in women who had tubectomy (50%) (Ranjit et al., 2018).
- Cigarette smoking and alcohol intake: "BV is the consequences of altering vaginal microflora by the chemical constituents present in cigarette smoke or these constituents probably reduces local immunity by reducing Langerhans cells in the cervical epithelium (Ranjit et al., 2018; Schorge et al., 2014; Tidbury et al., 2020). Further, smoking raises vaginal amines and helps an antiestrogen environment that prone women to BV (Ellington & Saccomano, 2020).Therefore, daily smokers have a higher rate of BV compared to occasional smokers (Ranjit et al., 2018). A study found that daily alcohol users had the highest rate of BV (38.5%) whereas 24% were nonalcoholic users and 16.7% were infrequent alcohol users (Ranjit et al., 2018). Smoking cigarettes and alcohol intake cause a decrease in *lactobacilli*, therefore, increases the prevalence of BV.
- Stress (Kenyon et al., 2013; Ranjit et al., 2018).
- Low vitamin-D levels (Kenyon et al., 2013).
- Infection with human immunodeficiency virus (HIV) (Tidbury et al., 2020).
- *Lack of male circumcision*: The higher rate of BV was reported in women who had uncircumcised partners perhaps as foreskin enables the transmission of microorganisms during intercourse. Further, male sexual partners probably help to carry anaerobic bacteria that cause BV (Kenyon et al., 2013; Mascarenhas et al., 2012).
- Dietary factors (Kenyon et al., 2013) (Fig. 3.1).

A study conducted in Brazil showed that tobacco, illegal drug use, and alcohol showed relation with BV ($P = 0.02$). Further, patients with BV stated that they had a higher number of sexual partners than those without the disease ($P = 0.01$) (Mascarenhas et al., 2012).

Another study reported that women between the age group 30 and 40 years, illiterate, unmarried and farmers were more predisposed to BV (Ranjit et al., 2018). The aforementioned various risk factors are preventable such as stress, dietary factors, preventing HIV infection, smoking and alcohol intake, sexual behavior, and excessive personal hygiene.

Pathophysiology

"*The female lower genital tract, comprising of vagina and ectocervix is a biological niche where several aerobe and anaerobe microorganisms in a dynamic balance exist*". This ecosystem relic is dynamic with fluctuations in composition and structure that are influenced by infections, sexual activity, menarche, age, menstrual cycle timing, medication, methods of contraception, pregnancy, and hygiene (Mastromarino, Vitali, & Mosca, 2013). The primary mechanisms by which *lactobacilli* employs their protective functions are: struggle with other microorganisms for vaginal epithelium adherence, immune system stimulation and nutrition; reducing vaginal pH by producing lactic acid (Mastromarino et al., 2013); *lactobacilli* produce H_2O_2, lactic acid antimicrobial substances and bacteriocin to provide a healthy ecology in the vagina and defeating pathogens growth and protect against infection caused by a microorganism (Masoudi, Miraj, & Kopaei, 2016; Mastromarino et al., 2013; Ranjit et al., 2018). This offers an indigenous defense action by preventing the growth of pathogenic and opportunistic organisms (Ranjit et al., 2018). Normal vaginal pH in adult women is 3.8–4.4 (Laue et al., 2018). Alkaline pH favors the growth of microorganisms leading to BV and these bacteria stick to the surface and form the epithelial cells clue cells (Khan, Shah, Gautam, & Patil, 2007).

Bacteria secrete enzymes that destroy the vaginal and cervical defensive layer of the epithelium (Masoudi et al., 2016). In women with a vaginal infection the percentage of H_2O_2 drops to 5% as normal vaginal flora is affected and lactobacilli decreases (Mastromarino et al., 2013). Diverse associations were observed in four clinical Amsel's criteria of BV with different bacterial species. *Lactobacillus amnionii* and *Eggerthella* were the only BV-related bacteria that were positively connected with all four clinical Amsel's criteria (Laue et al., 2018). Clusters of coccobacillus organism adherent to squamous vaginal epithelial cells giving them a granular appearance are called clue cells (Khan et al., 2007).

Clinical features

A substantial fraction of cases are asymptomatic, or had only foul-smelling homogenous vaginal discharge with no inflammatory complaints which were more noticeable after unprotected intercourse (Ranjit et al., 2018). Women usually complain of vaginal discharge, which is malodorous, and occasionally causes irritation (Bagnall & Rizzolo, 2017) which affects activity in society and sexual life. The overproduction of vaginal anaerobes regulates an increased production of amines (cadaverine, putrescine, and trimethylamine) that turn out to be volatile at alkaline pH, viz., during the menstrual cycle and after intercourse and contribute to the representative malodor of the vaginal discharge (Mastromarino et al., 2013). BV has a distinctive offensive "fishy smell" (Ellington & Saccomano, 2020). BV is often supposed with >4.5

vaginal pH. Other probable causes for increased pH include trichomoniasis, atrophic vaginitis, and desquamative inflammatory vaginitis (Ellington & Saccomano, 2020). A study reported 29.9% of patients had foul-smelling discharge whereas 14.3% of patients had a nonfoul discharge. The thin discharge was highly prevalent (39.8%) in the consistency of vaginal discharge (Ranjit et al., 2018).

A study showed most frequently observed clinical feature in BV patients was vaginal discharge (45%) 10% of patients showed dysmenorrhea, genital itching, and genital lesions (Mascarenhas et al., 2012).

Excessive abnormal vaginal discharge was noted in 30% of patients whereas 18.7% of patients reported scanty discharge (Ranjit et al., 2018).

Another study showed that vaginal discharge and malodour was the commonest symptoms of BV that was noted in all cases followed by itching and dysuria (Rao et al., 2016).

Bimanual examination

Demonstrates a thin, homogenous gray discharge that covers the vaginal wall with fishy odor sometimes is noticeable and bimanual examination is frequently normal, the vagina is not erythematous (Bagnall & Rizzolo, 2017).

Diagnosis

The diagnosis of BV includes Nugent score system (NSS), Amsel's criteria, Spiegel's criteria, anaerobic culture, Hays/Ison system, Schmidt's scoring system, gas–liquid chromatography, sialidase activity, and DNA probes for *G. vaginalis* (Muthusamy & Elangovan, 2016). Like the NSS, Spiegel's and Hays/Ison methods are Gram stain-based techniques that show results so similar to NSS and are substitutable with the NSS (Kenyon et al., 2013). Spiegel's and Nugent's criteria assess the vaginal discharge's normal flora in the Gram-stained smears (Rao et al., 2016). One additional diagnostic method BV, the Blue test when compared to NSS (Kenyon et al., 2013), detects "*sialidase activity and uses colour change technology*" and results are ready within 10 min but it does not detect *candida* and *trichomonas* species (Bagnall & Rizzolo, 2017). This test has 92–100% sensitivity and 98% specificity (Kenyon et al., 2013). In diagnosing BV, NSS, and Amsel's criteria are considered equally efficacious (Rao et al., 2016). Recently, *quantitative polymerase chain reaction (qPCR) assays* have been established to diagnose BV such as *A. vaginae.* Haahr et al., in their systemic review, included different qPCR panels in diverse settings, with 95% to 100% sensitivities and 90% to 95% specificities (Haahr et al., 2016).

Nugent scoring system (NSS)

Nugent scoring system was developed by Nugent (Mohammadzadeh, Dolatian, Jorjani, & Majd, 2014). It is a reproducible, reliable, standardized, and widely used method of assessing the presence or absence of BV (Kenyon et al., 2013). "*The diagnosis of BV is defined as a Nugent score of ≥7 of 10* (Kenyon et al., 2013). *A score of 0–3 is considered normal, 4–6 is intermediate, and*

*7–10 indicates bacterial vaginosis (*Bagnall & Rizzolo, 2017*). Large Gram-positive rods were taken as lactobacillus morphotypes; small Gram-negative to Gram-variable rods was considered as G. vaginalis and Bacteroides* spp. *morphotypes; curved Gram variable rods were considered as Mobiluncus* spp. *Morphotypes*" (Rao et al., 2016; Seth et al., 2017). A study showed that the 97% sensitivity and 98%, specificity of Nugent's score compared to Amsel's criteria (Coppolillo, Perazzi, Famiglietti, Eliseht, & Vay, 2003). Ranjit et al. found the rate of BV was 24.4% by Nugent's method (Ranjit et al., 2018) similar results in India were reported (Modak et al., 2011). In contrast, a prevalence rate of 48.6% by Nugent's scoring was also reported (Bitew et al., 2017).

Amsel's criteria

"*Amsel's composite criteria include the presence of thin homogeneous vaginal discharge, pH of the vagina being > 4.5, the presence of clue cells in a wet mount of the vaginal discharge and a positive whiff test (fishy odour of vaginal fluid after adding 10% KOH)*". According to Amsel's, if 3 of the 4 criteria are positive, the patient has bacterial vaginosis (Hillier et al., 2017; Mastromarino et al., 2013; Muthusamy & Elangovan, 2016; Ranjit et al., 2018). Rafiq et al. reported that the sensitivity and specificity, positive and negative predictive value was 85.3%, 95.6%, 86.4%, and 95.2%, respectively, of Amsel's clinical criteria (Rafiq, Nauman, Tariq, & Jalali, 2015). Amsel's criteria are as good as NSS as it is reliable, easy, cost-effective, fast, and can be used for precise and fast treatment in OPD (Nawani & Sujatha, 2013).

Mohammadzadeh et al. compared Amsel's criteria with those of the NSS and the Kappa coefficient was 0.8, confirming the reliability of Amsel's and NSS for BV (Mohammadzadeh, Dolatian, Jorjani, & Majd, 2014).

Rao et al. showed that in comparison with "Nugent's criteria the 78.72%, sensitivity, 92.35%, specificity, 75.51% PPV, and 93.54% NPV of Amsel's criteria. They concluded that Amsel's criteria and NSS are cost-effective, easy, simple, fast, and reliable, OPD procedures (Rao et al., 2016) (Table 3.1).

Vaginal pH determination

The pH of vaginal discharge has 97% sensitivity 26% specificity and it is the most sensitive among Amsel's criteria (Khan et al., 2007). A study found pH > 4.5 among women of BV (Mohammadzadeh, Dolatian, Jorjani, AlaviMajd, & Borumandnia, 2014).

TABLE 3.1 Amsel criteria and Nugent scoring for diagnosis of bacterial vaginosis.

Amsel criteria	**Nugent score**	
Thin homogeneous vaginal discharge	**Criteria**	**Score**
pH of the vagina being >4.5	Normal	0–3
Presence of clue cells in a wet mount of the vaginal discharge	Intermediate	4–6
A positive whiff test	Bacterial vaginosis	7–10

Whiff test

The whiff test has a 33.9% sensitivity and 86.9% specificity of (Khan et al., 2007).

Presence of clue cells

Khan et al. describe the 81% sensitivity and 91% specificity of more than 20% of clue cells on the wet mount for diagnosis of BV (Khan et al., 2007). Clue cells have 76.7% and 92.4% sensitivity and specificity, respectively; therefore, it is the most appreciated criterion for diagnosis of BV (Mohammadzadeh, Dolatian, Jorjani, AlaviMajd, & Borumandnia, 2014). Rafiq et al. reported the sensitivity, specificity, positive, and negative predictive value for clue cells was 89%, 79.6%, 58.5%, and 95.6%, respectively (Rafiq et al., 2015).

A study showed 47% were diagnosed to have BV by Amsel Criteria, 48% of patients had vaginal discharge, 45% and 47% had a clue and positive whiff test, respectively, and 44% had pH > 4.5, and 48% were diagnosed by NSS (Rao et al., 2016).

A cross-sectional comparative study reported Amsel's criteria detected BV in 35.38% whereas with Nugent score 14.61% were identified and 16.15% were identified by culture (Muthusamy & Elangovan, 2016).

Real-time PCR test

A longitudinal study reported a sensitivity of 91%, specificity of 97%, the positive predictive value of 91% and a negative predictive value of 97% for molecular test in comparison with modified Ison criteria (Breding, Selbing, Farnebäck, Hermelin, & Larsson, 2020).

Pap smear

According to CDC 2015 guidelines, vaginal swab cultures are not used (Ellington & Saccomano, 2020).The Pap smear should not be used to initiate treatment of BV as it is not a reliable test and has low sensitivity and specificity (Bagnall & Rizzolo, 2017).

Complications

Recurrent BV can cause psychological distress to women because of malodorous odor and avoid sexual relations. The most frequently BV associated gynecological complications are PID, STDs, AIDS, genital herpes (Donders, 2010), endometritis, salpingitis, and tubo-ovarian abscess (Laue et al., 2018).

During pregnancy, the preterm delivery and spontaneous abortion risk increase by more than two and nine times respectively in BV. A meta-analysis reported more than four times the risk of BV before 20 weeks of gestation and seven times when BV is identified before 16 weeks of gestation. However, the pathophysiological mechanisms remain uncertain (Subtil et al., 2018).

Clinical studies have reported that BV is linked to adverse pregnancy outcomes such as spontaneous miscarriages, postabortion endometritis, posthysterectomy vaginal cuff

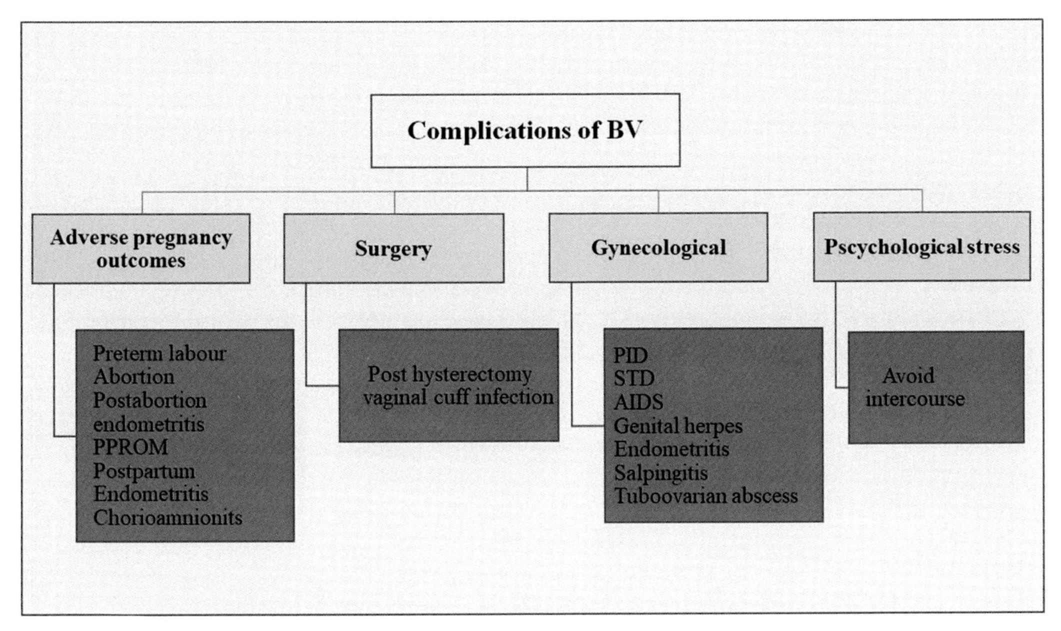

FIG. 3.2 Complications of bacterial vaginosis.

infection (Donders, 2010; Rao et al., 2016), preterm labour (El-Saied, Amer, Elbohoty, Saad, & Mansour, 2016; Homayouni et al., 2014), preterm prelabour rupture of the membrane (PPROM) (El-Saied et al., 2016; Homayouni et al., 2014),chorioamnionitis, postpartum infection (El-Saied et al., 2016; Rao et al., 2016), and postpartum endometritis (Homayouni et al., 2014; Laue et al., 2018). Approximately 30% of spontaneous preterm births are caused due to genital tract infection as micro-organisms from the vagina and cervix can enter through the uterine cavity and infect the membranes, placenta, and fetus (McDonald et al., 1997). BV probably facilitate the transmission of STIs such as *Chlamydia trachomatis*, *Neisseria gonorrhoeae*, *T. vaginalis* and herpes simplex virus type 2 (HSV-2) and HIV acquisition (Tidbury et al., 2020) (Fig. 3.2).

Management

Preventive measures

Asymptomatic women usually do not need treatment, unless they are pregnant, the recurrence of BV can be prevented by counseling women to stop douching, cessation of smoking, restraining from multiple sexual partners, consistent condom use, and use of combined oral contraceptives (Bagnall & Rizzolo, 2017).

Treatment

Though BV is a self-limiting infection, however, frequent recurrence and chronicity are reported and metronidazole, clindamycin and tinidazole are commonly used antibiotics (Donders, 2010; Donders et al., 2014). Metronidazole (MTZ) is the first-line antibiotic to treat

BV, has shown a 70%–80% success rate but within 3 months 30% recurrence rate was reported. MTZ is associated with several adverse effects includes nausea, metallic taste, transient neutropenia (Najafi et al., 2019), vomiting, dizziness, fatigue, constipation, swelling of the tongue, abdominal pain, vulvar itching, and uterine cramps, rash, and darkening of the urine (Abdali et al., 2015). It is harmless even during the first trimester and does not affect the foetus. Further, it is safe in breastfeeding women (Bagnall & Rizzolo, 2017). Clindamycin is the second choice in the treatment of BV and has some adverse effects like nausea, colitis, abdominal cramps, vomiting, diarrhea, and liver enzyme derangement (Najafi et al., 2019). Clindamycin is considered safe in both oral as well as vaginal forms for use in pregnant women. Clindamycin vaginal cream is less absorbable systemically than oral formula hence, useful to breastfeeding women (Bagnall & Rizzolo, 2017). *Recommended regimens are* (Donders, 2010*) (1) oral metronidazole 500 mg BID for 5 days; (2) Oral clindamycin 300 mg BD for 7 days; (3) 2% vaginal clindamycin cream once daily for 7 days; (4) Metronidazole 0.75% vaginal gel once daily for 5 days; (5) 2 g of metronidazole in a single dose; and (6) 2 g of tinidazole in a single dose* (Fig. 3.3).

Young women with chronic symptoms are switching toward complementary and alternative medicine (CAM) therapy because of such complications (Askari et al., 2020). According to the Centre for Disease Control (CDC) treatment for BV includes multidose oral and vaginal formulations of clindamycin and metronidazole (Hillier et al., 2017). A double-dummy, double-blind, randomized noninferiority phase III, a study in 577 patients reported 60.1% and 59.5% cure rate with single-dose secnidazole (2 g) regimen multiple-dose metronidazole regimen (500 mg BD for 7 days), respectively (Bohbot, Vicaut, Fagnen, & Brauman, 2010).

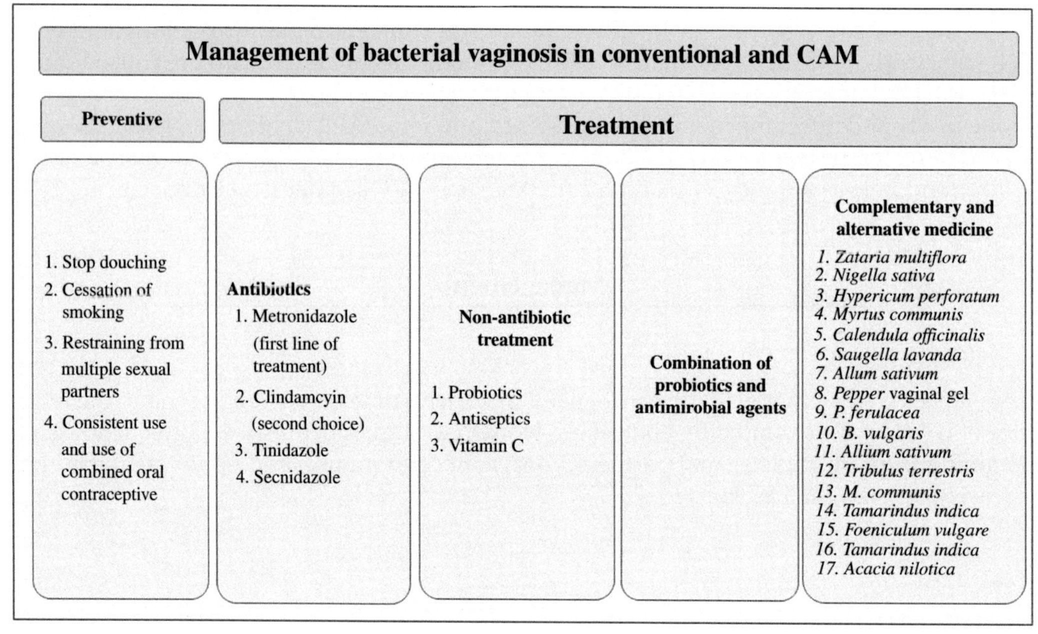

FIG. 3.3 Management of bacterial vaginosis.

- A parallel-group, single-blind, multicenter, randomized study in 540 patients with BV infections reported no significant differences in cure rates between the Clindesse™ and Cleocin treatment groups. They concluded that Clindesse™ vaginal cream had similar efficacy and safety as Cleocin vaginal cream to treat BV (Faro et al., 2005).
- In a multicenter, RCT reported that 46% of women who used oral tinidazole 2 g daily for 2 days and 64% of women who used 1 g for 5 days had a resolution from BV (Livengood et al., 2007).
- A single-dose clindamycin cream reported 64% clinical and 57% microbiologic cure rate in women with BV and a single-dose 1.3% metronidazole gel showed 37% and 18% clinical and microbiologic cure rate respectively at the 21-day visit (Faro et al., 2005).
- Hillier et al., reported therapeutic microbiologic, and clinical, cure rates were 40.3%, 40.3%, and 67.7%, for 2 g secnidazole and 21.9%, 23.4%, and 51.6% for 1 g secnidazole compared with 6.5%, 6.5%, and 17.7%, for placebo, respectively. Both doses were well-tolerated (Hillier et al., 2017).

Nonantibiotic treatment of BV

- *Probiotics*: "Probiotics exert beneficial health effects in the host and is a product containing viable defined microorganisms inadequate numbers that alter the microbiota in a compartment of the host" (Laue et al., 2018). Lactic acid is a natural acid produced that acts as a vaginal acidify agent, however, it had a short duration of action (El-Saied et al., 2016). Many studies have considered add on effect of probiotics with antimicrobial regimens as recurrent infections and bacterial resistance is reported after frequent use of antimicrobials. The probiotic containing *lactobacillus* is often used for BV treatment as it can reduce intravaginal pH, extracellular proteins, bacteriocin and creates specific molecules such as hydrogen peroxide. These molecules inhibit growth and kill pathogenic bacteria (Darmayanti et al., 2017). Vaginal suppositories of probiotics deposit the *lactobacillus* strains on the vaginal mucosa, whereas oral probiotics endure gastrointestinal transport and intensify the number of strains in the colon and feces. This promotes recolonization of the vagina through the proximity of the rectum and vagina (Tidbury et al., 2020). In several RCTs, the adjuvant use of probiotics after an initial course of antibiotics to avoid taking repetitive intake of antibiotics has been tested (Donders et al., 2014). Administration of *Lactobacillus rhamnosus* GR-1 and *L. reuteri* RC-14 with metronidazole improved the efficacy (Anukam et al., 2006).Estriol in combination with probiotics stimulates epithelial proliferation and maturation in the vagina. This helps to produce *lactobacilli* by stimulating glycogen production in the epithelial cells (Tidbury et al., 2020). Another study reported additional intake of yoghurt that contains probiotic strains ameriolated the symptoms of BV and recovery rate and also helped to recover the vaginal microbial pattern (Laue et al., 2018).

  Bradshaw et al. reported that BV recurrence does not reduce with a combination of oral metronidazole and vaginal clindamycin, or oral metronidazole with an extended course of vaginal *L. acidophilus* probiotic (Bradshaw et al., 2012).
- *Antiseptics*: For over half a century antiseptics have been used to treat vaginal infections. Antiseptics similar to antibiotics lead to recolonize indigenous lactobacilli and enable to

abolish the vaginal anaerobic microbiota related to bacterial vaginosis. Antiseptics use act through bacterial cell membrane disruption. A systematic search was conducted (1970–2010) reported noninferiority of polyhexamethylene biguanide and chlorhexidine in comparison to clindamycin or metronidazole. Another study reported that a single oral dose of metronidazole was more effective than a single vaginal douche with hydrogen peroxide. Further, they concluded that disinfectants and antiseptics used for BV has been needs future studies for the long-term efficacy and safety for vaginal use (Verstraelen, Verhelst, Roelens, & Temmerman, 2012).

- *Vitamin C*: A experiment reported that the BV recurrences were reduced from 32% to 16% of patients over 6 months use of monthly vitamin C vaginal application for 6 days after menstruation group (Krasnopolsky et al., 2013). Another RCT reported similar rates of efficacy and adverse effect of vaginal vitamin C suppository and vaginal metronidazole to treat BV (El-Saied et al., 2016).

Efficacy of a combination of probiotics and antimicrobial agents

A systematic review included 692 women of reproductive age from five RCTs that involved a patient who underwent treatment of BV. The authors reported that antimicrobial and probiotic combined treatment decreases 50% recurrence risk of BV compared to antimicrobial treatment alone (Darmayanti et al., 2017).

Complementary and alternative medicine: The encumbrance that inflicts on women's quality of life with recurrent need extra care to minimize it. Hence, better ways to prevent recurrence that does not require repeated courses of antibiotics are needed such as complementary and alternative treatment.

- A systematic review of 13 trials found that *Zataria multiflora, Nigella sativa, Hypericum perforatum* vaginal gel 3%, *Myrtus communis* L. plus metronidazole or *Berberis vulgaris* plus metronidazole, *Hypercum perforatum* were as effective as the standard drug in the improvement of both clinical symptoms and laboratory parameters. Further, other herbal medicines including *Calendula officinalis, Saugella lavanda*, garlic, *pepper* vaginal gel, *P. ferulacea, B. vulgaris,* and *Allium sativum* was found to be as effective as metronidazole with fewer side effects (Najafi et al., 2019).
- One study reported that therapeutic composition made of *Tribulus terrestris* is beneficial to treat bacterial, fungal, and viral infections, predominantly in gynecologic vaginal infections in a local dosage form such as suppository (Alexis, 2004).
- A comparative, double-blind reported that *Z. multiflora* cream had a similar effect as vaginal or oral metronidazole against both clinical symptoms and laboratory parameters (Abdali et al., 2015; Najafi et al., 2019).
- An RCT reported that a combination of *M. communis* L. or *B. vulgaris* in metronidazole base enhanced the efficiency of BV therapy, with no relapse in patients in the test groups however, 30% of patients in the reported relapse during 3 weeks of follow-up in the metronidazole group (Masoudi et al., 2016).
- Another RCT result showed that the therapeutic effect was 70% and 48.3% after treatment with garlic and metronidazole respectively ($P < 0.001$). Further, metronidazole was related to more complications in terms of side effects than garlic with significant

differences between the two treatment groups ($P = 0.032$) (Mohammadzadeh, Dolatian, Jorjani, AlaviMajd, & Borumandnia, 2014).

- The result of a study showed thatherbal suppository of *T. terrestris*, *M. communis*, *Tamarindus indica*, and *Foeniculum vulgare* had a similar effect as metronidazole in the treatment of BV (Baery et al., 2018).
- A randomized controlled study showed that response to *Prangos ferulacea* vaginal cream plus oral metronidazole for a week was 94% and 88% according to Amsel's clinical criteria and Nugent microscopic criteria respectively. Whereas placebo vaginal cream plus metronidazole was 94% and 85% according to Amsel's criteria and Nugent criteria, respectively. Further, the authors concluded that *Prangos ferulacea* vaginal cream hastened the recovery from BV (Motlagh et al., 2018).
- A comparative double-dummy, double-blind, randomized study showed that *H. perforatum* vaginal gel (3%) was as effective as metronidazole vaginal gel (0.75%) against preventing BV recurrence. (Charandabi, Mohammadzadeh, Khalili, & Javadzadeh, 2014).
- Another study assessed the effect of *N. sativa* on BV, patients with BV were divided into two groups that received either Phytovagex vaginal suppository 1% (containing *N. sativa*) once a day or metronidazole (250 mg) orally twice daily for a week. The treatment success rate was 74.2% and 69.2% for Phytovagex and metronidazole, respectively (Jafarnezhad, Mask, Rakhshandeh, & Shakeri, 2017).
- A randomized comparative study of *C. officinalis* cream (5 g) vs metronidazole in reproductive age with BV for a week intravaginally. The results showed that after a week of intervention in both groups all women were free of symptoms proving their efficacy in BV without any efficacy (Pazhohideh et al., 2018).
- Jahufer and Begum conducted an RCT to evaluate the efficacy of *Acacia nilotica* bark in BV that included 45 patients (2:1). The test group was administered orally, *A. nilotica* decoction (30 g) BID for 1 month whereas in the standard group tab. Metronidazole (400 mg) BID for 7 days. The author concluded that both groups were similar in efficacy with no statistical significance ($P = 1.000$) (Jahufer & Begum, 2014) (Table 2 and Table 3).

The aforementioned studies show that clinical trials have proven the efficacy of herbal medicine in BV. However, phase III and postmarketing clinical trials are recommended and make patents so that these herbal medicines are available worldwide.

Conclusion

BV is the commonest vaginal infection seen in the Gynecological OPDs that causes abnormal vaginal discharge and if not treated, it is allied to serious complications. Its prevalence among reproductive-aged women globally ranges from 23% to 29%. They are various risk factors that can be avoided to prevent BV. The gold standard test to the diagnosis of BV includes NSS, and Amsel's criteria though are tests are available. The management includes antibiotics, nonantibiotics and complementary and alternative treatment with their own merits and demerits. The positive effect of herbal medicines

TABLE 3.2 Clinical trials of complementary and alternative medicine.

First author, Year, Place	Patient age (Y)	Diagnosis method	Type of intervention	Control group	Duration of treatment	Outcome	Adverse event	References
Abdali, 2015, Iran	Reproductive age	Nugent microscopic criteria	*Z. multiflora* cream + placebo tablet ($n = 70$	Oral metronidazole tablet + placebo cream ($n = 70$)	7 days	There was no significant difference between intervention and control groups ($P > 0.05$)	Irritation	Abdali et al. (2015)
Masoudi, 2016, Iran	18–40	Amsel clinical criteria	Vaginal gel *Berberis vulgaris* 5% with metronidazole base ($n = 40$) and Vaginal gel *Myrtus communis* 2% with metronidazole ($n = 40$)	Metronidazole vaginal gel 0.75% on bacterial ($n = 40$)	5 days	Cure rate in both intervention groups was higher than metronidazole alone ($P < 0.001$) there was no significant difference between to intervention groups ($P = 0.18$)	Not mentioned	Masoudi et al. (2016)
Mohammadzadeh, 2014, Iran	18–44	Amsel clinical criteria, Nugent microscopic criteria	Garlic tablet ($n = 60$)	Metronidazole tablet ($n = 60$)	7 days	Treatment success in garlic group (63.3%) was higher than metronidazole (48.3%), however, nonsignificant ($P = 0.141$)	More frequent side effects in metronidazole group than the intervention group	Mohammadzadeh, Dolatian, Jorjani, AlaviMajd, & Borumandnia (2014)
Baery, 2018, Iran	18–50	Amsel clinical criteria	vaginal suppository of Forzejeh (*Tribulus terrestris* + Myrtus commnuis + *Foeniculum vulgare* + *Tamarindus indica*) ($n = 64$)	Vaginal suppository of metronidazole ($n = 63$)	7 days	Fozejeh was as effective as metronidazole concerning the amount and odor of discharge, Amsel criteria score, pelvic pain and cervical inflammation	No serious side effect	Baery et al. (2018)

Azadpour Motlagh, 2018, Iran	15–49	Amsel clinical criteria, Nugent microscopic criteria	oral metronidazole + Prangos ferulacea vaginal cream ($n = 50$)	Oral metronidazole + placebo vaginal cream ($n = 50$)	7 days	The two groups after treatment had similar results in terms of patients' complaints, Amsel clinical criteria score and microscopic criteria Nugnt	Nausea, and metallic taste (Overall 16%)	Motlagh et al. (2018)
Mohammad, 2014, Iran	18–49	Amsel clinical criteria	*Hypericum perforatum* + placebo ($n = 82$)	Metronidazole + placebo ($n = 80$)	5 days	There was no significant difference between intervention (82% improvement) and control groups (85% improvement) ($P = 0.574$) except less itching in intervention group ($P = 0.018$)	Vaginal irritation at first day (13%), nausea, vomiting, dizziness and vaginal dryness	Charandabi et al. (2014)
Jafarnezhad, 2017, Iran	15–49	Amsel clinical criteria	Phytovagex vaginal suppository (*Nigella sativa*) + placebo tab	Metronidazole oral tablet + placebo vaginal suppository ($n = 31$)	7 days	Treatment success rate was 74.2% for Phytovagex and 69.2% for metronidazole group ($P = 0.68$)	Mild burning (12.9%)	Jafarnezhad et al. (2017)
Pazhohideh, 2018, Iran	18–45	Amsel clinical criteria	*Calendula officinalis* cream ($n = 40$)	Metronidazole vaginal cream ($n = 40$)	7 days	There was no significant difference between the intervention and control groups. All patients cured	none	Pazhohideh et al. (2018)
Jahufer R, 2014, India	18–45	Amsel clinical criteria	*Acacia arabica* (chal babool decoction) ($n = 30$)	Tablet Metronidazole ($n = 15$)	Test drug for 1 month and cotrol group for 7 days	Both groups were similar in efficacy with no statistical significance ($P = 1.000$)	None	Jahufer & Begum (2014)

TABLE 3.3 Herbal plants with activities and phytoconstituents.

S no	Drug name	Actions	Phytoconstituent	References
1.	*Zataria multiflora*	Carvacrol is antiseptic and antifungal and thymol is antiseptic and antiworm. These compounds have antimicrobial properties	saponins, caffeic acid, resin, tannin, resonates, 2.6% volatile oil, thymol, and carvacrol	Abdali et al. (2015)
2.	*Myrtus communis*	Astringent, antidiarrhea and hair strengthening and growing effects. The plant extract can inhibit the growth of bacteria like *Staphylococcus aureus*, *Pseudomonas aeruginosa*, and *Escherichia coli*	Essential oil in leaves consists of tannins, flavonoids, vitamin C without cardiac glycosides and alkaloids	Masoudi et al. (2016)
3.	*Berberis vulgaris*	Appetizer and antiinflammatory, antibacterial, antihistaminic, antimicrobial, antifungal	Isoquanils like Berberin, Berbamine, and Palmatine	Masoudi et al. (2016)
4.	*Tribulus terrestris*	Antimicrobial activity, useful for treating bacterial, fungal and viral infections, particularly in gynaecologic infections		Alexis (2004); Baery et al. (2018)
5.	*Tamarindus indica*	Antimicrobial activity		Baery et al. (2018)
6.	*Foeniculum vulgare*	Antimicrobial activity		Baery et al. (2018)
7.	*Prangosferulacea*	Antibacterial, antiinflammatory, antioxidant, antimicrobial properties, Prangos ferulacea is used for the treatment of joint inflammation 14 vascular occlusion, hemorrhage, vaginal itching and infection, neutralization of toxins, uterine infections, and removal of dead fetus.	Monoterin, Sescoueeterin, Coumarin, Flavondeid, Tannin, Salpounin, and Alkaloid	Motlagh et al. (2018)
8.	*Hypericum perforatum*	Antibacterial and anti microbial properties		Charandabi et al. (2014)

might be probably considered as potential candidates to combat bacterial vaginosis. Further, phase III and postmarketing trials in large sample sizes are recommended to prove the efficacy of herbal medicines.

References

Abdali, K., Jahed, L., Amooee, S., Zarshenas, M., Tabatabaee, H., Bekhradi, R., et al. (2015). Comparison of the effect of vaginal *Zataria multiflora* cream and oral metronidazole pill on results of treatments for vaginal infections including trichomoniasis and bacterial vaginosis in women of reproductive age. *BioMed Research International*, *2015*, 1–7. https://doi.org/10.1155/2015/683640.

Alexis B. Treatment of vulvovaginitis with spirostanol enriched extract from *Tribulus terrestris*. United States Patent US 6,818,231;1-10. 2004.

Anukam, K., Osazuwa, E., Ahonkhai, I., Ngwu, M., Osemene, G., Bruce, A. W., et al., & Reid, G. (2006). Augmentation of antimicrobial metronidazole therapy of bacterial vaginosis with oral probiotic *Lactobacillus rhamnosus* GR-1 and *Lactobacillus reuteri* RC-14: Randomized, double-blind, placebo controlled trial. *Microbes and Infection, 8*(6), 1450–1454. https://doi.org/10.1016/j.micinf.2006.01.003.

Askari, S. F., Jahromi, B. N., Dehghanian, A., Zarei, A., Tansaz, M., Badr, P., et al. (2020). Effect of a novel herbal vaginal suppository containing myrtle and oak gall in the treatment of vaginitis: A randomized clinical trial. *DARU Journal of Pharmaceutical Sciences, 28*(2), 603–614. https://doi.org/10.1007/s40199-020-00365-6.

Baery, N., Nejad, A. G., Amin, M., Mahroozade, S., Mokaberinejad, R., Bioos, S., et al. (2018). Effect of vaginal suppository on bacterial vaginitis based on Persian medicine (Iranian traditional medicine): A randomised double blind clinical study. *Journal of Obstetrics and Gynaecology, 38*(8), 1110–1114. https://doi.org/10.1080/01443615.2018.1445706.

Bagnall, P., & Rizzolo, D. (2017). Bacterial vaginosis: A practical review. *JAAPA, 30*(12), 15–21. https://doi.org/10.1097/01.JAA.0000526770.60197.fa.

Bhat, T. A., & Begum, W. (2017). Efficacy of *Tamarindus indicus, Melia azadirach* and *Santalum album* in syndromic management of abnormal vaginal discharge: A single-blind randomised controlled trial. *Journal of Complementary and Integrative Medicine, 15*(2), 1–8. https://doi.org/10.1515/jcim-2015-0023.

Bhilwar, M., Lal, P., Sharma, N., Bhalla, P., & Kumar, A. (2015). Prevalence of reproductive tract infections and their determinants in married women residing in an urban slum of North-East Delhi, India. *Journal of Natural Science, Biology and Medicine, 6*, S29–S34. https://doi.org/10.4103/0976-9668.166059.

Bitew, A., Abebaw, Y., Bekele, D., & Mihret, A. (2017). Prevalence of bacterial vaginosis and associated risk factors among women complaining of genital tract infection. *International Journal of Microbiology, 2017*, 1–8. https://doi.org/10.1155/2017/4919404.

Bohbot, J. M., Vicaut, E., Fagnen, D., & Brauman, M. (2010). Treatment of bacterial vaginosis: A multicenter, double-blind, double-dummy, randomised phase III study comparing secnidazole and metronidazole. *Infectious Diseases in Obstetrics and Gynecology, 2010*. https://doi.org/10.1155/2010/705692, 705692.

Bradshaw, C. S., Pirotta, M., Guingand, D. D., Hocking, J. S., Morton, A. N., Garland, S. M., et al. (2012). Efficacy of oral metronidazole with vaginal clindamycin or vaginal probiotic for bacterial vaginosis: Randomised placebo-controlled double-blind trial. *PLoS One, 7*(4). https://doi.org/10.1371/journal.pone.0034540, e34540.

Breding, K., Selbing, A., Farnebäck, M., Hermelin, A., & Larsson, P. (2020). Diagnosis of bacterial vaginosis using a novel molecular real-time PCR test. *Journal of Womens Health and Gynecology, 7*, 1–7.

Charandabi, S. M. A., Mohammadzadeh, Z., Khalili, F. A., & Javadzadeh, Y. (2014). Effect of *Hypericum perforatum* L. compared with metronidazole in bacterial vaginosis: A double-blind randomized trial. Asian Pac. *Journal of Tropical Biomedicine, 4*(11), 896–902. https://doi.org/10.12980/APJTB.4.201414B160.

Chaudhary, V., Prakesh, V., Agarwal, K., Agrawal, V. K., Singh, A., & Pandey, S. (2012). Clinico-microbiological profile of women with vaginal discharge in a tertiary care hospital of northern India. *International Journal of Medical Science and Public Health, 1*, 75–80.

Chiaffarino, F., Parazzini, F., De Besi, P., & Lavezzari, M. (2004). Risk factors for bacterial vaginosis. *European Journal of Obstetrics, Gynecology, and Reproductive Biology, 117*(2), 222–226. https://doi.org/10.1016/j.ejogrb.2004.05.012.

Coppolillo, E. F., Perazzi, B. E., Famiglietti, A. M., Eliseht, M. G., & Vay, C. A. (2003). Diagnosis of bacterial vaginosis during pregnancy. *Journal of Lower Genital Tract Disease, 2*, 117–121. https://doi.org/10.1097/00128360-200304000-00008.

Darmayanti, A. T., Murti, B., & Susilawati, T. N. (2017). The effectiveness of adding probiotic to antimicrobial agents for the treatment of bacterial vaginosis: A systematic review. *Indonesian Journal of Medicine, 2*(3), 161–168. https://doi.org/10.26911/theijmed.2017.02.03.03.

Donders, G. (2010). Diagnosis and management of bacterial vaginosis and other types of abnormal vaginal bacterial flora: A review. *Obstetrical & Gynecological Survey, 65*(7), 462–473. https://doi.org/10.1097/OGX.0b013e3181e09621.

Donders, G. G., Zodzika, J., & Rezeberga, D. (2014). Treatment of bacterial vaginosis: What we have and what we miss. *Expert Opinion on Pharmacotherapy, 15*(5), 645–657. https://doi.org/10.1517/14656566.2014.881800.

Ellington, K., & Saccomano, S. J. (2020). Recurrent bacterial vaginosis. *The Nurse Practitioner, 45*(10), 27–32. https://doi.org/10.1097/01.NPR.0000696904.36628.0a.

El-Saied, N., Amer, M., Elbohoty, A., Saad, M., & Mansour, M. (2016). Efficacy of vitamin C vaginal suppository in treatment of bacterial vaginosis a randomized controlled trial. *Journal of Gynecology Research*, *2*(1), 103. https://doi.org/10.15744/2454-3284.2.103.

Faro, S., Skokos, C. K., & Clindesse Investigators Group. (2005). The efficacy and safety of a single dose of Clindesse vaginal cream versus a seven-dose regimen of Cleocin vaginal cream in patients with bacterial vaginosis. *Infectious Diseases in Obstetrics and Gynecology*, *13*(3), 155–160. https://doi.org/10.1080/10647440500148321.

Fethers, K. A., Fairley, C. K., Hocking, J. S., Gurrin, L. C., & Bradshaw, C. S. (2008). Sexual risk factors and bacterial vaginosis: A systematic review and meta-analysis. *Clinical Infectious Diseases*, *47*, 1426–1435. https://doi.org/10.1086/592974.

Garba, D. J., Zabaze, S. S., Tabitha, V. S., James, G., & Makshwar, K. (2014). Microbiological diagnosis of bacterial vaginosis in pregnant women in a resource limited setting in North Central Nigeria. *American Journal of Life Sciences*, *2*(6), 356–360. https://doi.org/10.11648/j.ajls.20140206.15.

Haahr, T., Ersbøll, A. S., Karlsen, M. A., Svare, J., Sneider, K., Hee, L., et al. (2016). Treatment of bacterial vaginosis in pregnancy in order to reduce the risk of spontaneous preterm delivery- a clinical recommendation. *Acta Obstetricia et Gynecologica Scandinavica*, *95*, 850–860. https://doi.org/10.1111/aogs.12933.

Hay, P. (2017). Bacterial vaginosis. *Research*, *6*, 1761. https://doi.org/10.12688/f1000research.11417.1 (F1000 Faculty Rev).

Hillier, S. L., Nyirjesy, P., Waldbaum, A. S., Schwebke, J. R., Morgan, F. G., Adetoro, N. A., et al. (2017). Secnidazole treatment of bacterial vaginosis: A randomized controlled trial. *Obstetrics and Gynecology*, *130*(2), 379–386. https://doi.org/10.1097/AOG.0000000000002135.

Homayouni, A., Bastani, P., Ziyadi, S., Charandabi, S. M. A., Ghalibaf, M., Mortazavian, A. M., et al. (2014). Effects of probiotics on the recurrence of bacterial vaginosis: A review. *Journal of Lower Genital Tract Disease*, *18*(1), 79–86. https://doi.org/10.1097/LGT.0b013e31829156ec.

Jafarnezhad, F., Mask, M. K., Rakhshandeh, H., & Shakeri, M. T. (2017). Comparison of the percentage of medical success for Phytovagex vaginal suppository and metronidazole oral tablet in women with bacterial vaginosis. *The Iranian Journal of Obstetrics, Gynecology and Infertility*, *20*(3), 29–39.

Jahufer, R., & Begum, W. (2014). Efficacy of bark of *Acacia arabica* in management of bacterial vaginosis: A randomized controlled trial. *International Journal of Current Research and Review*, *6*(1), 79–88.

Kenyon, C., Colebunders, R., & Crucitti, T. (2013). The global epidemiology of bacterial vaginosis: A systematic review. *American Journal of Obstetrics and Gynecology*, *209*(6), 505–523. https://doi.org/10.1016/j.ajog.2013.05.006.

Khan, K. J., Shah, R., Gautam, M., & Patil, S. (2007). Clue cells. *Indian Journal of Sexually Transmitted Diseases*, *28*(2), 108–109.

Krasnopolsky, V. N., Prilepskaya, V. N., Polatti, F., Zarochentseva, N. V., Bayramova, G. R., Caserini, M., et al. (2013). Efficacy of vitamin C vaginal tablets as prophylaxis for recurrent bacterial vaginosis: A randomised, double-blind, placebo-controlled clinical trial. *Journal of Clinical Medical Research*, *5*(4), 309–315. https://doi.org/10.4021/jocmr1489w.

Laue, C., Papazova, E., Liesegang, A., Pannenbeckers, A., Arendarski, P., Linnerth, B., et al. (2018). Effect of a yoghurt drink containing *Lactobacillus* strains on bacterial vaginosis in women-a double-blind, randomised, controlled clinical pilot trial. *Beneficial Microbes*, *9*(1), 35–50. https://doi.org/10.3920/BM2017.0018.

Livengood, C. H., Ferris, D. G., Wiesenfeld, H. C., Hillier, S. L., Soper, D. E., Nyirjesy, P., et al. (2007). Effectiveness of two tinidazole regimens in treatment of bacterial vaginosis: A randomized controlled trial. *Obstetrics and Gynecology*, *110*(2), 302–309. https://doi.org/10.1097/01.AOG.0000275282.60506.3d.

Madhivanan, P., Krupp, K., Chandrasekaran, V., Karat, C., Arun, A., Cohen, C., et al. (2008). Prevalence and correlates of bacterial vaginosis among young women of reproductive age in Mysore, India. *Indian Journal of Medical Microbiology*, *26*(2), 132–137. https://doi.org/10.4103/0255-0857.40526.

Manandhar, R., Sharma, J., Pokharel, B. M., Shrestha, B., & Pradhan, N. (2005). Bacterial vaginosis in Tribhuvan University Teaching Hospital. *Journal of Institute of Medicine Nepal*, *27*(2), 14–17.

Mascarenhas, R. E., Machado, M. S., Costa e Silva, B. F., Pimentel, R. F., Ferreira, T. T., Leoni, F. M., et al. (2012). Prevalence and risk factors for bacterial vaginosis and other vulvovaginitis in a population of sexually active adolescents from Salvador, Bahia, Brazil. *Infectious Diseases in Obstetrics and Gynecology*, *2012*, 1–8. https://doi.org/10.1155/2012/378640.

Masoudi, M., Miraj, S., & Kopaei, M. R. (2016). Comparison of the effects of *Myrtus communis* L, *Berberis vulgaris* and metronidazole vaginal gel alone for the treatment of bacterial vaginosis. *Journal of Clinical and Diagnostic Research*, *10*(3), QC04-07. https://doi.org/10.7860/JCDR/2016/17211.7392.

Mastromarino, P., Vitali, B., & Mosca, L. (2013). Bacterial vaginosis: A review on clinical trials with probiotics. *The New Microbiologica*, *36*(3), 229–238 (PMID: 23912864).

McDonald, H. M., O'Loughlin, J. A., Vigneswaran, R., Jolley, P. T., Harvey, J. A., Bof, A., et al. (1997). Impact of metronidazole therapy on preterm birth in women with bacterial vaginosis flora (*Gardnerella vaginalis*): A randomised, placebo controlled trial. *British Journal of Obstetrics and Gynaecology*, *104*(12), 1391–1397. https://doi.org/10.1111/j.1471-0528.1997.tb11009.x.

Modak, T., Arora, P., Agnes, C., Ray, R., Goswami, S., Ghosh, P., et al. (2011). Diagnosis of bacterial vaginosis in cases of abnormal vaginal discharge: Comparison of clinical and microbiological criteria. *Journal of Infection in Developing Countries*, *5*(5), 353–360. https://doi.org/10.3855/jidc.1153.

Mohammadzadeh, F., Dolatian, M., Jorjani, M., AlaviMajd, H., & Borumandnia, N. (2014). Comparing the therapeutic effects of garlic tablet and oral metronidazole on bacterial vaginosis: A randomized controlled clinical trial. *Iranian Red Crescent Medical Journal*, *16*(7), 1–6. https://doi.org/10.5812/ircmj.19118.

Mohammadzadeh, F., Dolatian, M., Jorjani, M., & Majd, H. A. (2014). Diagnostic value of Amsel's clinical criteria for diagnosis of bacterial vaginosis. *Global Journal of Health Science*, *7*(3), 8–14. https://doi.org/10.5539/gjhs.v7n3p8.

Motlagh, A. A., Dolatian, M., Mojab, F., Nasiri, M., Ezatpour, B., Sahranavard, Y., et al. (2018). The effect of *Prangos ferulacea* vaginal cream on accelerating the recovery of bacterial vaginosis: A randomized controlled clinical trial. *International Journal of Community Based Nursing & Midwifery*, *6*(2), 100–110. 29607339.

Muthusamy, S., & Elangovan, S. (2016). Comparison of Amsel's criteria, Nugent score and culture for the diagnosis of bacterial vaginosis. *National Journal of Laboratory Medicine*, *5*(1), 37–40. https://doi.org/10.7860/NJLM/2016/17330:2095.

Najafi, M. N., Rezaee, R., Najafi, N. N., Mirzaee, F., Burykina, T. I., Lupuliasa, D., et al. (2019). Herbal medicines against bacterial vaginosis in women of reproductive age: A systematic review. *Farmácia*, *67*(6), 931–940.

Nawani, M., & Sujatha, R. (2013). Diagnosis and prevalence of bacterial vaginosis in a teritiary care Centre at Kanpur. *Journal of Evolution of Medical and Dental Sciences*, *2*(22), 3959–3963.

Pazhohideh, Z., Mohammadi, S., Bahrami, N., Mojab, F., Abedi, P., & Maraghi, E. (2018). The effect of *Calendula officinalis* versus metronidazole on bacterial vaginosis in women: A double-blind randomized controlled trial. *Journal of Advanced Pharmaceutical Technology & Research*, *9*(1), 15–19. https://doi.org/10.4103/japtr.JAPTR_305_17.

Peebles, K., Velloza, J., Balkus, J. E., McClelland, R. S., & Barnabas, R. V. (2019). High global burden and costs of bacterial vaginosis: A systematic review and meta-analysis. *Sexually Transmitted Diseases*, *46*(5), 304–311. https://doi.org/10.1097/OLQ.0000000000000972.

Rafiq, S., Nauman, N., Tariq, A., & Jalali, S. (2015). Diagnosis of bacterial vaginosis in females with vaginal discharge using Amsel's clinical criteria and Nugent scoring. *Journal of Rawalpindi Medical College.*, *19*(3), 230–234.

Ranjit, E., Raghubanshi, B. R., Maskey, S., & Parajuli, P. (2018). Prevalence of bacterial vaginosis and its association with risk factors among nonpregnant women: A hospital based study. *International Journal of Microbiology*, *2018*, 1–9. https://doi.org/10.1155/2018/8349601.

Rao, S. R., Pindi, K. G., Rani, U., Sasikala, G., & Kawle, V. (2016). Diagnosis of bacterial vaginosis: Amsel's criteria vs Nugent's scoring. *Scholars Journal of Applied Medical Sciences*, *4*(6C), 2027–2031. https://doi.org/10.21276/sjams.2016.4.6.32.

Schorge, J. O., Schaffer, J. O., Halvorson, L. M., Hoffman, B. I., Bradshaw, K. D., & Cunningham, F. G. (2014). *Williams Gynaecology* (pp. 51–52). New York: McGraw-Hill.

Seth, A. R., Chaitra, S., Vaishnavi, S., & Sharath Chandra, G. R. (2017). Prevalence of bacterial vaginosis in females in the reproductive age group in Kadur, Karnataka, India. *International Journal of Reproduction, Contraception, Obstetrics and Gynecology*, *6*(11), 4863–4865.

Subtil, D., Brabant, G., Tilloy, E., Devos, P., Canis, F., Fruchart, A., et al. (2018). Early clindamycin for bacterial vaginosis in pregnancy (PREMEVA): A multicentre, double-blind, randomised controlled trial. *The Lancet.*, *392*(10160), 2171–2179. https://doi.org/10.1016/S0140-6736(18)31617-9.

Tidbury, F. D., Langhart, A., Weidlinger, S., & Stute, P. (2020). Non-antibiotic treatment of bacterial vaginosis-a systematic review. *Archives of Gynecology and Obstetrics*, *303*(1), 37–45. https://doi.org/10.1007/s00404-020-05821-x.

Verstraelen, H., Verhelst, R., Roelens, K., & Temmerman, M. (2012). Antiseptics and disinfectants for the treatment of bacterial vaginosis: A systematic review. *BMC Infectious Diseases*, *12*(1), 1–8. https://doi.org/10.1186/1471-2334-12-148.

Vodstrcil, L. A., Hocking, J. S., Law, M., Walker, S., Tabrizi, S. N., Fairley, C. K., et al. (2013). Hormonal contraception is associated with a reduced risk of bacterial vaginosis: A systematic review and meta-analysis. *PLoS One*, *8*(9), 1–16. https://doi.org/10.1371/journal.pone.0073055.

CHAPTER

4

Computer-aided knee joint MR image segmentation—An overview

Punit Kumar Singh[a] *and Sudhakar Singh*[b]

[a]Department of BioEngineering, Integral University Lucknow, Lucknow, Uttar Pradesh, India
[b]Department of Biomedical Engineering, Lovely Professional University, Phagwara, Punjab, India

Introduction

Knee osteoarthritis (OA) is a constant and dynamic infection portrayed by basic changes in the ligament, bone, synovium, and other joint structures (Felson, 2013). Knee OA is one of the main sources of handicap in the United States, which results in a lessening in personal satisfaction just as an enormous money-related weight on medicinal service frameworks and society. Knee OA is pervasive and expensive. It was assessed that symptomatic knee OA harasses more than 9.9 million US grown-ups in 2010 (Jevsever, 2013). Meanwhile, the evaluated normal lifetime treatment cost for every individual determined to have knee OA was $140,300 (Losina et al., 2015). There is developing comprehension about knee OA following quite a while of research, despite the fact that the pathogenesis of knee OA and the improvement system are as yet obscure. One of the principal impacts of knee OA is the debasement of the articular ligament, causing agony and loss of joint portability (Heidari, 2011).

Distinctive clinical imaging modalities could be utilized to explore and quantitatively measure the knee joint ligament, for example, X-beam, magnetic reverberation imaging (MRI), and computed tomography (CT) (Chan & Vese, 2001). Among them, X-beam and CT have comparative reactions on delicate tissues and ligaments and just can gauge ligament thickness through the separation between bones (Braun & Gold, 2012). An innovation of CT, eluded as knee arthrography, could conquer the above drawback and show the articular ligament plainly (Ghelman, 1985). Be that as it may, the preinfusion, which is imperative in knee arthrography, may cause agony and uneasiness for certain individuals other than potential entanglements (Newberg, Munn, & Robbins, 1985). X-ray is the main imaging methodology for noninvasive appraisal of the articular ligament, and the ligament decay can be broken

https://doi.org/10.1016/B978-0-323-99031-8.00011-9

down successfully (Eckstein, Burstein, & Link, 2006; Eckstein, Cicuttini, Raynauld, Waterton, & Peterfy, 2006). X-ray utilizes ground-breaking attractive field and radio recurrence heartbeats to create pictures of organs, delicate tissues, bones, and for all intents and purposes all inward body structures (Edelman & Warach, 1993). A normal knee MRI picture may contain several cuts (e.g., 160 cuts for every sweep in the OAI dataset (Heimann & Meinzer, 2009; Heimann, Morrison, Styner, Niethammer, & Warfield, 2010) contingent upon the example rate. It costs hours for experienced radiologists to break down a solitary output. The huge work cost makes the conclusion costly, wasteful, and hard to imitate. The personal computer (PC)-supported division advancements of knee MRI are in earnest need (Boesen et al., 2017). Programmed/self-loader division is troublesome because of the low differentiation, clamor, and particularly the flimsy structure of the ligament. Programmed volume and thickness estimations dependent on the manual mark is the initial move toward programmed examination (Kass, Witkin, & Terzopoulos, 1988). The self-loader division strategies perform knee MRI division with the intercession of specialists/radiologists through human-PC connection (Duryea et al., 2007). Self-loader division could be practiced utilizing an assortment of calculations, for example, dynamic shapes (Solloway, Hutchinson, Waterton, & Taylor, 1997), locale developing (Eckstein, Burstein, & Link, 2006; Eckstein, Cicuttini, et al., 2006; Eckstein et al., 1996; Stammberger, Eckstein, Englmeier, & Reiser, 1999), ray casting (Pakin, Tamez-Pena, Totterman, & Parker, 2002), live wire (Dodin, Martel-Pelletier, Pelletier, & Abram, 2011; Dodin, Pelletier, Martel-Pelletier, & Abram, 2010), watershed (Folkesson, Dam, Olsen, Pettersen, & Christiansen, 2005; Folkesson, Olsen, Pettersen, Dam, & Christiansen, 2005), and so on. Scientists have led the studies in related regions, for example, cerebrum MRI picture division (Balafar, Ramli, Saripan, & Mashohor, 2010; Iglesias & Sabuncu, 2015), 3D clinical picture division (Heimann et al., 2010), multimap book picture division, and so on. Be that as it may, as far as we could possibly know, there is no viable review for knee MRI division yet. This paper depicts the knee MRI division standards and gives a thorough audit of PC-helped knee MRI division strategies (Tang, Millington, Acton, Crandall, & Hurwitz, 2006). Additionally, we think about the current strategies and favorable circumstances/weaknesses and point out the future research bearing. The remainder of this paper is sorted out as follows: "Knee MR image segmentation" section condenses the current knee joint MRI division strategies, "Configurations with mathematical forms" section portrays the volume and thickness estimation techniques, "Classification processes" section gives a concise perspective on existing benchmarks and datasets just as the exhibition assessment measurements, and "Conclusion and future trends" section closes the whole paper and calls attention to the future pattern.

Knee MR image segmentation

Overview

The precise division assumes a basic job in PC-helped knee OA finding, and various techniques were created in late decades. Contingent upon the degree of robotization, the current techniques could be classified into manual division, self-loader division, and programmed division. In the methodology of manual division, the radiologists sketch the forms of various tissues on the MR pictures in a cut-by-cut way. The manual division is work serious and tedious. It could take 3–4 h for a very much prepared radiologist to complete a single output. What is more, manual division experiences low reproducibility.

Self-loader techniques are proposed to address the detriments of the manual strategy by presenting PC calculations. Self-loader techniques, for example, dynamic shape and locale developing (Bae et al., 2009; Dalal et al., 2007; Liu, Wang, Zhang, Gao, & Shen, 2015; Piplani, Disler, McCauley, Holmes, & Cousins, 1996; Wang, Donoghue, & Rueckert, 2013; Wang et al., 2016), give a human-PC interface that clients (radiologists/specialists) can enter their expert information (tourist spots) into the framework. At this point, the calculation will produce the tissue limits dependent on the given tourist spots naturally. Self-loader division strategies profit from the expert information and stay away from serious work. PC-aided diagnosis (CAD) framework-prepared self-loader division programming is broadly applied in the facility as a result of the high exactness.

The previous decades have seen the fast advancement of programmed knee joint division. Scientists in both the scholarly community and medication expect that by utilizing PCs with the related programming, the frameworks can fragment MRI pictures and quantize the ligament thickness and volume with least or no human mediation. Completely programmed division can be accomplished by various methodologies, for example, pixel/voxel grouping (Dalvi, Abugharbieh, Wilson, & Wilson, 2007; Dam, Lillholm, Marques, & Nielsen, 2015; Folkesson, Dam, et al., 2005; Folkesson, Dam, Olsen, Pettersen, & Christiansen, 2007; Kapur, Beardsley, Gibson, Grimson, & Wells, 1998; Prasoon et al., 2013), deformable model (Prasoon et al., 2012; Prasoon et al., 2013; Prasoon, Petersen, et al., 2013), and diagram-based techniques (Ababneh, Prescott, & Gurcan, 2011; Bae et al., 2009; Lee, Gumus, Moon, Kwoh, & Bae, 2014). Completely programmed knee joint division experiences the low difference of the MR picture and the dainty structure of the ligament. Most best-in-class techniques share the comparative method. They start at the bone division, which is less intricate, and afterward sketch the limits of the ligament dependent on the consequence of bone division. The ligament volume and thickness are the chief factors in OA analysis and progress assessment. The assignment should be possible by just tallying the number of pixels when the limits of the ligament are resolved. Nonetheless, the assessment of thickness is progressively intricate. A few examinations characterize the thickness of the two-dimensional (2D) MR picture cut (Solloway et al., 1997), whereas others characterize the thickness of the three-dimensional (3D) space (Cohen et al., 1999). 3D Euclidean distance transformation (EDT) (Stammberger et al., 1999) is generally used to decide the separation between the ligament surface and the bone-cartilage interface (BCI). There are likewise investigations of moved 3D ligament structures into thickness guides and discover neighborhood ligament misfortune among the guides (Kauffmann et al., 2003).

Bones are the greatest and the most notable structures in the knee joint. Different structures, for example, ligaments, are joined to the bones. The division of bones is more straightforward than ligament division because of higher differences and much more clear edges. Numerous techniques perform ligament division dependent on bone division results, whereas different strategies explain both bone division and ligament division simultaneously. There are additionally specialists who create start-to-finish frameworks to illuminate the knee MR picture division issue in full. The sample test image is shown in Fig. 4.1.

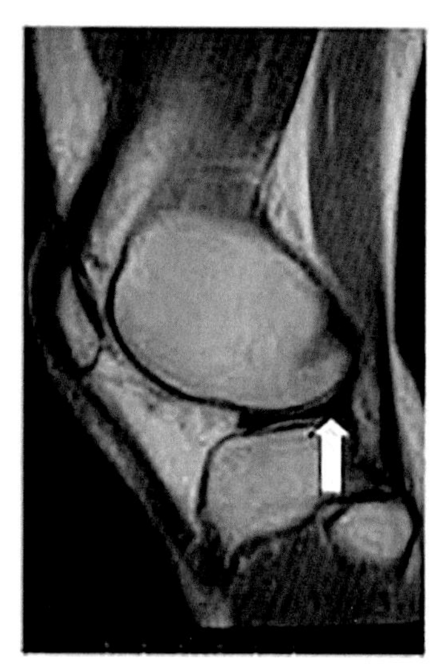
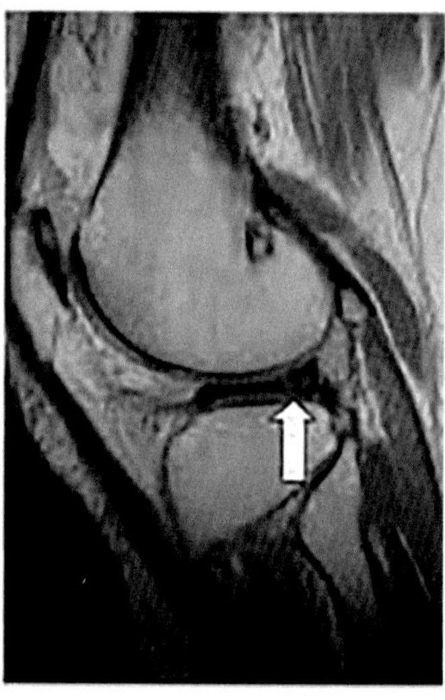

FIG. 4.1 Sample images of the knee. *Source: Hind Medical College, Lucknow. No permission required.*

Deformable models

Deformable models (Bae et al., 2009; Chan & Vese, 2001; Felson, 2013; McInerney & Terzopoulos, 1996), one of the most seriously read model-based methodologies for PC-helped clinical picture examination, can suit the regularly noteworthy fluctuation of natural structures after some time and across various people. This subsection overviews the uses of deformable models to knee joint MRI division.

Active contours

Dynamic shape models (Bae et al., 2009; Chan et al., 1991; Chan & Vese, 2001; Eckstein, Burstein, & Link, 2006; Felson, 2013; Solloway et al., 1997; Stammberger, Hohe, Englmeier, Reiser, & Eckstein, 2000) are well known for self-loader clinical picture division. A functioning shape model is characterized by a lot of focuses vi, where $I=0, 1, \ldots, n-1$; the interior versatile vitality *Ei*; and the outside vitality *Ee*. The inner vitality controls the disfigurements made to the snake, and the outside vitality term drives the bend to the article shape. The calculation looks for the best division by limiting the vitality work. A lot of dynamic form models were presented for knee MRI division, for example, B-spline (Cohen et al., 1999), Bezier spline, geodesic snake, GVF snake, and snake. The presentation of dynamic form models relies especially upon the choice of focuses. Be that as it may, there existed scarcely any writing about how to choose the underlying focuses in the field of knee MRI examination. Existing techniques began from the focuses inputted by human administrators. Dynamic shape models profit from the cooperation of human specialists, just as the continuous direction and rectification. Be that as it may, the reliance on human interest thwarted the advancement of the calculations.

The exemplary snake models thought about both neighborhood highlights and worldwide shapes limited the edge vitality to distort the form over the most noteworthy neighboring picture edge. The advancement of these models was nonraised, and it could fall into a nearby least. Another impediment of the exemplary snake models was their troubles in fitting inward limits. By presenting the slope vector stream as another outer power, GVF snake, the model could tackle the affectability to in-statement and the fitting issue on curved shape limit. Tang et al. (Duryea et al., 2007; Hakky, Jarraya, Ratzlaff, Guermazi, & Duryea, 2015) used GVF snake for knee MRI division, and the trial results indicated that the technique accomplished better reproducibility, contrasting with past investigations. Lorigo et al. (Williams et al., 2010; Yin et al., 2010) used the GAC snake model for knee MRI division and accomplished exact outcomes. The GAC snake model depended on the connection between dynamic shapes and the calculation of geodesics or insignificant separation bends. The calculation was hearty to in-statement and could deal with the topological fluctuations of the bend. Nonetheless, the calculation experienced issues when the edges were obscure or with a low difference. Techniques dependent on the dynamic shape model divided knee MRI in a cut-by-cut way. The wide utilization of cut choice methods could accelerate the division procedure, diminish the work cost, and improve the presentation. The strategy in Duryea et al. (2007) played out a 2D division on each cut of the MRI. A cut close to the focal point of the ligament plate was chosen, and a seed point on the bone-ligament edge was stamped.

Dynamic shapes had been broadly applied to knee MRI division. High accuracy could be accomplished by making an insignificant number of client inputs. Dynamic forms additionally permitted an accomplished client to direct, address, and approve the ongoing division. In any case, the dynamic shapes had constraints in the union, and the improvement issue drives vulnerability and poor security of the division. Table 4.1 presents the current knee MRI division strategies dependent on the dynamic shape model and records the shortcoming of the techniques.

TABLE 4.1 Current knee MRI segmentation strategies.

Methodology	References	Goal	Failure
Strong geodesic contour	Lorigo, Faugeras, Grimson, Keriven, and Kikinis (1998)	Bone	The SGC snake model was prone to visibility on the edges
CB-spline; spline at Bezier	Cohen et al. (1999), Lynch et al. (2000), Carballido-Gamio, Bauer, Lee, Krause, and Majumdar (2005), and Carballido-Gamio et al. (2008)	Cartilage	The methods were sensitive to initialization; the optimization was nonconvex; the convergence was not guaranteed
Directional snake of gradient vector flow	Tang et al. (2006)	Cartilage	The gradient vector flow snake model was unable to accommodate long, fine concave shapes
Effective highlighter	Duryea et al. (2007), Kauffmann et al. (2003), and Brem et al. (2009)	Cartilage	Initialization sensitive; optimization is nonconvex, and the algorithm quickly collapses into local minimum

Configurations with mathematical forms

The configurations with mathematical forms (CF) describe the shape of target objects using a series of n landmarks and learning from a training set of labeled images the appropriate ranges of shape variations. The appearance pattern can be applied to the new images using various algorithms, where the most commonly studied are the build action form (BAF) (Cootes, Edwards, & Taylor, 1999; Williams et al., 2010) and the based on positive appearance (BPA). Active appearance models are more descriptive compared to the active contour models and allow the algorithms to identify target points without operator interference. They could converge to surface boundaries based on form constraints, where the models of the active contour are powerless. All BPA and BAF algorithms use the same reference mathematical description of the structure of the target object. Nevertheless, the active shape model seeks to match a set of model points to an image, and BPA seeks to match both the model point position and the object texture representation to an image. There are two key distinctions between the two algorithms (Davies, Twining, Cootes, Waterton, & Taylor, 2002a, 2002b). First, BAF uses only the image texture in specific regions around the landmarks, whereas BPA uses the area as a whole; first, BAF minimizes the distance between the model points and the corresponding points contained in the image, whereas BPA seeks to minimize the disparity between the synthesized image and the goal image. BAF is usually easier, so it more precisely locates the position of the function point compared with BPA. BAF is usually easier, so it more precisely locates the position of the function point compared with BPA.

Solloway et al. (1997) used 2D BAF for slice-by-slice segmentation of bones and cartilages. In addition to modeling variance in shape, the approach modeled the gray-level presence of the objects of interest by analyzing tiny patches of photographs around each landmark. They believed that their model of local appearance could minimize sharp edge reliance. 2D models could not model 3D systems because they neglected the interaction between the slices. In comparison, the effects of 2D techniques often need more analysis. Compared to 2D models, 3D models draw more scrutiny. BAF and BPA are constructed using samples labeled with the MRI. Owing to the differences between the interobjects, identifying the relation between the dense landmarks is a critical step for both model creation and segmentation. Most research studies (Williams, Holmes, et al., 2010; Williams, Taylor, Waterton, & Holmes, 2004; Yin et al., 2010) used the minimal summary period (MSP) (Davies et al., 2002a, 2002b) and its correspondence variations (Cootes, Twining, Petrovic, Schestowitz, & Taylor, 2005). In comparison to MSP, Schmid and Magnenat-Thalmann (2008) recommended a Dalal et al. (2007) and believed that their approach did not need any unique topology of form and that a previous correspondence was sufficient to refinish.

System initialization is commonly applicable to authentication technologies. Fripp et al. (2005) submitted that the system was prone to initialization (Solloway et al., 1997). They added a stable, affine registration to automatically configure the application. The implementation of registration could boost the robustness of to configure. In most research, main component analysis (MCA) has been applied as an integral component of BAF and BPA algorithms to identify the main directions of shape variance in a training collection of example shapes. In addition to principal component analysis capturing global shape variations, Markov random fields (MRFs) are used to capture regional shape variations and local deformations (Seim et al., 2010).

TABLE 4.2 Summary of the latest BPA and BAF-related research.

Authors	Design	Goals	Failure
Solloway et al. (1997)	BAF(2D)	Bone/cartilage	1. The software neglected 3D structures; 2. We required postprocessing
Fripp, Crozier, Warfield, and Ourselin (2007, 2010), Fripp et al. (2005), Schmid and Magnenat-Thalmann (2008), and Schmid, Kim, and Magnenat-Thalmann (2011)	BAF(3D)	Bone	1. It can be very time-consuming to search for initial model pose parameters; 2. The initialization was focused on landmarks which were identified by hand
Gilles and Magnenat-Thalmann (2010) and Seim et al. (2010)	BAF(3D)	Bone/cartilage	1. These models can hit an optimal locale and heavily rely on their initial position; 2. The segmentation of the cartilage mostly relies on the predetermined "thickness" parameter
Lee et al. (2014), Williams, Holmes, et al. (2010), and Williams, Vincent, et al. (2010)	BPA	Bone/cartilage	Variance outside of such spaces cannot be accurately identified if no further relaxing step is used

BAF and BPA can well model the structure of the knee joint, which can be extended to the segmentation of the bone cartilage. Because of the cartilage's thin structure, these models do well on bone segmentation than on cartilage segmentation. BPAs and BAFs depend heavily on the used datasets and are thus not easy to replicate. Table 4.2 summarizes the latest BPA and BAF-related research.

Classification processes

Both polygon mesh classification-based methods and region-growing methods follow the same concept of learning from training samples one or more classifiers to differentiate foreground from context. Classification-based approaches are stable, and the increasing scale of training data may benefit. In comparison, polygon mesh-based labeling approaches are the most efficient and simple way to obtain full-automatic MRI segmentation of the knees.

Polygon mesh classification

In recent years, a great deal of work has been dedicated to approaches focused on polygon mesh classification (PMC). There are clear benefits of such approaches. First, the conceptual paradigm is descriptive, and the computation mechanism is straightforward; second, to attain an improved efficiency, a range of technologies, local and regional, should be combined; and third, by efficiently using significant training samples, such approaches could achieve an improved efficiency. One of the main disadvantages is oversegmentation of the thickness of the cartilage. In comparison, the approaches based on PMC are unreliable compared with other approaches because of a large number of polygon meshes. Folkesson, Dam, et al. (2005) published the first study on PMC-based knee-based MRI segmentation, which was also called the first fully automated knee-joint MRI segmentation (Öztürk & Albayrak, 2016). The system included a two-class k-nearest neighbor (kNN) classifier for the difference between the foreground (articular cartilage) and other tissues, and a three-class kNN classifier allocated each voxel to the bone (tibial and femoral), cartilage, and background.

A variety of features were used, which included the 3-jet (first-, second-, and third-order derivatives on xx, yy, and zz) (Florack, 1993), the Hessian matrix, the tensor structure (ST) (Florack, 1993), and so on. They contended that the most important apps were Hessian and ST.

The initial algorithm's 2.5-h segmentation cycle had impeded its application. More PMC-based approaches have been suggested in recent years to complement the rapid progress of computational power. To achieve detailed segmentation of the cartilage, a two-stage classification system was used. For highly accurate noise detection, they used kNN as the first classification. They then using a nonlinear SVM to make the final call on unexplained polygon meshes. Huang et al. (2015) developed the algorithm based on Fulkerson's methods using a probabilistic version of the kNN classifier to integrate the results of the classification into a Bayesian context. This work's focus is to add multiatlas registration to produce the spatial prior. A thorough analysis of the experimental approach and the findings are summed up. In addition, this approach was used as the basis for the identification of disease regions using knee MRI.

The polygonal mesh classifier relies on the local characteristic and location of each polygonal mesh. Researchers are trying to add the global function to the polygonal mesh classification to improve performance. Liu et al. (2015) used the random forest as a polygonal mesh classifier, mixing characteristics of the presence and meaning. They believed that the cartilage was usually small and thus, the findings of the preliminary segmentation were too inaccurate to provide sufficient background details. To solve the problem, they implemented a method of multiatlas registration. Adams and Bischof (1994), Dam et al. (2015), and Prasoon et al. (2012) suggested a segmentation system in a multistructure environment, integrating rigid multiatlas registration with voxel classification. By optimizing structured shared information (NMI) using optimization of L-BFGS-B (Fortran subroutines for large-scale restricted optimization), they reported two provided scans. Using a basic sampling scheme, they solved the unbalance between various systems. k-nearest neighbors were used as a polygonal mesh classifier in the training process, and the floating forward feature selection was used for feature selection based on the overlap value of the Dice number. The studies made use of materials from 1907.

Polygonal mesh classification-based methods are the most common methods considering not only the local presentation but also the interaction between slices. The new practice is to merge the global characteristic with the process of polygonal mesh classification. Another important challenge to tackle is how to reduce the reliance on position information. The descriptive PMC-dependent segmentation algorithms are listed in Table 4.3.

TABLE 4.3 PMC-based methods for knee joint segmentation.

Authors
Folkesson, Dam, et al. (2005)
Folkesson, Olsen, et al. (2005)
Prasoon et al. (2012)
Shan, Charles, and Niethammer (2012)
Wang, Donoghue, and Rueckert (2013)
Liu et al. (2015)
Dam et al. (2015)
Prasoon, Petersen, et al. (2013)

Region growing

Region growing is one of the most common methods for the segmentation of photographs. One or more seed points have to be chosen as the first stage of area development. The regions then expanded from the seed points to the neighboring points based on the membership criteria of a region. For example, the parameter may be pixel size, texture grayscale, color, and so on. Region growing methods are categorized as methods based on classification because the above process could be solved as a problem of binary classification. Methods in these regions have also been researched widely in other fields. In knee MRI segmentation, most can area-based methods are semiautomatic because the seed points have to be manually selected. It introduced an automated system for picking seed points. A candidate with a 3D Gaussian was created in this process, and a classifier decided whether it was cartilage. The candidates are chosen as seeds were categorized into cartilage. The outcome of the polygonal mesh classification was then used as the criterion for that area. Region growing methods have difficulty in managing the thin cartilage structure; thus, it is commonly applicable to bone segmentation, and a few researchers use it to segment cartilages.

Graph-based approaches

Segmentation based on graphs is commonly used in multisurface and multiobject segmentation activities. By using the graph model, the methods model the initial segmentation and then refine the process by reducing unique cost functions. Many approaches based on graphs require effects of presegmentation as the initialization. Presegmentation tests are automatically built on methods (Bae et al., 2009; Öztürk & Albayrak, 2016; Wang, Donoghue, & Rueckert, 2013). An efficient layered graph image segmentation of multiple objects and surfaces has been proposed. The algorithm took a sample of presegmentation as the reference and then performed the process of segmentation using a graph model. The final segmentation was achieved by cost feature optimization.

A level-set-based algorithm for automatic presegmentation of the bone surface was used as initialization (Millington, Li, Wu, Hurwitz, & Sonka, 2005); then the tacit surface was transformed to an actual, structured triangular mesh. Finally, the mesh was used to create a network around the presectioned bone surface in a narrow band, and a multisurface matrix scan algorithm was used to obtain the exact cartilage and bone surfaces simultaneously. Li, Millington, Wu, Chen, and Sonka (2005) and Tang et al. (2006) used a graph-construction scheme based on triangulated surface meshes derived from a topological presegmentation and used an effective graph-cut algorithm to optimize globally. Seeds were hand-painted to signify cartilage and noncartilage. The seeds were propagated slice by slice, and the cartilage regions were automatically segmented by a graph-cut algorithm. The optimal segmentation follows global expectations. The presegmentation can be automatically performed.

The graph-based approaches produce excellent results as commonly used models on knee MRI segmentation. Graph-based methods are less effective than deformable models but more effective than PMC-based methods. One of the main limitations to graph-based approaches is that these approaches need to initialize findings from presegmentation. Graph-based approaches perform well for segmentation refining, but assembling an end-to-end method is hard.

Segmentation of the dual atlas

Graph-based approaches are less effective than deformable models but better than VC-based approaches. One of the main disadvantages to graph-based approaches is that configuration of such approaches requires effects to presegmentation. Graph-based approaches perform well in minimizing the segmentation, but an end-to-end approach is difficult to build. Knee MRI segmentation is a common issue in dual-atlas segmentation (DAS). DAS problems may be overcome by disseminating the sticker from one branded slice to another. This method can also be seen as recording unlabeled slices on one or more labeled slices. Atlas-related approaches (Glocker et al., 2007; Lynch et al., 2000; Stammberger et al., 1999) have been widely researched since the last century. Segmentation was accomplished by registration aimed at updating the atlas in such a way as to increase the conditional posterior density of the trained atlas. Writers have employed linear primal/dual programming to speed up the registration process. Registration is used in some studies to allow for the initial segmentation. According to Lee et al. (2014), a nonrigid identification system registered all training cases to a target picture and picked the best fit atlases. A locally weighted voting (LWV) algorithm was implemented to fuse the atlas information and produce the initial segmentation. Atlas-dependent approaches and other approaches may be mixed classification-based and mathematical model-dependent. Kauffmann et al. (2003) used 3D imaging techniques to remove human involvement and subjectivity from the process. Atlas segmentation is one of the most commonly used DAS processes. The principle is easy to understand, and there are a number of common approaches. Registration phase efficiency however is time-consuming, so they require at least one labeled piece (Table 4.4).

Quantitative analysis

Cartilage volume and thickness are important considerations in the treatment of OA and in the estimation of development. The methods of measurement play a vital role both in clinical trials and in medical studies. This section describes existing techniques for calculating the

TABLE 4.4 Analysis of the different approaches.

Method	Pros	Cons
CMF and BPA	1. Might be fullyautomatic methods; 2. The designers were able to accommodate missing limits well	1. The techniques have cartilage modeling difficulties; 2. The outcome depends primarily on the representativeness of the samples for the preparation
Effective schematic	1. High complexity of the computations; 2. With expert engagement, the methods could achieve accurate results	1. An effective schematic-based automated framework is hard to build; 2. Convergence problem
PMC classification	1 Model concise; 2. Strong precision; 3. Through offering more testing samples, the output could be increased	1. Oversegmentation and postprocessing needs; 2. The sophistication of the computations is high
Graph-based	1. Strong precision; 2. It may be necessary to achieve economic optimization	1. Strong computing complexity and requiring heavy storage; 2. In the majority of cases, initialization is required
Atlas-based	1. Strong precision; 2. The method and results are based on intuition	1. It needs to have current labeled slices; 2. Schwierigkeiten in the treatment of combinations

volume and thickness of the cartilage. The literature reveals a detailed study of the volume estimation processes. By supplying the inner and outer surfaces/boundaries of the cartilage, the calculation of the volume of the cartilage may be achieved directly by counting pixels or polygonal meshes; thus, it is considered a question solved in most studies. The estimation of thickness compared with the calculation of volume is more complicated. The definition of thickness is most commonly known as the shortest distance between the brain-computer interface base points and their respective points. Researchers, therefore, use different thickness concepts in the studies. The thickness estimation may be performed in both 2D and 3D spaces, and the methods can be specified for locating the related points. In some early studies and Heuer, Sommers, Reid, and Bottlang (2001), the thickness was described as the distance between the brain-computer interface base points and the corresponding intersection points on the surface of the cartilage. Solloway et al. (1997) proposed a 2D form known as "M-norm." They measured the medial axis between the inner and outer surfaces and distributed a number of points around the medial axis in equal measure. The usual was cast against both inner and outer boundaries at each point, and the thickness was preserved as the distance between the points of intersection. Dalal et al. (2007) and Tang et al. (2006) determined a regular vector centered on a plain composed of the point and its nearest points, rather than a single point. The form known as "T-norm" could prevent the potential instability. For 3D spaces, 3D Euclidean distance transformation (Lee et al., 2014; Stammberger et al., 1999) and the Laplace equation (Huang et al., 2015) are the most commonly used methods. A few heuristic analysis methods are often used to locate the nearest point, such as scanning the inner cartilage in the usual direction and "Spans and Ridge" (Williams, Holmes, et al., 2010; Williams et al., 2006; Williams et al., 2003; Yin et al., 2010). Some researchers move the initial coordination system to

TABLE 4.5 Methods common for calculating thickness.

Authors	2D/3D	Method	Weakness
Solloway et al. (1997)	2D	M-Norm	1. The distance set in 2D space was not able to correctly represent the true thickness; 2. The solution is unpredictable
Tang et al. (2006)	2D	T-Norm	The distance set in 2D space was not able to correctly represent the true thickness
Cohen et al. (1999)	3D	Casting light	When the brain-computer interface is not smooth, the system is unstable
Stammberger et al. (1999, 2000), Carballido-Gamio et al. (2005, 2008), and Fripp et al. (2010)	3D	3D mapping of the Euclidean distance	High-cost computation
Williams, Taylor, Gao, and Waterton (2003) and Williams et al. (2004)	3D	"Spans and Reefs"	When the brain-computer interface is not smooth, the system is unstable
Shan et al. (2012), Shan, Zach, Charles, and Niethammer (2014), and Huang et al. (2015)	3D	3D equation of a Laplace	High-cost computation
Kauffmann et al. (2003)	3D	Diction map	High-cost computation

a brain computer interface-based coordination system in which the map of thickness can be determined simply. The standard techniques for calculating thickness are summarized in Table 4.5.

Different artifacts and parameters may be used to determine the efficiency of MR image segmentation methods for the knee. Some studies use cartilage volume and cartilage thickness to determine efficiency. The reproducibility of tests-retests is a key criterion for assessment for semiautomatic segmentation methods. The ground realities in knee MR picture segmentation usually emerge from radiologist-sketched outcomes. The findings of manual segmentation are used as the ground realities for semiautomatic and automated segmentation tests. Technical users use semiautomatic segmentation tests to test the efficiency of automated processes as well (Table 4.6).

TABLE 4.6 Metrics for segmentation of the knee MR picture.

Data	Index	Truth in the field
OAI (Eckstein, Wirth, & Nevitt, 2012)	4796 applicants, the longest 108 months of follow-up as of Dec 2016	No subsequent segmentation
SKI10 (Heimann et al., 2010)	100 samples of instruction, 50 check experiments	Segmentation and validation
PROMISE12 (Litjens et al., 2014)	49 workout samples, 29 research tests	Segmentation, but not validation

Conclusion and future trends

The segmentation of computer-aided knee MRIs has tremendous promise both in clinical diagnosis and in medical studies. The central problem in the segmentation of knee MRI is specifically identifying the bone and cartilage boundary. For bone segmentation, the current literature confirms that bone segmentation can be performed well with the deformable model-based approaches. The semiautomatic approaches based on region widening often yield satisfactory results. Classification-based approaches based on voxel/pixel could perform bone segmentation very well. However, because of the higher numerical efficiency, researchers favor deformable structures for bone segmentation. Segmentation of the articular cartilage still requires further study, especially the fully automatic segmentation of the cartilage.

References

Ababneh, S. Y., Prescott, J. W., & Gurcan, M. N. (2011). Automatic graph-cut based segmentation of bones from knee magnetic resonance images for osteoarthritis research. *Medical Image Analysis, 15*, 438–448.

Adams, R., & Bischof, L. (1994). Seeded region growing. *IEEE Transactions on Pattern Analysis and Machine Intelligence, 16*, 641–647.

Bae, K., Shim, H., Tao, C., Chang, S., Wang, J., Boudreau, R., et al. (2009). Intra-and inter-observer reproducibility of volume measurement of knee cartilage segmented from the OAI MR image set using a novel semi-automated segmentation method. *Osteoarthritis and Cartilage, 17*, 1589–1597.

Balafar, M. A., Ramli, A. R., Saripan, M. I., & Mashohor, S. (2010). Review of brain MRI image segmentation methods. *Artificial Intelligence Review, 33*, 261–274.

Boesen, M., Ellegaard, K., Henriksen, M., Gudbergsen, H., Hansen, P., Bliddal, H., et al. (2017). Osteoarthritis year in review 2016: Imaging. *Osteoarthritis and Cartilage, 25*, 216–226.

Braun, H. J., & Gold, G. E. (2012). Diagnosis of osteoarthritis: Imaging. *Bone, 51*, 278–288.

Brem, M., Lang, P., Neumann, G., Schlechtweg, P., Schneider, E., Jackson, R., et al. (2009). Magnetic resonance image segmentation using semi-automated software for quantification of knee articular cartilage—Initial evaluation of a technique for paired scans. *Skeletal Radiology, 38*, 505–511.

Carballido-Gamio, J., Bauer, J., Lee, K.-Y., Krause, S., & Majumdar, S. (2005). Combined image processing techniques for characterization of MRI cartilage of the knee. *Conference Proceedings: Annual International Conference of the IEEE Engineering in Medicine and Biology Society*, e6.

Carballido-Gamio, J., Bauer, J. S., Stahl, R., Lee, K.-Y., Krause, S., Link, T. M., et al. (2008). Inter-subject comparison of MRI knee cartilage thickness. *Medical Image Analysis, 12*, 120–135.

Chan, W. P., Lang, P., Stevens, M. P., Sack, K., Majumdar, S., Stoller, D. W., et al. (1991). Osteoarthritis of the knee: Comparison of radiography, CT, and MR imaging to assess extent and severity. *AJR. American Journal of Roentgenology, 157*, 799–806.

Chan, T. F., & Vese, L. A. (2001). Active contours without edges. *IEEE Transactions on Image Processing, 10*, 266–277.

Cohen, Z. A., McCarthy, D. M., Kwak, S. D., Legrand, P., Fogarasi, F., Ciaccio, E. J., et al. (1999). Knee cartilage topography, thickness, and contact areas from MRI: In-vitro calibration and in-vivo measurements. *Osteoarthritis and Cartilage, 7*, 95–109.

Cootes, T. F., Edwards, G. J., & Taylor, C. J. (1999). Comparing active shape models with active appearance models. In *BMVC* (pp. 173–182).

Cootes, T. F., Twining, C. J., Petrovic, V. S., Schestowitz, R., & Taylor, C. J. (2005). Groupwise construction of appearance models using piece-wise affine deformations. In *BMVC* (pp. 879–888).

Dalal, P., Munsell, B. C., Wang, S., Tang, J., Oliver, K., Ninomiya, H., et al. (2007). A fast 3D correspondence method for statistical shape modeling. In *Computer vision and pattern recognition, 2007. CVPR'07. IEEE conference on* (pp. 1–8).

Dalvi, R., Abugharbieh, R., Wilson, D., & Wilson, D. R. (2007). Multi-contrast MR for enhanced bone imaging and segmentation. In *2007 29th annual international conference of the IEEE engineering in medicine and biology society* (pp. 5620–5623).

Dam, E. B., Lillholm, M., Marques, J., & Nielsen, M. (2015). Automatic segmentation of high-and low-field knee MRIs using knee image quantification with data from the osteoarthritis initiative. *Journal of Medical Imaging, 2*, 024001.

Davies, R. H., Twining, C. J., Cootes, T. F., Waterton, J. C., & Taylor, C. J. (2002a). 3D statistical shape models using direct optimisation of description length. In *European conference on computer vision* (pp. 3–20).

Davies, R. H., Twining, C. J., Cootes, T. F., Waterton, J. C., & Taylor, C. J. (2002b). A minimum description length approach to statistical shape modeling. *IEEE Transactions on Medical Imaging, 21*, 525–537.

Dodin, P., Martel-Pelletier, J., Pelletier, J.-P., & Abram, F. (2011). A fully automated human knee 3D MRI bone segmentation using the ray casting technique. *Medical & Biological Engineering & Computing, 49*, 1413–1424.

Dodin, P., Pelletier, J.-P., Martel-Pelletier, J., & Abram, F. (2010). Automatic human knee cartilage segmentation from 3-D magnetic resonance images. *IEEE Transactions on Biomedical Engineering, 57*, 2699–2711.

Duryea, J., Neumann, G., Brem, M., Koh, W., Noorbakhsh, F., Jackson, R., et al. (2007). Novel fast semi-automated software to segment cartilage for knee MR acquisitions. *Osteoarthritis and Cartilage, 15*, 487–492.

Eckstein, F., Burstein, D., & Link, T. M. (2006). Quantitative MRI of cartilage and bone: Degenerative changes in osteoarthritis. *NMR in Biomedicine, 19*, 822–854.

Eckstein, F., Cicuttini, F., Raynauld, J.-P., Waterton, J., & Peterfy, C. (2006). Magnetic resonance imaging (MRI) of articular cartilage in knee osteoarthritis (OA): Morphological assessment. *Osteoarthritis and Cartilage, 14*, 46–75.

Eckstein, F., Gavazzeni, A., Sittek, H., Haubner, M., Lösch, A., Milz, S., et al. (1996). Determination of knee joint cartilage thickness using three-dimensional magnetic resonance chondro—Crassometry (3D MR—CCM). *Magnetic Resonance in Medicine, 36*, 256–265.

Eckstein, F., Wirth, W., & Nevitt, M. C. (2012). Recent advances in osteoarthritis imaging-the osteoarthritis initiative. *Nature Reviews Rheumatology, 8*, 622–630.

Edelman, R. R., & Warach, S. (1993). Magnetic resonance imaging. *New England Journal of Medicine, 328*, 708–716.

Felson, D. T. (2013). Osteoarthritis as a disease of mechanics. *Osteoarthritis and Cartilage, 21*, 10–15.

Florack, L. M. J. (1993). *The syntactical structure of scalar images*. Universiteit Utrecht, Faculteit Geneeskunde.

Folkesson, J., Dam, E., Olsen, O. F., Pettersen, P., & Christiansen, C. (2005). Automatic segmentation of the articular cartilage in knee MRI using a hierarchical multi-class classification scheme. In *International conference on medical image computing and computer-assisted intervention* (pp. 327–334).

Folkesson, J., Dam, E. B., Olsen, O. F., Pettersen, P. C., & Christiansen, C. (2007). Segmenting articular cartilage automatically using a voxel classification approach. *IEEE Transactions on Medical Imaging, 26*, 106–115.

Folkesson, J., Olsen, O. F., Pettersen, P., Dam, E., & Christiansen, C. (2005). Combining binary classifiers for automatic cartilage segmentation in knee MRI. In *International workshop on computer vision for biomedical image applications* (pp. 230–239).

Fripp, J., Bourgeat, P., Mewes, A. J., Warfield, S. K., Crozier, S., & Ourselin, S. (2005). 3D statistical shape models to embed spatial relationship information. In *International workshop on computer vision for biomedical image applications* (pp. 51–60).

Fripp, J., Crozier, S., Warfield, S. K., & Ourselin, S. (2007). Automatic segmentation of the bone and extraction of the bone–cartilage interface from magnetic resonance images of the knee. *Physics in Medicine and Biology, 52*, 1617.

Fripp, J., Crozier, S., Warfield, S. K., & Ourselin, S. (2010). Automatic segmentation and quantitative analysis of the articular cartilages from magnetic resonance images of the knee. *IEEE Transactions on Medical Imaging, 29*, 55–64.

Ghelman, B. (1985). Meniscal tears of the knee: Evaluation by high-resolution CT combined with arthrography. *Radiology, 157*, 23–27.

Gilles, B., & Magnenat-Thalmann, N. (2010). Musculoskeletal MRI segmentation using multi-resolution simplex meshes with medial representations. *Medical Image Analysis, 14*, 291–302.

Glocker, B., Komodakis, N., Paragios, N., Glaser, C., Tziritas, G., & Navab, N. (2007). Primal/dual linear programming and statistical atlases for cartilage segmentation. In *International conference on medical image computing and computer-assisted intervention* (pp. 536–543).

Hakky, M., Jarraya, M., Ratzlaff, C., Guermazi, A., & Duryea, J. (2015). Validity and responsiveness of a new measure of knee osteophytes for osteoarthritis studies: Data from the osteoarthritis initiative. *Osteoarthritis and Cartilage, 23*, 2199–2205.

Heidari, B. (2011). Knee osteoarthritis prevalence, risk factors, pathogenesis, and features: Part I. *Caspian Journal of Internal Medicine, 2*, 205–212.

Heimann, T., & Meinzer, H.-P. (2009). Statistical shape models for 3D medical image segmentation: A review. *Medical Image Analysis*, *13*, 543–563.

Heimann, T., Morrison, B. J., Styner, M. A., Niethammer, M., & Warfield, S. (2010). Segmentation of knee images: A grand challenge. In *Proc. MICCAI workshop on medical image analysis for the clinic* (pp. 207–214).

Heuer, F., Sommers, M., Reid, J., & Bottlang, M. (2001). *Estimation of cartilage thickness from joint surface scans: Comparative analysis of computational methods. Vol. 50* (pp. 569–570). ASME-Publications-Bed.

Huang, C., Shan, L., Charles, H. C., Wirth, W., Niethammer, M., & Zhu, H. (2015). Diseased region detection of longitudinal knee magnetic resonance imaging data. *IEEE Transactions on Medical Imaging*, *34*, 1914–1927.

Iglesias, J. E., & Sabuncu, M. R. (2015). Multi-atlas segmentation of biomedical images: A survey. *Medical Image Analysis*, *24*, 205–219.

Jevsever, D. S. (2013). Treatment of osteoarthritis of the knee: Evidence-based guideline. *Journal of the American Academy of Orthopaedic Surgeons*, *21*, 571–576.

Kapur, T., Beardsley, P., Gibson, S., Grimson, W., & Wells, W. (1998). Model-based segmentation of clinical knee MRI. In *Proc. IEEE Int'l workshop on model-based 3D image analysis* (pp. 97–106).

Kass, M., Witkin, A., & Terzopoulos, D. (1988). Snakes: Active contour models. *International Journal of Computer Vision*, *1*, 321–331.

Kauffmann, C., Gravel, P., Godbout, B., Gravel, A., Beaudoin, G., Raynauld, J.-P., et al. (2003). Computer-aided method for quantification of cartilage thickness and volume changes using MRI: Validation study using a synthetic model. *IEEE Transactions on Biomedical Engineering*, *50*, 978–988.

Lee, J.-G., Gumus, S., Moon, C. H., Kwoh, C. K., & Bae, K. T. (2014). Fully automated segmentation of cartilage from the MR images of knee using a multi-atlas and local structural analysis method. *Medical Physics*, *41*, 092303.

Li, K., Millington, S., Wu, X., Chen, D. Z., & Sonka, M. (2005). Simultaneous segmentation of multiple closed surfaces using optimal graph searching. In *Biennial international conference on information processing in medical imaging* (pp. 406–417).

Litjens, G., Toth, R., van de Ven, W., Hoeks, C., Kerkstra, S., van Ginneken, B., et al. (2014). Evaluation of prostate segmentation algorithms for MRI: The PROMISE12 challenge. *Medical Image Analysis*, *18*, 359–373.

Liu, Q., Wang, Q., Zhang, L., Gao, Y., & Shen, D. (2015). Multi-atlas context forests for knee MR image segmentation. In *International workshop on machine learning in medical imaging* (pp. 186–193).

Lorigo, L. M., Faugeras, O., Grimson, W. E. L., Keriven, R., & Kikinis, R. (1998). Segmentation of bone in clinical knee MRI using texture-based geodesic active contours. In *International conference on medical image computing and computer-assisted intervention* (pp. 1195–1204).

Losina, E., Paltiel, A. D., Weinstein, A. M., Yelin, E., Hunter, D. J., Chen, S. P., et al. (2015). Lifetime medical costs of knee osteoarthritis management in the United States: Impact of extending indications for total knee arthroplasty. *Arthritis Care & Research*, *67*, 203–215.

Lynch, J. A., Zaim, S., Zhao, J., Stork, A., Peterfy, C. G., & Genant, H. K. (2000). Cartilage segmentation of 3D MRI scans of the osteoarthritic knee combining user knowledge and active contours. In *Vol. 3979. Proceedings of the SPIE, Medical imaging 2000* (pp. 925–935). SPIE.

McInerney, T., & Terzopoulos, D. (1996). Deformable models in medical image analysis: A survey. *Medical Image Analysis*, *1*, 91–108.

Millington, S., Li, K., Wu, X., Hurwitz, S., & Sonka, M. (2005). Automated simultaneous 3D segmentation of multiple cartilage surfaces using optimal graph searching on MRI images. *Osteoarthritis and Cartilage*, *13*, S130.

Newberg, A., Munn, C., & Robbins, A. (1985). Complications of arthrography. *Radiology*, *155*, 605–606.

Öztürk, C. N., & Albayrak, S. (2016). Automatic segmentation of cartilage in high-field magnetic resonance images of the knee joint with an improved voxel-classification-driven region-growing algorithm using vicinity-correlated subsampling. *Computers in Biology and Medicine*, *72*, 90–107.

Pakin, S. K., Tamez-Pena, J. G., Totterman, S., & Parker, K. J. (2002). Segmentation, surface extraction, and thickness computation of articular cartilage. In *Vol. 4684. Proceedings of the SPIE, Medical imaging 2002* (pp. 155–166). SPIE.

Piplani, M. A., Disler, D. G., McCauley, T. R., Holmes, T. J., & Cousins, J. P. (1996). Articular cartilage volume in the knee: Semiautomated determination from three-dimensional reformations of MR images. *Radiology*, *198*, 855–859.

Prasoon, A., Igel, C., Loog, M., Lauze, F., Dam, E., & Nielsen, M. (2012). Cascaded classifier for large-scale data applied to automatic segmentation of articular cartilage. In *SPIE medical imaging*. 83144V-83144V-9.

Prasoon, A., Igel, C., Loog, M., Lauze, F., Dam, E. B., & Nielsen, M. (2013). Femoral cartilage segmentation in Knee MRI scans using two stage voxel classification. In *2013 35th annual international conference of the IEEE engineering in medicine and biology society (EMBC)* (pp. 5469–5472).

Prasoon, A., Petersen, K., Igel, C., Lauze, F., Dam, E., & Nielsen, M. (2013). Deep feature learning for knee cartilage segmentation using a triplanar convolutional neural network. In *International conference on medical image computing and computer-assisted intervention* (pp. 246–253).

Schmid, J., Kim, J., & Magnenat-Thalmann, N. (2011). Robust statistical shape models for MRI bone segmentation in presence of small field of view. *Medical Image Analysis*, *15*, 155–168.

Schmid, J., & Magnenat-Thalmann, N. (2008). MRI bone segmentation using deformable models and shape priors. In *International conference on medical image computing and computer-assisted intervention* (pp. 119–126).

Seim, H., Kainmueller, D., Lamecker, H., Bindernagel, M., Malinowski, J., & Zachow, S. (2010). Model-based auto-segmentation of knee bones and cartilage in MRI data. In *Medical image analysis for the clinic: A grand challenge, Beijing*.

Shan, L., Charles, C., & Niethammer, M. (2012). Automatic multi-atlas-based cartilage segmentation from knee MR images. In *2012 9th IEEE international symposium on biomedical imaging (ISBI)* (pp. 1028–1031).

Shan, L., Zach, C., Charles, C., & Niethammer, M. (2014). Automatic atlas-based three-label cartilage segmentation from MR knee images. *Medical Image Analysis*, *18*, 1233–1246.

Solloway, S., Hutchinson, C. E., Waterton, J. C., & Taylor, C. J. (1997). The use of active shape models for making thickness measurements of articular cartilage from MR images. *Magnetic Resonance in Medicine*, *37*, 943–952.

Stammberger, T., Eckstein, F., Englmeier, K.-H., & Reiser, M. (1999). Determination of 3 D cartilage thickness data from MR imaging: Computational method and reproducibility in the living. *Magnetic Resonance in Medicine*, *41*, 529–536.

Stammberger, T., Hohe, J., Englmeier, K. H., Reiser, M., & Eckstein, F. (2000). Elastic registration of 3D cartilage surfaces from MR image data for detecting local changes in cartilage thickness. *Magnetic Resonance in Medicine*, *44*, 592–601.

Tang, J., Millington, S., Acton, S. T., Crandall, J., & Hurwitz, S. (2006). Surface extraction and thickness measurement of the articular cartilage from MR images using directional gradient vector flow snakes. *IEEE Transactions on Biomedical Engineering*, *53*, 896–907.

Wang, Z., Donoghue, C., & Rueckert, D. (2013). Patch-based segmentation without registration: Application to knee MRI. In *International workshop on machine learning in medical imaging* (pp. 98–105).

Wang, P., He, X., Li, Y., Zhu, X., Chen, W., & Qiu, M. (2016). Automatic knee cartilage segmentation using multi-feature support vector machine and elastic region growing for magnetic resonance images. *Journal of Medical Imaging and Health Informatics*, *6*, 948–956.

Williams, T. G., Holmes, A. P., Waterton, J. C., Maciewicz, R. A., Hutchinson, C. E., Moots, R. J., et al. (2010). Anatomically corresponded regional analysis of cartilage in asymptomatic and osteoarthritic knees by statistical shape modelling of the bone. *IEEE Transactions on Medical Imaging*, *29*, 1541–1559.

Williams, T. G., Holmes, A. P., Waterton, J. C., Maciewicz, R. A., Nash, A. F., & Taylor, C. J. (2006). Regional quantitative analysis of knee cartilage in a population study using MRI and model based correspondences. In *Biomedical imaging: Nano to macro, 2006. 3rd IEEE international symposium on* (pp. 311–314).

Williams, T. G., Taylor, C. J., Gao, Z., & Waterton, J. C. (2003). Corresponding articular cartilage thickness measurements in the knee joint by modelling the underlying bone (commercial in confidence). In *Biennial international conference on information processing in medical imaging* (pp. 126–135).

Williams, T. G., Taylor, C. J., Waterton, J. C., & Holmes, A. (2004). Population analysis of knee cartilage thickness maps using model based correspondence. In *Biomedical imaging: Nano to macro, 2004. IEEE international symposium on* (pp. 193–196).

Williams, T. G., Vincent, G., Bowes, M., Cootes, T., Balamoody, S., Hutchinson, C., et al. (2010). Automatic segmentation of bones and inter-image anatomical correspondence by volumetric statistical modelling of knee MRI. In *2010 IEEE international symposium on biomedical imaging: From nano to macro* (pp. 432–435).

Yin, Y., Zhang, X., Williams, R., Wu, X., Anderson, D. D., & Sonka, M. (2010). LOGISMOS—layered optimal graph image segmentation of multiple objects and surfaces: Cartilage segmentation in the knee joint. *IEEE Transactions on Medical Imaging*, *29*, 2023–2037.

CHAPTER

5

Computational approach to assess mucormycosis: A systematic review

Debanjan Mitra[a], *Benjir Nachhmin*[a], *Arindam Ganguly*[b], *and Pradeep Mohapatra*[a]

[a]Department of Microbiology, Raiganj University, Raiganj, Uttar Dinajpur, West Bengal, India
[b]Department of Microbiology, Bankura Sammilani College, Bankura, West Bengal, India

Introduction

The current COVID-19 situation has pushed the world into a dangerous situation. The death march continues. Scientists are still busy searching for suitable drugs for the treatment of SARS-CoV-2. Researchers worldwide have pursued many chemical lead compounds as drug molecules for COVID-19 (Cavasotto & Di Filippo, 2021; Mitra, Paul, Thatoi, & Mohapatra, 2021; Shah, Modi, & Sagar, 2020). Nowadays, bioactive compounds are being used to cure various diseases (Mitra, Dey, Biswas, & Mohapatra., 2021; Seo & Choi, 2021), and this case is no exception. Many researchers have already proposed several bioactive compounds from different biological sources to treat SARS-CoV-2 infection (Mitra & Bose, 2021; Mitra & Mohapatra., 2021; Mitra et al., 2021; Verma et al., 2021). Study of the virus genome and proteins is also underway (Mitra, Pal, & Mohapatra, 2020; Nawaz, Fournier-Viger, Shojaee, & Fujita, 2021; Valerio et al., 2021; Wang & Jiang, 2021).

According to WHO, in February of 2022 there were 434,154,739 confirmed cases of infection by this deadly novel coronavirus, with 5,960,972 deaths reported worldwide. Currently, additional alarming situations are appearing among those who have recovered from COVID-19. Fatal mucormycosis infection has often been observed among hospitalized patients recovering from COVID-19 in India. Those who have received supplementary oxygen therapy or mechanical ventilation are being affected most often.

Mucormycosis is a type of angioinvasive fungal disease caused by members of Zygomycetes. Mucormycosis can result in a fast-growing, acute, and occasionally fatal disease caused by different fungi of the Mucorales order, commonly found in the soil or

Computational Intelligence in Healthcare Applications
https://doi.org/10.1016/B978-0-323-99031-8.00004-1

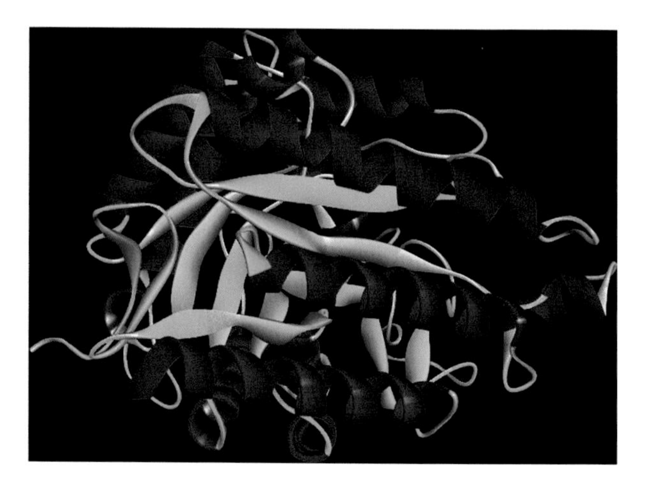

FIG. 5.1 Crystal structure of glycoside hydrolase from *Rhizomucor miehei. No permission required.*

environment. *Rhizopus arrhizus* is a widespread cause of mucormycosis throughout the world (Jeong et al., 2019; Prakash et al., 2019). Infections due to *Rhizopus microsporus, Rhizomucor miehei* (Fig. 5.1), and *Rhizopus homothallicus* are the main fungi causes of rapid mucormycosis in India (Pandey et al., 2018; Prakash et al., 2019). A rhinofacial mucormycosis was reported in India that was caused by *Rhizomucor variabilis* (Hemashettar et al., 2011). It was identified by multilocus DNA sequence data. The *Apophysomyces* species ranks second after *Rhizopus* sp. in India, whereas the *Lichtheimia* sp. have been found to cause mucormycosis in developed countries (García et al., 2021; Kachuei et al., 2021). Recently, it was reported that mucormycosis occurred due to *Saksenaea* sp. in a 34-year-old man. Mucormycosis with pulmonary disease was reported to be caused by *Cunninghamella bertholletiae* (Hu et al., 2021; Malini & Pasupathi, 2021; Yamamoto et al., 2021).

Mucormycosis occurs from subjection to mucormyete molds. These organisms are present in the soil, piles of compost, leaves, rotting wood, etc. A patient can be exposed to mucormycosis by breathing the affected mold spores from the air. In turn, they may develop the infection in the sinuses, central nervous system, face, eyes, and lungs. The mucormycosis infection can also be passed to human skin via a cut or burn. In that case, the wound or burn ends up becoming the main region of infection. Immunocompetent patients with severe traumatic injury also exhibit cutaneous mucormycosis infection due to *Rhizopus microspores* (El Deeb, Al Soub, Almaslamani, Al Khuwaiter, & Taj-Aldeen, 2005).

There are also some rare modes of transmission of mucormycosis. A healthy person can also have mucormycosis infection due to oroantral communication (Nilesh, Malik, & Belgaumi, 2015). Recent investigations found that self-extraction of teeth can be a possible source of mucormycosis (Agarwal et al., 2020).

In 1855, Friedrich Küchenmeister used the term mucormycosis for the first time. In 1876, Fürbringer identified mucormycosis in human lungs. In 1884, *Mucor corymbifera* and *Mucor rhizopodiformis* were identified as responsible for the development of the disease in rabbits by Lichtheim (McBride, Corson, & Dammin, 1960). In 1972, the first outbreak of the mucormycosis epidemic was reported in an amphibian collection in Germany. *Mucor amphibiorum* was responsible for that outbreak (Creeper, Main, Berger, Huntress, &

TABLE 5.1 Different types of mucormycosis and their symptoms.

Types of mucormycosis	Symptoms
Rhino-cerebral mucormycosis	Headache, fever, reddish and swollen skin over nose and sinuses, swelling of eyes, large dark spot on nose, pain in facial area
Pulmonary (lung) mucormycosis	Fever, chest pain, shortness of breath, bloody dark fluid appears with coughing
Gastrointestinal mucormycosis	Abdominal distension, bloody dark vomit, diffuse abdominal pain
Renal mucormycosis	Flank pain, fever, pyuria, hematuria
Cutaneous mucormycosis	Reddish and swollen skin, pain, black skin
Disseminated mucormycosis	Infection spreads through the bloodstream

No permission required.

Boardman, 1998). In Spain, mucormycosis increased significantly within the period 2007–15. The mortality rate was also high (47.4%) in Spain (Guinea et al., 2017).

Currently, six types of mucormycosis have been identified based on the infection area of fungus in the human body. The divisions of mucormycosis were described based on their symptoms. Those types are rhino-cerebral mucormycosis, pulmonary (lung) mucormycosis, gastrointestinal mucormycosis, renal mucormycosis, cutaneous mucormycosis, and disseminated mucormycosis. The symptoms of each type are different due to the location of the infected area (Table 5.1).

Computational approach in mucormycosis

Computational studies of diseases are currently playing a very important role in medicine. Analysis of genes and proteins can reveal their nature and also increase knowledge in the field of health care (Lucassen & Houlston, 2014; Mitra & Das Mohapatra, 2021a, 2021b). Pan-genomic studies of mucormycosis have revealed the host-pathogen interaction. A study also revealed that the infection developed a dysbiotic microbiome with low α-diversity, dominated by *staphylococci* (Shelburne et al., 2015).

Immunoinformatics studies of mucormycosis indicate that T-cell immunotherapy can increase immunity in patients (Castillo-Caro et al., 2015). A genomic and transcriptomics study has been carried out on 30 species of fungi for host-fungal determination. This study showed platelet-derived growth factor (PDGF) receptor B signaling in response to divergent pathogenic fungi and inhibition of this receptor reduced the damage caused by the fungi (Chibucos et al., 2016). Computational modeling of mucormycosis has provided knowledge of innate immune effectors and early immune response. The mathematical model pointed up the role of proinflammatory waves for phagocytic assignment and the significance of the inhibition of spore germination for shielding from fungal infection (Inglesfield et al., 2018). An RNAi-based functional genomics study allowed the identification of new virulence factors in those species responsible for mucormycosis through a forward genetic approach. Knock-out mutations on those genes also showed their effect on virulence (Trieu et al., 2017). Investigation of ex-sRNAs from *Rhizopus delemar* revealed that it has a rich reservoir of secreted ex-sRNAs and also elucidated their virulence mechanism (Liu, Bruni, Taylor, Zhang, & Wang, 2018). Whole-genome sequencing of *Mucor circinelloides* has helped to investigate how patients are infected by invasive wound mucormycosis. This study also showed that patients can be infected by two different strains at the same time (Garcia-Hermoso et al., 2018).

Computational studies are also playing a principal role in drug discovery (Mitra & Bose, 2021; Mitra, Dey, et al., 2021; Mitra & Mohapatra, 2021; Mitra, Paul, et al., 2021). RNA-dependent RNA polymerase from *Rhizopus oryzae* was taken as a target in drug discovery. Sofosbuvir showed a very high binding affinity with the target and appears to be a lead compound (Elfiky, 2019). A bioactive compound named 1,8 cineole that is present in the essential oils of eucalyptus leaves showed promise against the β-glucan synthase enzyme of *Candida albicans*. ADMET properties of ligands were also studied (Sharma & Kaur, 2021).

Mucormycosis association with COVID-19

During the second wave of COVID-19, mucormycosis badly hit people in India who had recently recovered from COVID-19 (Gandra, Ram, & Levitz, 2021; Nambiar, Varma, & Damdoum, 2021). Through May 25, 2021, a total of 45,432 cases were reported and more than 4300 persons had lost their lives. Mehta and Pandey reported a case of a 60-year-old patient who had been admitted due to COVID-19, but was also was infected with rhino-orbital mucormycosis during the treatment time (Mehta & Pandey, 2020).

Another case of rhino-cerebral mucormycosis associated with COVID-19 was reported in a 41-year-old man with a history of diabetes mellitus. COVID-19 was detected by RT-PCR. The mucormycosis infection had effects on nasal passages, the oral cavity, the sinuses, and the brain (Alekseyev, Didenko, & Chaudhry, 2021). A similar case was reported in the northwestern part of India. A rare and fatal gastrointestinal mucormycosis with COVID-19 was reported in an 86-year-old male patient. The patient died within 36 h after being admitted to the hospital (Nehara et al., 2021). COVID-19 associated mucormycosis was also reported in a 55-year-old male patient who also suffered from diabetes (do Monte Junior et al., 2020). A group of researchers has thoroughly investigated the mucormycosis cases, which represent an epidemic condition within a pandemic due to the association with COVID-19. In this case study, conducted over 3 months, 47 patients were identified who had mucormycosis with COVID-19. Among them, 10 patients had diabetes and 2 patients had already taken the first dose of the COVID-19 vaccine. A case report suggested that the imprudent application of corticosteroids and broad-spectrum antibiotics with poor glycemic limitations increased the high infection and mortality rate (Patel et al., 2021). A post-COVID-19 sino-orbital mucormycosis infection was reported by Maini et al. (2021). The patient developed symptoms of mucormycosis 18 days after recovering from COVID-19. Amphotericin B and fluconazole were used to treat that patient (Maini et al., 2021). A COVID-19 patient was diagnosed with paranasal sinus mucormycosis. That patient had to undergo emergency sinus surgery (Selarka et al., 2021). Two 36-year-old persons suffered from rhino-orbital-cerebral mucormycosis and

COVID-19 at the same time. Oxygen support was needed in both patients (Dallalzadeh, Ozzello, Liu, Kikkawa, & Korn, 2021; Saldanha, Reddy, & Vincent, 2021).

Diseases act as risk factors of mucormycosis

Mucormycosis is not only associated with COVID-19, but also with other diseases with which the chances of infection increase. Researchers have already reported many case studies that indicate patients with those diseases have a greater chance of acquiring mucormycosis infection. Those diseases (Fig. 5.2) which act as risk factors of mucormycosis include:

Diabetes

Diabetes is the most reported disease found to be associated with mucormycosis. Hopkins and Treloar (1997) reported that a young woman with diabetic ketoacidosis was also infected with mucormycosis. Even after appropriate treatment with surgical debridement and use of amphotericin B, the central nervous system of that patient was invaded by the fungus (Hopkins & Treloar, 1997). A survey of rhino-orbital-cerebral mucormycosis was reported with diabetes mellitus patients. In this survey, 23 men and 12 women were studied. Among them, 5 patients had type I diabetes mellitus, 29 had type II diabetes mellitus, and one had secondary diabetes (Bhansali et al., 2004). Patients suffering from type I diabetes mellitus with rhino-orbital-cerebral mucormycosis have improved through the combined approach of amphotericin B and surgery (Bhadada et al., 2005). Rammaert, Lanternier, Poirée, Kania,

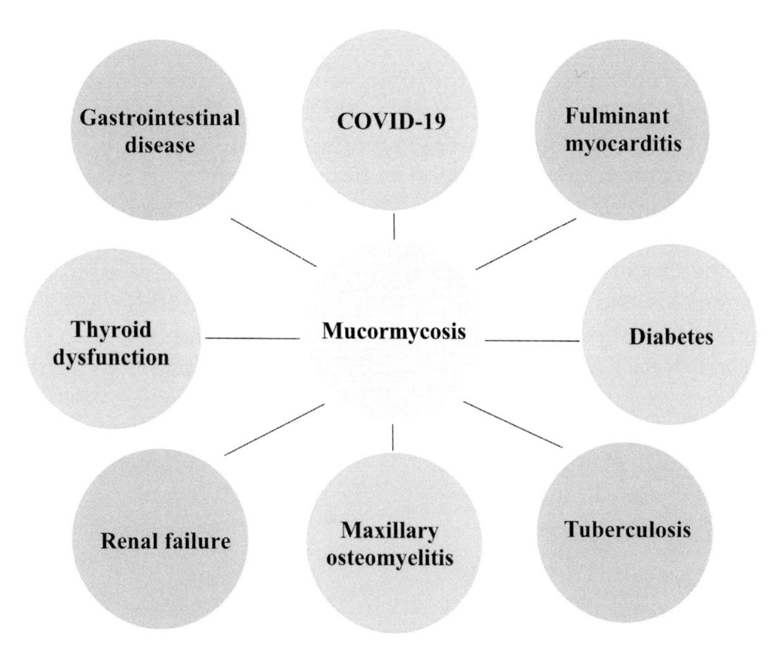

FIG. 5.2 Diseases that act as risk factors of mucormycosis. *No permission required.*

and Lortholary (2012) provided a review of mucormycosis cases and highlighted the absolute necessity of prompt and rapid treatment to reduce the mortality rate. They also reported that ketoacidosis increases the risk of mucormycosis (Rammaert et al., 2012). A rare pulmonary mucormycosis showed high mortality and morbidity in patients with diabetes mellitus. Diabetes acts as a high risk factor of mucormycosis due to impaired innate immunity (Iqbal, Irfan, Jabeen, Kazmi, & Tariq, 2017). A survey of 418 patients from 1986 to 2016 in Mexico showed 72% of patients had diabetes mellitus. The mortality rate was reported as 51%. *Rhizopus* sp. were the most responsible microorganisms for mucormycosis of those patients. Of these patients, 89% used amphotericin B and 90% of patients needed surgery (Corzo-León, Chora-Hernández, Rodríguez-Zulueta, & Walsh, 2018). Lipatov et al. (2018) reported a rare case of uncontrolled diabetes of a 62-year-old male patient. The patient had 30 years of smoking history, and pulmonary mucormycosis affected the patient very badly (Lipatov et al., 2018). Similar pulmonary mucormycosis was also reported in a 52-year-old male patient who had type II diabetes. That case report suggested difficulty in diagnosis and the significance of histological examination in identifying the mucormycosis (Mekki et al., 2020). Pathophysiology of mucormycosis showed how diabetic patients are badly infected by fungus, creating serious life-threatening situations (Afroze, Korlepara, Rao, & Madala, 2017). Overall reports from those previous investigations suggest that mainly 50- to 65-year-old patients with diabetes bear a higher risk of being infected by mucormycosis. Type II diabetic patients suffer more with mucormycosis as compared to type I. Mortality rates are also very high (almost 60%–70%) in diabetic patients who have suffered from mucormycosis. Recovery is generally completed within 12 weeks with early diagnosis and proper treatment.

Tuberculosis

Tuberculosis is a serious infection of the lungs caused by *Mycobacterium tuberculosis*. General signs and symptoms include fever, cough, night sweats, chest pain, and loss of appetite, weight loss, and fatigue. There are many case reports suggesting that tuberculosis also acts as a risk factor of mucormycosis. A coinfection of pulmonary mucormycosis and tuberculosis was reported in a 52-year-old patient. The patient was an insulin-dependent diabetic person. Computed tomography showed badly affected lung conditions. Amphotericin B was used in the treatment of the mucormycosis (Jiménez-Zarazúa et al., 2019). A similar type case, i.e., pulmonary tuberculosis and mucormycosis, was reported in 2019 in a 30-year-old diabetic patient. The patient had shortness of breath, chest pain, low fever, and cough (Aggarwal, Chander, Janmeja, & Katyal, 2015). The use of a high amount of insulin in the diabetic patient showed a coinfection of pulmonary tuberculosis and mucormycosis. A 54-year-old patient diagnosed as diabetic for 15 years faced this same type of coinfection (Ramesh, Kaur, Deepak, & Kumar, 2020). Mortality rates are also high due to coinfection of mucormycosis, tuberculosis, and diabetes.

Maxillary osteomyelitis

Maxillary osteomyelitis is a rare condition of bone inflammation generally caused by odontogenic bacteria. In 2011, Pandey et al. reported on a group of patients who were coinfected with mucormycosis and maxillary osteomyelitis. The patients had an average age of 50 years. During the treatment, necrotic gray-colored bone in the right maxillary molar region was reported in all those patients (Pandey et al., 2018). In 2021, many similar cases of coinfections were reported. A 52-year-old patient who also had this type of coinfection had a history of tooth pain and extensive tissue necrosis (Selvamani, Donoghue, Bharani, & Madhushankari, 2015). An immunocompromised 62-year-old female patient faced tooth extraction due to rhino-maxillary osteomyelitis. After a certain time, she was admitted to the hospital with mucormycosis and incursive surgical care was needed to control the infection (Kumar, Singh, Pandey, Agrawal, & Singh, 2015). However, the mortality rate in such cases of coinfection remains low if proper care is given in due time. An informatics-based review indicated that 71% of males were infected with osteomyelitis and arthritis with mucormycosis. Most of the patients were immunocompromised. The data was taken from EMBASE and PUBMED databases from 1978 to 2014 (Taj-Aldeen et al., 2017).

Renal disease

Very few cases have been reported regarding renal failure and mucormycosis coinfections. A 48-year-old patient admitted to the hospital reported a history of hematuria. A medical report had suggested the renal failure of that patient. During treatment, the patient also suffered from mucormycosis (Sheth, Talwalkar, Desai, & Acharya, 1981). Amphotericin B is excreted by the kidney and can be nephrotoxic in nature, so the use of this drug in renal failure patients to treat mucormycosis has some risks. Treatments of this type of patient are very difficult. Gandhi et al. (2005) had discussions about renal dysfunction such as renal failure, chronic kidney disease, etc. They found that the use of intravenous liposomal amphotericin is more effective than amphotericin B for the treatment of mucormycosis with renal dysfunction. Currently, two liposomal amphotericins, i.e., AmBisome and Fungisome, are available on the market (Gandhi et al., 2005). A 20-year-old female renal transplant recipient had idiopathic intracranial hypertension followed by rhinocerebral mucormycosis. This type of condition creates immunosuppression, which leads to life-threatening complications (Jha, Gude, Chennamsetty, & Kotari, 2013). The diagnosis of renal mucormycosis is found by use of renal histology sections of nephrectomized kidneys and renal biopsy. Removal of infected tissues and use of posaconazole can save a patient's life (Gupta & Gupta, 2012). A clinical study with 1900 renal transplant patients for 6 years concluded that fungal infections during renal transplants can increase mortality and morbidity (Patel et al., 2017).

Thyroid dysfunction

Thyroid involvement with mucormycosis is rarely reported. A 52-year-old male patient was admitted to the hospital due to throat pain and nonproductive cough. Organ transplants had also been reported in the patient's history. The thyroid was seeded during a fungal infection. Later it created laryngeal nerve paralysis and the patient died (Prasad et al., 2018). Another case report of a 79-year-old male patient with mucormycosis of the thyroid gland was reported. Thyroid nodule and dysphonia occurred due to chronic lymphocytic leukemia.

Fungal infection was identified by biopsy test, and surgery was needed to save the life of that patient (Mascarella et al., 2019).

Gastrointestinal disease

Gastrointestinal disease is a disorder of the digestive system. In 2012, Goel et al. reported on an immunocompetent child with gastrointestinal mucormycosis. Later the patient responded to aggressive treatment with surgical debridement and the use of polymyxin E antifungal agents for successful management of the disease (Goel et al., 2013). A 55-year-old woman was reported to have mucormycosis and gastric disorder. Inflammation of gastric mucosa was revealed through a biopsy test (Naqvi, Nadeem Yousaf, Chaudhary, & Mills, 2020). Amphotericin B and posaconazole were used for fungal treatment. Trzaska, Correia, Villegas, May, and Voelz (2015) suggested a treatment of gastric mucormycosis by pH manipulation. Acetic acid appears highly effective against mucormycosis. They suggested that dilute acetic acid can be used in the treatment of mucormycosis (Trzaska et al., 2015).

Heart diseases and cancer

Some other diseases were also reported to have coinfection with mucormycosis. Patients with those diseases can easily be infected by this type of fungal pathogen. A 40-year-old woman with aplastic anemia treated with some immunosuppressive agents and steroids was admitted to the hospital with fulminant myocarditis. It was a special clinical condition that occurs due to cardiac inflammation. Patients generally die due to ventricular arrhythmias and cardiogenic shock. During treatment, she was infected by mucormycosis. She died after 90 min of cardiopulmonary resuscitation in the hospital (Basti, Taylor, Tschopp, & Sztajzel, 2004). In 2019, a 19-year-old man was admitted to the hospital due to cardiac pain and dry cough. Computed tomography and echocardiography of that patient showed fungal infection in the heart due to *Mucor* sp. Amphotericin B and surgery were needed in the treatment of that patient (Krishnappa, Naganur, Palanisamy, & Kasinadhuni, 2019). In 1958, Hutter published a review that indicated that cancer patients also are infected massively by fungal disease (Hutter, 1959). Ben-Ami, Luna, Lewis, Walsh, and Kontoyiannis (2009) reported a case of serious coinfection of pulmonary mucormycosis in cancer patients. A total of 20 patients were studied in that survey. Between them, 95 patients were found with hematologic malignancies and mucormycosis (Ben-Ami et al., 2009).

Statistical analysis for identification of high-risk factors

Statistical analysis and biological factors are shown to be related rather than opposing, in that statistical analysis of a dataset is dependent on experimental design, no quantity of statistical refinement can save a poorly designed study, and good design of experiments is essential (Lovell, 2013). A two-tailed test is a statistical procedure that evaluates whether a sample is more or less than a range of values by using a two-sided critical area of distribution. It is utilized in statistical significance testing and null hypothesis testing. Here, the study of statistical analysis is divided into two parts. First, we collected a number of patients from previous reports and applied statistical analysis on the sample number to identify the highest

TABLE 5.2 Name of disease with reported patient numbers and rate of mortality from previous reports.

Disease	No. of patients	Mortality rate (%)
Diabetes	131	73.2
Tuberculosis	3	60
Thyroid disease	2	50
Maxillary osteomyelitis	3	52
Myocarditis	1	0
Chronic kidney disease	2	0
Idiopathic intracranial hypertension	1	0

risk factor. Second, statistical analysis was done on patients' age of the highest risk factor for further study (see Table 5.2).

The different numbers of patients with diabetes, tuberculosis, thyroid disease, maxillary osteomyelitis, myocarditis, renal disease, and idiopathic intracranial hypertension were identified. The highest number of mucormycosis patients was found in the case of diabetes. The two-tailed t-test was applied to this dataset. The degree of freedom was 6. The P-value was .3135, which indicates the difference is considered not to be statistically significant due to the high differences in the number of patients. But this nonsignificant data indicates that diabetic patients are highly reported with mucormycosis coinfection. Mortality rates are highest in the case of coinfection with diabetes (Fig. 5.3), so diabetes appears to be the strongest risk factor for mucormycosis.

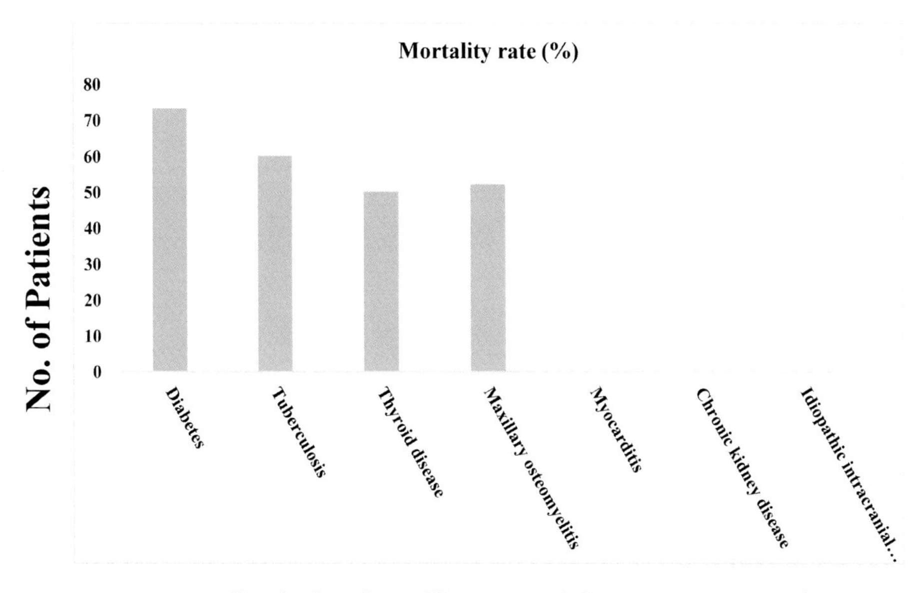

FIG. 5.3 Mortality rate of different diseases with mucormycosis coinfection. *No permission required.*

The total case report of diabetes and mucormycosis coinfection was taken into consideration to check the effect of the patient's age on risk. Researchers reported a total of 142 mucormycosis patients with diabetes. Among them, 93 were male and 49 were female patients (Fig. 5.4). A two-tailed *t*-test on that case study showed a *P*-value of .0011, which indicates that the data is extremely significant. The mean of that test was 18.53. The 95% confidence interval of this difference was found on a range of 11.86–28.71. The standard error was a low value of 3.441, and the standard deviation was 9.11. The significant distribution of patients concerning their age revealed that the chance of coinfection of diabetes patients by mucormycosis is high at ages 41–50.

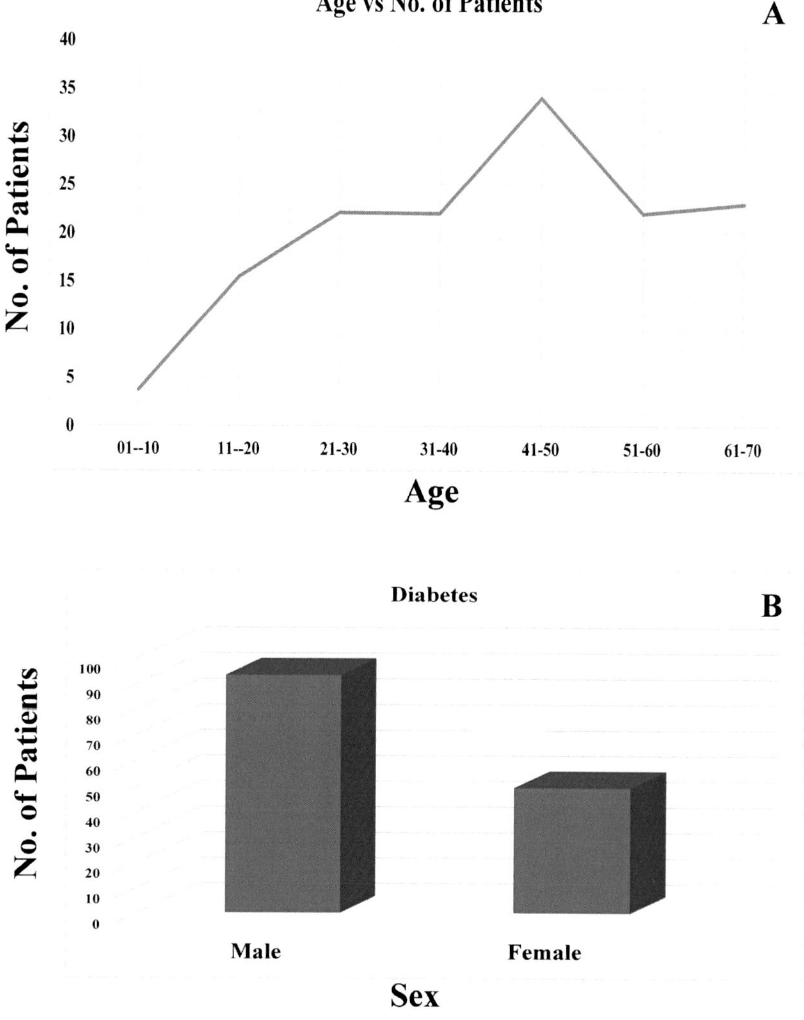

FIG. 5.4 (A) Graphical representation of number of patients vs their age. (B) Bar plot represents that males are more affected than females by mucormycosis. *No permission required.*

Conclusion

Currently, mucormycosis has become a terrible epidemic. Apart from India, post-COVID-19 fungal infections have also been reported in the United States and Oman, but in very few numbers. Right now, no country matches the sheer number of black fungus cases being reported in India. Artificial intelligence and computational studies are playing a vital role in diagnosing and treating this disease. Through this, we have obtained much information about mucormycosis that may help in the treatment of this disease. The computational method of diagnosis for mucormycosis has shown a new direction. With its help, it has been possible to diagnose this disease very easily. Patients with diabetes and tuberculosis are most likely to develop the disease. Graphical representations of previous reports indicate diabetic patients have a greater chance of being infected by mucormycosis. The survival rate in the case of diabetes and mucormycosis coinfection is very low. The two-tailed t-test also revealed that diabetic patients are mainly infected at age 41–50. Timely diagnosis and treatment can reduce mortality in most cases. Amphotericin B has been widely used in the treatment of mucormycosis (except for renal mucormycosis). However, some other drugs are also used in some specific cases. Computational methods and in silico studies will play a crucial role in the development of vaccines.

References

Afroze, S. N., Korlepara, R., Rao, G. V., & Madala, J. (2017). Mucormycosis in a diabetic patient: A case report with an insight into its pathophysiology. *Contemporary Clinical Dentistry*, *8*(4), 662–666. https://doi.org/10.4103/ccd.ccd_558_17.

Agarwal, S., Anand, A., Ranjan, P., Meena, V. P., Ray, A., Dutta, R., et al. (2020). Case of mucormycosis of mandible after self-extraction of teeth incidentally detected to have chronic granulomatous disease: Case report and literature review. *Medical Mycology Case Reports*, *28*, 55–59. https://doi.org/10.1016/j.mmcr.2020.03.005.

Aggarwal, D., Chander, J., Janmeja, A. K., & Katyal, R. (2015). Pulmonary tuberculosis and mucormycosis co-infection in a diabetic patient. *Lung India*, *32*(1), 53–55. https://doi.org/10.4103/0970-2113.148452.

Alekseyev, K., Didenko, L., & Chaudhry, B. (2021). Rhinocerebral mucormycosis and COVID-19 pneumonia. *Journal of Medical Cases*, *12*(3), 85–89. https://doi.org/10.14740/jmc3637.

Basti, A., Taylor, S., Tschopp, M., & Sztajzel, J. (2004). Fatal fulminant myocarditis caused by disseminated mucormycosis. *Heart (British Cardiac Society).*, *90*(10). https://doi.org/10.1136/hrt.2004.038273, e60.

Ben-Ami, R., Luna, M., Lewis, R. E., Walsh, T. J., & Kontoyiannis, D. P. (2009). A clinicopathological study of pulmonary mucormycosis in cancer patients: Extensive angioinvasion but limited inflammatory response. *Journal of Infection*, *59*(2), 134–138. https://doi.org/10.1016/j.jinf.2009.06.002.

Bhadada, S., Bhansali, A., Reddy, K. S. S., Bhat, R. V., Khandelwal, N., & Gupta, A. K. (2005). Rhino-orbital-cerebral mucormycosis in type 1 diabetes mellitus. *Indian Journal of Pediatrics*, *72*(8), 671–674. https://doi.org/10.1007/BF02724075.

Bhansali, A., Bhadada, S., Sharma, A., Suresh, V., Gupta, A., Singh, P., et al. (2004). Presentation and outcome of rhino-orbital-cerebral mucormycosis in patients with diabetes. *Postgraduate Medical Journal*, *80*(949), 670–674. https://doi.org/10.1136/pgmj.2003.016030.

Castillo-Caro, P., Wright, K. E., Bose, S., Hazrat, Y., Bollard, C., & Cruz, C. R. (2015). Developing T cell based immunotherapies for mucormycosis post HSCT. *Cytotherapy*, *S76*. https://doi.org/10.1016/j.jcyt.2015.03.574.

Cavasotto, C. N., & Di Filippo, J. I. (2021). In silico drug repurposing for COVID-19: Targeting SARS-CoV-2 proteins through docking and consensus ranking. *Molecular Informatics*, *40*(1). https://doi.org/10.1002/minf.202000115.

Chibucos, M. C., Soliman, S., Gebremariam, T., Lee, H., Daugherty, S., Orvis, J., et al. (2016). An integrated genomic and transcriptomic survey of mucormycosis-causing fungi. *Nature Communications*, *7*. https://doi.org/10.1038/ncomms12218.

Corzo-León, D. E., Chora-Hernández, L. D., Rodríguez-Zulueta, A. P., & Walsh, T. J. (2018). Diabetes mellitus as the major risk factor for mucormycosis in Mexico: Epidemiology, diagnosis, and outcomes of reported cases. *Medical Mycology*, *56*(1), 29–43. https://doi.org/10.1093/mmy/myx017.

Creeper, J. H., Main, D. C., Berger, L., Huntress, S., & Boardman, W. (1998). An outbreak of mucormycosis in slender tree frogs (Litoria adelensis) and white-lipped tree frogs (Litoria infrafrenata). *Australian Veterinary Journal*, *76*(11), 761–762. https://doi.org/10.1111/j.1751-0813.1998.tb12312.x.

Dallalzadeh, L. O., Ozzello, D. J., Liu, C. Y., Kikkawa, D. O., & Korn, B. S. (2021). Secondary infection with rhino-orbital cerebral mucormycosis associated with COVID-19. *Orbit (London)*. https://doi.org/10.1080/01676830.2021.1903044.

do Monte Junior, E. S., dos Santos, M. E. L., Ribeiro, I. B., de Luz, G. O., Baba, E. R., Hirsch, B. S., et al. (2020). Rare and fatal gastrointestinal Mucormycosis (Zygomycosis) in a COVID-19 patient: A case report. *Clinical Endoscopy*, 746–749. https://doi.org/10.5946/ce.2020.180.

El Deeb, Y., Al Soub, H., Almaslamani, M., Al Khuwaiter, J., & Taj-Aldeen, S. J. (2005). Post-traumatic cutaneous mucormycosis in an immunocompetent patient. *Annals of Saudi Medicine*, *25*(4), 343–345. https://doi.org/10.5144/0256-4947.2005.343.

Elfiky, A. A. (2019). The antiviral Sofosbuvir against mucormycosis: An in silico perspective. *Future Virology*, *14*(11), 739–744. https://doi.org/10.2217/fvl-2019-0076.

Gandhi, B. V., Bahadur, M. M., Dodeja, H., Aggrwal, V., Thamba, A., & Mali, M. (2005). Systemic fungal infections in renal diseases. *Journal of Postgraduate Medicine*, *51*(5), S30–S36.

Gandra, S., Ram, S., & Levitz, S. M. (2021). The "black fungus" in India: The emerging syndemic of COVID-19–associated mucormycosis. *Annals of Internal Medicine*. https://doi.org/10.7326/m21-2354.

García, O. F., Guerrero-Torres, L., Roman-Montes, C. M., Rangel-Cordero, A., Martínez-Gamboa, A., Ponce-de-León, A., et al. (2021). Isolation of Rhizopus microsporus and Lichtheimia corymbifera from tracheal aspirates of two immunocompetent critically ill patients with COVID-19. *Medical Mycology Case Reports*. https://doi.org/10.1016/j.mmcr.2021.07.001.

Garcia-Hermoso, D., Criscuolo, A., Lee, S. C., Legrand, M., Chaouat, M., Denis, B., et al. (2018). Outbreak of invasive wound mucormycosis in a burn unit due to multiple strains of Mucor circinelloides f. circinelloides resolved by whole-genome sequencing. *mBio*, *9*(2). https://doi.org/10.1128/mBio.00573-18.

Goel, P., Jain, V., Sengar, M., Mohta, A., Das, P., & Bansal, P. (2013). Gastrointestinal mucormycosis: A success story and appraisal of concepts. *Journal of Infection and Public Health*, *6*(1), 58–61. https://doi.org/10.1016/j.jiph.2012.08.004.

Guinea, J., Escribano, P., Vena, A., Muñoz, P., Martínez-Jiménez, M. D. C., Padilla, B., et al. (2017). Increasing incidence of mucormycosis in a large Spanish hospital from 2007 to 2015: Epidemiology and microbiological characterization of the isolates. *PLoS One*, *12*(6). https://doi.org/10.1371/journal.pone.0179136.

Gupta, K. L., & Gupta, A. (2012). Mucormycosis and acute kidney injury. *Journal of Nephropathology*, *1*(3), 155–159. https://doi.org/10.5812/nephropathol.8111.

Hemashettar, B. M., Patil, R. N., O'Donnell, K., Chaturvedi, V., Ren, P., & Padhye, A. A. (2011). Chronic rhinofacial mucormycosis caused by Mucor irregularis (Rhizomucor variabilis) in India. *Journal of Clinical Microbiology*, *49*(6), 2372–2375. https://doi.org/10.1128/JCM.02326-10.

Hopkins, M. A., & Treloar, D. M. (1997). Mucormycosis in diabetes. *American Journal of Critical Care*, *6*(5), 363–367. https://doi.org/10.4037/ajcc1997.6.5.363.

Hu, Z., Wang, L., Zou, L., Chen, Z., Yi, Y., Meng, Q., et al. (2021). Coinfection pulmonary mucormycosis and aspergillosis with disseminated mucormycosis involving gastrointestinalin in an acute B-lymphoblastic leukemia patient. *Brazilian Journal of Microbiology*. https://doi.org/10.1007/s42770-021-00554-8.

Hutter, R. V. P. (1959). Phycomycetous infection (mucormycosis) in cancer patients: A complication of therapy. *Cancer*, *12*(2), 330–350. https://doi.org/10.1002/1097-0142(195903/04)12:2<330::AID-CNCR2820120217>3.0.CO;2-F.

Inglesfield, S., Jasiulewicz, A., Hopwood, M., Tyrrell, J., Youlden, G., Mazon-Moya, M., et al. (2018). Robust phagocyte recruitment controls the opportunistic fungal pathogen Mucor circinelloides in innate granulomas in vivo. *mBio*, *9*(2). https://doi.org/10.1128/mbio.02010-17.

Iqbal, N., Irfan, M., Jabeen, K., Kazmi, M. M., & Tariq, M. U. (2017). Chronic pulmonary mucormycosis: An emerging fungal infection in diabetes mellitus. *Journal of Thoracic Disease*, *9*(2), E121–E125. https://doi.org/10.21037/jtd.2017.02.31.

Jeong, W., Keighley, C., Wolfe, R., Lee, W. L., Slavin, M. A., Kong, D. C. M., et al. (2019). The epidemiology and clinical manifestations of mucormycosis: A systematic review and meta-analysis of case reports. *Clinical Microbiology and Infection*, *25*(1), 26–34. https://doi.org/10.1016/j.cmi.2018.07.011.

Jha, R., Gude, D., Chennamsetty, S., & Kotari, H. (2013). Intracranial hypertension: An unusual presentation of mucormycosis in a kidney transplant recipient. *Indian Journal of Nephrology*, *23*(2), 130–132. https://doi.org/10.4103/0971-4065.109437.

Jiménez-Zarazúa, O., Vélez-Ramírez, L. N., Alcocer-León, M., Utrilla-Álvarez, J. D., Martínez-Rivera, M. A., Flores-Saldaña, G. A., et al. (2019). A case of concomitant pulmonary tuberculosis and mucormycosis in an insulin-dependent diabetic patient. *Journal of Clinical Tuberculosis and Other Mycobacterial Diseases*, *16*.

Kachuei, R., Badali, H., Vaezi, A., Jafari, N. J., Ahmadikia, K., Kord, M., et al. (2021). Fatal necrotising cutaneous mucormycosis due to novel Saksenaea species: A case study. *Journal of Wound Care*, *30*(6), 465–468. https://doi.org/10.12968/jowc.2021.30.6.465.

Krishnappa, D., Naganur, S., Palanisamy, D., & Kasinadhuni, G. (2019). Cardiac mucormycosis: A case report. *European Heart Journal—Case Reports*, *3*(3). https://doi.org/10.1093/ehjcr/ytz130.

Kumar, N., Singh, A. K., Pandey, S., Agrawal, S., & Singh, S. (2015). Rhinomaxillary osteomyelitis due to mucormycosis in an immunocompromised geriatric patient. *Egyptian Journal of Oral & Maxillofacial Surgery*, 66–70. https://doi.org/10.1097/01.OMX.0000464798.32075.d8.

Lipatov, K., Patel, C., Lat, T., Shakespeare, A., Wang, B., & Prakash, G. (2018). Pulmonary mucormycosis in a patient with uncontrolled diabetes. *Federal Practitioner*, *35*(1).

Liu, M., Bruni, G. O., Taylor, C. M., Zhang, Z., & Wang, P. (2018). Comparative genome-wide analysis of extracellular small RNAs from the mucormycosis pathogen Rhizopus delemar. *Scientific Reports*, *8*(1). https://doi.org/10.1038/s41598-018-23611-z.

Lovell, D. P. (2013). Biological importance and statistical significance. *Journal of Agricultural and Food Chemistry*, *61*(35), 8340–8348. https://doi.org/10.1021/jf401124y.

Lucassen, A., & Houlston, R. S. (2014). The challenges of genome analysis in the health care setting. *Genes*, *5*(3), 576–585. https://doi.org/10.3390/genes5030576.

Maini, A., Tomar, G., Khanna, D., Kini, Y., Mehta, H., & Bhagyasree, V. (2021). Sino-orbital mucormycosis in a COVID-19 patient: A case report. *International Journal of Surgery Case Reports*, *82*. https://doi.org/10.1016/j.ijscr.2021.105957, 105957.

Malini, S., & Pasupathi, A. (2021). Mucormycosis: A rare fungal infection in patients affected with covid 19. *International Journal of Pharmaceutical Research*, 976–979.

Mascarella, M. A., Schweitzer, L., Alreefi, M., Silver, J., Caglar, D., Loo, V. G., et al. (2019). The infectious thyroid nodule: A case report of mucormycosis associated with ibrutinib therapy. *Journal of Otolaryngology—Head and Neck Surgery*, *48*(1). https://doi.org/10.1186/s40463-019-0376-1.

McBride, R. A., Corson, J. M., & Dammin, G. J. (1960). Mucormycosis. Two cases of disseminated disease with cultural identification of rhizopus; review of literature. *The American Journal of Medicine*, *28*(5), 832–846. https://doi.org/10.1016/0002-9343(60)90138-8.

Mehta, S., & Pandey, A. (2020). Rhino-orbital mucormycosis associated with COVID-19. *Cureus*. https://doi.org/10.7759/cureus.10726.

Mekki, S. O., Hassan, A. A., Falemban, A., Alkotani, N., Alsharif, S. M., Haron, A., et al. (2020). Pulmonary mucormycosis: A case report of a rare infection with potential diagnostic problems. *Case Reports in Pathology*, *2020*, 5845394.

Mitra, D., & Bose, A. (2021). Remarkable effect of natural compounds that have therapeutic effect to stop COVID-19. In *Recent advances in pharmaceutical sciences* (pp. 115–126). Innovare Academic Sciences Pvt. Ltd.

Mitra, D., & Das Mohapatra, P. K. (2021a). Discovery of novel cyclic salt bridge in thermophilic bacterial protease and study of its sequence and structure. *Applied Biochemistry and Biotechnology*, *193*(6), 1688–1700. https://doi.org/10.1007/s12010-021-03547-3.

Mitra, D., & Das Mohapatra, P. K. (2021b). Cold adaptation strategy of psychrophilic bacteria: An in-silico analysis of isocitrate dehydrogenase. *Systems Microbiology and Biomanufacturing*, *1–11*. https://doi.org/10.1007/s43393-021-00041-z.

Mitra, D., Dey, A., Biswas, I., & Mohapatra. (2021). Bioactive compounds as a potential inhibitor of colorectal cancer; an insilico study of gallic acid and pyrogallol. *Annals of Colorectal Research*, *9*(1), 32–39.

Mitra, D., & Mohapatra. (2021). Inhibition of SARS-CoV-2 protein by bioactive compounds of edible mushroom; a bioinformatics insight. *International Journal of Advances in Science, Engineering and Technology*, *9*(2), 84–88.

Mitra, D., Pal, A. K., & Mohapatra, P. K. D. (2020). *In-silico study of SARS-CoV-2 and SARS with special reference to intra-protein interactions, a plausible explanation for stability, divergency and severity of SARS-CoV-2*. Research Square (Preprint).

Mitra, D., Paul, M., Thatoi, H., & Mohapatra, P. K. D. (2021). Study of potentiality of dexamethasone and its derivatives against Covid-19. *Journal of Biomolecular Structure & Dynamics*. https://doi.org/10.1080/07391102.2021.1942210.

Mitra, D., Verma, D., Mahakur, B., Kamboj, A., Srivastava, R., Gupta, S., et al. (2021). Molecular docking and simulation studies of natural compounds of Vitex negundo L. against papain-like protease (PLpro) of SARS CoV-2 (coronavirus) to conquer the pandemic situation in the world. *Journal of Biomolecular Structure & Dynamics*. https://doi.org/10.1080/07391102.2021.1873185.

Nambiar, M., Varma, S. R., & Damdoum, M. (2021). Post-Covid alliance-mucormycosis, a fatal sequel to the pandemic in India. *Saudi Journal of Biological Sciences*. https://doi.org/10.1016/j.sjbs.2021.07.004.

Naqvi, H. A., Nadeem Yousaf, M., Chaudhary, F. S., & Mills, L. (2020). Gastric Mucormycosis: An infection of fungal invasion into the gastric mucosa in immunocompromised patients. *Case Reports in Gastrointestinal Medicine*, 1–7. https://doi.org/10.1155/2020/8876125.

Nawaz, M. S., Fournier-Viger, P., Shojaee, A., & Fujita, H. (2021). Using artificial intelligence techniques for COVID-19 genome analysis. *Applied Intelligence*, *51*(5), 3086–3103. https://doi.org/10.1007/s10489-021-02193-w.

Nehara, H. R., Puri, I., Singhal, V., Ih, S., Bishnoi, B. R., & Sirohi, P. (2021). Rhinocerebral mucormycosis in COVID-19 patient with diabetes a deadly trio: Case series from the north-western part of India. *Indian Journal of Medical Microbiology*, *39*(3), 380–383. https://doi.org/10.1016/j.ijmmb.2021.05.009.

Nilesh, K., Malik, N. A., & Belgaumi, U. (2015). Mucormycosis in a healthy elderly patient presenting as oro-antral fistula: Report of a rare incidence. *Journal of Clinical and Experimental Dentistry*, *7*(2), e333–e335. https://doi.org/10.4317/jced.52064.

Pandey, M., Singh, G., Agarwal, R., Dabas, Y., Jyotsna, V. P., Kumar, R., et al. (2018). Emerging rhizopus microsporus infections in India. *Journal of Clinical Microbiology*, *56*(6). https://doi.org/10.1128/JCM.00433-18.

Patel, A., Agarwal, R., Rudramurthy, S. M., Shevkani, M., Xess, I., Sharma, R., et al. (2021). Multicenter epidemiologic study of coronavirus disease-associated mucormycosis, India. *Emerging Infectious Diseases*, *27*(9), 2349–2359. https://doi.org/10.3201/eid2709.210934.

Patel, M. H., Patel, R. D., Vanikar, A. V., Kanodia, K. V., Suthar, K. S., Nigam, L. K., et al. (2017). Invasive fungal infections in renal transplant patients: A single center study. *Renal Failure*, *39*(1), 294–298. https://doi.org/10.1080/0886022X.2016.1268537.

Prakash, H., Ghosh, A. K., Rudramurthy, S. M., Singh, P., Xess, I., Savio, J., et al. (2019). A prospective multicenter study on mucormycosis in India: Epidemiology, diagnosis, and treatment. *Medical Mycology*, *57*(4), 395–402. https://doi.org/10.1093/mmy/myy060.

Prasad, N., Manjunath, R., Bhadauria, D., Marak, R. S. K., Sharma, R., Agarwal, V., et al. (2018). Mucormycosis of the thyroid gland: A cataclysmic event in renal allograft recipient. *Indian Journal of Nephrology*, *28*(3), 232. https://doi.org/10.4103/ijn.IJN_192_17.

Ramesh, P., Kaur, G., Deepak, D., & Kumar, P. (2020). Disseminated pulmonary mucormycosis with concomitant tuberculosis infection in a diabetic patient. *International Journal of Mycobacteriology*, *9*(1), 95–97. https://doi.org/10.4103/ijmy.ijmy_186_19.

Rammaert, B., Lanternier, F., Poirée, S., Kania, R., & Lortholary, O. (2012). Diabetes and mucormycosis: A complex interplay. *Diabetes & Metabolism*, *38*(3), 193–204. https://doi.org/10.1016/j.diabet.2012.01.002.

Saldanha, M., Reddy, R., & Vincent, M. J. (2021). Of the article: Paranasal mucormycosis in COVID-19 patient. *Indian Journal of Otolaryngology and Head & Neck Surgery*, 1–4.

Selarka, L., Sharma, S., Saini, D., Sharma, S., Batra, A., Waghmare, V. T., et al. (2021). Mucormycosis and COVID-19: An epidemic within a pandemic in India. *Mycoses*. https://doi.org/10.1111/myc.13353.

Selvamani, M., Donoghue, M., Bharani, S., & Madhushankari, G. S. (2015). Mucormycosis causing maxillary osteomyelitis. *Journal of Natural Science, Biology and Medicine*, *6*(2), 456–459. https://doi.org/10.4103/0976-9668.160039.

Seo, D. J., & Choi, C. (2021). Antiviral bioactive compounds of mushrooms and their antiviral mechanisms: A review. *Viruses*, *13*(2). https://doi.org/10.3390/v13020350.

Shah, B., Modi, P., & Sagar, S. R. (2020). In silico studies on therapeutic agents for COVID-19: Drug repurposing approach. *Life Sciences*, *252*. https://doi.org/10.1016/j.lfs.2020.117652.

Sharma, A. D., & Kaur, I. (2021). Targeting β-glucan synthase for mucormycosis "the 'black fungus" maiming covid patients in India: Computational insights. *Journal of Drug Delivery and Therapeutics*, *11*(3-S), 9–14. https://doi.org/10.22270/jddt.v11i3-s.4873.

Shelburne, S. A., Ajami, N. J., Chibucos, M. C., Beird, H. C., Tarrand, J., Galloway-Peña, J., et al. (2015). Implementation of a pan-genomic approach to investigate holobiont-infecting microbe interaction: A case report of a leukemic patient with invasive mucormycosis. *PLoS One*, *10*(11). https://doi.org/10.1371/journal.pone.0139851.

Sheth, S. M., Talwalkar, N. C., Desai, A. P., & Acharya, V. N. (1981). Rhinocerebral mucormycosis in a case of renal failure. *Journal of Postgraduate Medicine*, *27*(3).

Taj-Aldeen, S. J., Gamaletsou, M. N., Rammaert, B., Sipsas, N. V., Zeller, V., Roilides, E., et al. (2017). Bone and joint infections caused by mucormycetes: A challenging osteoarticular mycosis of the twenty-first century. *Medical Mycology*, *55*(7), 691–704. https://doi.org/10.1093/mmy/myw136.

Trieu, T. A., Navarro-Mendoza, M. I., Pérez-Arques, C., Sanchis, M., Capilla, J., Navarro-Rodriguez, P., et al. (2017). RNAi-based functional genomics identifies new virulence determinants in mucormycosis. *PLoS Pathogens*, *13*(1). https://doi.org/10.1371/journal.ppat.1006150.

Trzaska, W. J., Correia, J. N., Villegas, M. T., May, R. C., & Voelz, K. (2015). pH manipulation as a novel strategy for treating mucormycosis. *Antimicrobial Agents and Chemotherapy*, *59*(11), 6968–6974. https://doi.org/10.1128/AAC.01366-15.

Valerio, L., Ferrazzi, P., Sacco, C., Ruf, W., Kucher, N., Konstantinides, S. V., et al. (2021). Course of D-dimer and C-reactive protein levels in survivors and nonsurvivors with COVID-19 pneumonia: A retrospective analysis of 577 patients. *Thrombosis and Haemostasis*, *121*(1), 98–101. https://doi.org/10.1055/s-0040-1721317.

Verma, D., Mitra, D., Paul, M., Chaudhary, P., Kamboj, A., Thatoi, H., et al. (2021). Potential inhibitors of SARS-CoV-2 (COVID 19) proteases PLpro and Mpro/ 3CLpro: Molecular docking and simulation studies of three pertinent medicinal plant natural components. *Current Research in Pharmacology and Drug Discovery*, *2*. https://doi.org/10.1016/j.crphar.2021.100038, 100038.

Wang, B., & Jiang, L. (2021). Principal component analysis applications in COVID-19 genome sequence studies. *Cognitive Computation*. https://doi.org/10.1007/s12559-020-09790-w.

Yamamoto, K., Mawatari, M., Fujiya, Y., Kutsuna, S., Takeshita, N., Hayakawa, K., et al. (2021). Survival case of rhinocerebral and pulmonary mucormycosis due to Cunninghamella bertholletiae during chemotherapy for acute myeloid leukemia: A case report. *Infection*, *49*(1), 165–170. https://doi.org/10.1007/s15010-020-01491-8.

CHAPTER

6

A review of diabetes management tools and applications

Hossain Shahriar[a], *Sweta Sneha*[b], *Yesake Abaye*[b], *Talha Hashmi*[b], *Shakaria Wilson*[b], *and Usen Usen*[b]

[a]Department of Information Technology, Kennesaw State University, Kennesaw, GA, United States [b]Department of Information Systems and Security, Kennesaw State University, Kennesaw, GA, United States

Introduction

Diabetes mellitus is a huge health burden that affects the overall healthcare system in the United States. According to the Centers for Disease Control and Prevention (CDC), it is estimated that 10.5% of the population (34 million people) in the United States suffer from diabetes. The number of people diagnosed with diabetes continues to rise globally affecting the overall healthcare systems of the world. Diabetes is a difficult disease to manage due to the multiple chronic health complications associated with it. The high prevalence of diabetes and its subsequent complications make it a common comorbid condition in hospitalized patients. In addition to the primary disease itself, diabetes, in the long term, can cause many other conditions that affect different parts of the body. 98% of diabetic patients suffer from complications stemming from having the disease (CDC, 2020).

Some of the common chronic complications and comorbidities associated with diabetes include kidney disease, hypertension, cardiac disease, neuropathy, and retinopathy. These chronic comorbidities increase mortality and morbidity rates among the diabetes patient population. It also severely hampers improvement in clinical patient outcomes and contributes to skyrocketing healthcare costs in our system. It is estimated that diabetes directly costs the US healthcare system around $237 billion dollars annually (CDC, 2020). Readmission is the repeat hospitalization of patients after being discharged from the hospital. Hospital readmissions cost around $26 billion dollars annually (CDC, 2020). The CDC estimates that

https://doi.org/10.1016/B978-0-323-99031-8.00023-5

the average healthcare expenditures of diabetic patients are more than twice that of non-diabetics.

Management and treatment of diabetes usually consists of diet, exercise, oral medications, and injectable agents, including insulin therapy. The main goal of treatment is controlling blood sugars to a desirable level. This is achieved by patients regularly checking their blood sugar levels and ensuring they are within the goal treatment levels, prescribed by their doctor. It requires patients to be meticulously committed to regularly checking their blood sugars and making necessary adjustments in their treatment regimen. This could entail, injecting additional doses of insulin, change in dietary intake, or exercise level. Since diabetics are often also affected with associated health comorbidities, treatment becomes increasingly complex. It can increase the incidence of non-compliance to therapy due to the nature of managing multiple diseases. Also, poor adherence to medications and treatment follow-ups is persistent in this group due to the increased demands of managing multiple conditions. The incorporation of technology in the management of diabetes has the potential to alleviate this burden and improve patient outcomes and reduce healthcare costs.

Background

The evolution of technology in the past few decades has drastically changed and improved the methods in which diseases can be monitored, controlled, and ultimately cured. One disease in particular which is greatly impacted by this change is diabetes. For the majority of history, diabetes has been a deadly disease for which there has been no cure or treatment. From the time it was noticed by the ancient Egyptians more than 3000 years ago to when the Greeks actually named it "diabetes," its record is clearly stated. Early treatments ranged from exercising (horseback riding was recommended) to consumption of rancid animal food (Higuera, 2020). It was only in the late 1800s that the importance of the pancreas and blood sugar was related to diabetes. This discovery helped with the development of insulin, and the first administration of this treatment in 1922 (Higuera, 2020). The understanding of diabetes further progressed as blood sugar could be tested, and insulin given accordingly. The previously deadly type 1 diabetes had suddenly become treatable, and its treatment has become more developed and efficient as time has passed.

The rapid advancement in the digital consumer technology field in recent years, has led to the introduction of different connected technology tools and devices for use in administering healthcare. Digital health is a broad scope health care delivery method utilizing different technologies. It can be categorized in different models such as telehealth, mobile health, and telemonitoring (Eiland et al., 2018). These technologies range in use from mobile apps in general health to gadgets used as medical devices. Digital health is expected to revolutionize the way health care is delivered by simplifying the current processes of diagnosing, treating, and preventing diseases. The greatest beneficiaries are patients who have chronic medical conditions such as diabetes and hypertension. It has enormous potential to improve clinical outcomes by increasing access to care and assisting healthcare providers to deliver quality care remotely.

Digital health initiatives employ tools and devices such as mobile apps, medical software, smart watch, medical wearables, and monitoring systems that monitor and collect health data. Wearable devices in real time, continually capture health data and vital signs such as heart rate, blood glucose, and blood pressure. The devices are integrated via a global network connection system, utilizing Wi-Fi and Bluetooth technology that provide real-time observations and assessments for further response (Eiland et al., 2018). Some of these digital devices offer built-in interaction systems that allow regular clinician and patient communications. The vast amount of health data collected via these digital health tools can be integrated into patients' electronic health records and analyzed for further interventions and future research.

The amount of technology, from different health applications to smart watches, which track blood sugar and diabetes have vastly escalated in the past decade alone. Diabetes patients must on a regular basis monitor their glucose levels using diabetes management accessories; glucose sensors need to be replaced on a regular basis. The goal of this paper is to explain the history of diabetes management, research these new innovations, and explain the ways in which they might benefit diabetes patients.

Review of tools

Diabetes is prevalent throughout many communities. The symptoms that lead to diabetes may be ignored because some people do not pay attention to their bodies. Three major symptoms of diabetes are increased thirst, increased urination, and increased hunger, which are often overlooked (Diabetes, 2020). Diabetes has been researched for many years and is proven to be a precursor for many other diseases. It is important to monitor glucose levels to prevent other illnesses in the future. It is a major health concern during the pandemic. People with diabetes are more likely to contract COVID-19 and have harsher symptoms than those without diabetes. It is very important to monitor your health and fix problems that may lead to diabetes such as diet, high blood pressure, and low-income communities. Not only is it important to educate all communities about diabetes, it is also beneficial to inform them of the many applications and tools that can be used to help them throughout their journey. The tools and applications are intended to manage health through diet, exercise, and glucose levels which aid in the prevention of becoming seriously ill. In the next subsections, we provide details of some of the common tools and apps. In the next subsections, we discuss some of the popular diabetes management tools and applications currently available in the market.

Diabeo system

The Diabeo System is a medical software device and application system that is approved in Europe for managing insulin-dependent diabetic patients (Franc et al., 2020). The Diabeo System combines a smartphone application downloadable by the patients and a web portal for healthcare providers that can access data in real-time. The overview of the diabeo system as illustrated in Fig. 6.1 shows patients enter relevant health data such as glucose measurements, daily diet of carbohydrate intake, and exercise activity levels via the phone application. The Diabeo software utilizes algorithms with calculators that automatically calculate

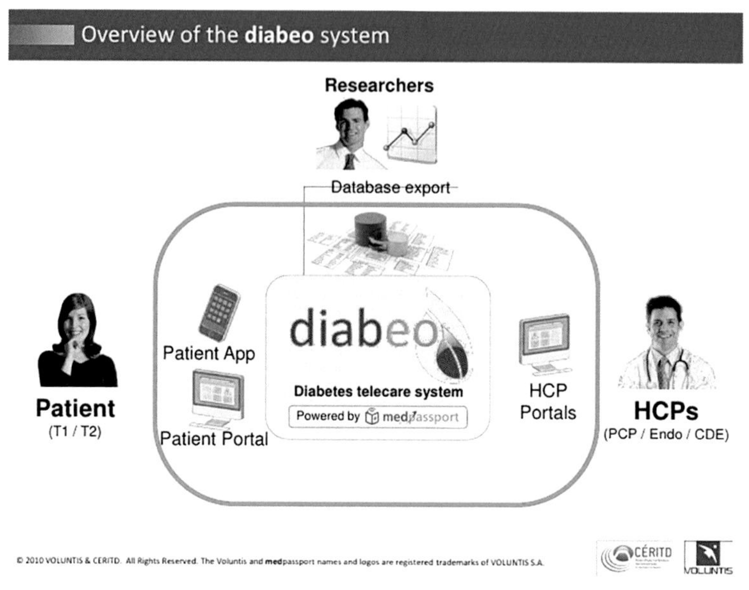

FIG. 6.1 Diabeo System for Management of Diabetes.

insulin doses based on the protocol prescribed by the healthcare provider. The dose calculator automatically adjusts and calculates insulin doses by adjusting for the reported health data by patients reported in real-time. The clinicians can access all the relevant data using the web portal and can make modifications to the protocol remotely. Additionally, a clinical decision support system provides recommendations and feedback for the providers based on the available data from the Diabeo system. The web portal also allows communication between patients and their healthcare provider and other staff on their care team including nutritionists.

A multicenter, European randomized trial assessing the effectiveness of the Diabeo System over a 6-month period, showed significant benefits including better blood sugar control in chronic, uncontrolled, insulin-dependent diabetic patients (Franc et al., 2020). The study also highlighted those benefits seen in controlling blood glucose levels was even greater when patients used the Diabeo app more than once daily. Another study evaluating Diabeo software, concluded that Diabeo software significantly improved blood sugar control in chronic, poorly controlled diabetic patients without requiring additional medical resources and at a lower overall cost than standard care (Charpentier et al., 2011).

Dexcom G6 glucose monitoring system

Dexcom is a wearable medical software system device that is approved by the U.S. Food and Drug Administration (FDA) for the monitoring of blood sugar in diabetic patients (dexcom.com, accessed 2021 Dexcom, 2021). It utilizes a microsensor application to measure glucose levels continuously. The system consists of a microsensor application that is inserted underneath the skin via an applicator with a transmitter (dexcom.com, accessed 2021). This

FIG. 6.2 Dexcom G6 Glucose Monitoring System.

microsensor device transmits the blood sugar data wirelessly to a display monitor or a smart device such as a smart phone or watch (Fig. 6.2). The Dexcom G6 system continually measures blood sugar levels and reports results in real time. It generates blood glucose measurement reading every 5 min, 24/7 and up to 288 readings a day (dexcom.com, accessed 2021). It simplifies the glucose monitoring process for patients by eliminating multiple painful and expensive fingerstick checks at home. The system has interoperable capabilities and can transmit data to healthcare providers and be integrated with other diabetes management systems such as insulin pumps, insulin dosing systems, apps, and other electronic devices (dexcom.com, accessed 2021). It also assists insulin-dependent diabetic patients with insulin dosing when integrated with automated systems that administer insulin when blood glucose is out of range. This helps control blood sugar by assuring a timely administration of insulin via alerts from the Dexcom monitoring system.

An observational study by Roze et al. (2020) discovered that Dexcom G6 glucose monitoring system provided significant improvements in patient clinical outcomes with a lifetime cost benefit compared to self-monitoring of blood sugar by patients. In addition, the Dexcom G6 glucose monitoring system has been shown to reduce the incidence of hypoglycemia—low blood sugar in insulin-dependent diabetes patients. The HypoDE study, a randomized controlled trial, found a clinically significant reduction in the number of hypoglycemic events in type 1 diabetes patients that utilized the Dexcom G6 system for managing their diabetes (Heinemann et al., 2018).

Insulia app

Insulia is a prescription medical application for smart digital devices intended for patients with type 2 diabetes (insulia.com, accessed 2021). Insulia is available for download by patients on the App Store and Google Play once prescribed by a healthcare provider. The app has a simple interface for entering and reading values by the user, sample view of the app can be seen below in Fig. 6.3. The application system consists of a web portal for the healthcare provider and a mobile app for the patient for management of insulin administration (insulia.com, accessed 2021). Its specific goal is to deliver the appropriate dose of insulin in real-time based on the blood sugar level. This is achieved by the mobile app's synchronized continuous insulin dosing recommendations based on clinical algorithms built into the system. The application also delivers educational and nutrition coaching alerts based on blood glucose values. The web portal is used by the healthcare provider to prescribe the system and configure a patient-targeted therapy to control blood sugar. Data are continuously shared with the clinicians and the prescriber is able to assess response to therapy in real-time and make modifications to the treatment remotely via the web portal.

Eversense CGM system

The Eversense CGM is an approved U.S. Food and Drug Administration (FDA) system for use in people 18 years and above for the measurement of blood glucose levels in diabetes patients. It is made up of an implanted sensor coated with a fluorescent-like chemical that detects the glucose level, the glucose trend. It also comes with the capacity to send alerts for

FIG. 6.3 Insulia insulin dosing application.

detection and prediction of episodes of hypoglycemia (low blood glucose) and hyperglycemia (high blood glucose). This sensor is implanted under the skin by a qualified healthcare provider during a procedure and it will regularly measure glucose levels in adult diabetes patients for up to 90 days before replacement. The implanted sensor utilizes a novel light-based technology to measure the glucose levels of a patient and then send its information to a mobile app to alert wearers if their glucose levels are too high or too low. The implanted sensor when exposed to a patient's blood sugar will produce an amount of light that will be measured by the sensor. Transmission of the sensor blood sugar reading is done every 5 min using a compatible mobile device such as smartphone or tablet that is running a device-specific mobile app with a reminder for the user to calibrate two times daily. It has the option for the data to be uploaded to Senseonics' data management system. However, the listed procedures and medications below should not be carried out or given to individuals who have Eversense CGM System sensor implanted in them.

- Electrocautery: When an electrocautery; a heated electrode that is applied to tissues to stop the flow of blood, is used near an inserted sensor of the Eversense CGM, it may damage the sensor.
- MRI: The Smart Transmitter from the Eversense CGM system must be removed before undergoing any MRI procedure as it is unsafe and incompatible with magnetic resonance imaging.
- Diathermy: The procedure where heat is generated through electrical pulses should not be carried out on people that the Eversense CGM sensor has been inserted on as the heat energy generated during this procedure can pass through sensor and damage healthy tissues in the inserted area.
- Lithotripsy: It is recommended that people who have inserted Eversense CGM sensor should not undergo lithotripsy procedure as the effect of the interactions between the ultrasound shock waves emanating from the lithotripsy treatment and the sensor are unknown.
- Antibiotics such as tetracycline could incorrectly lower the Eversense CGM sensor sugar readings. So, it is advisable that patients taking tetracycline medication should not depend on the sensor sugar readings.
- Dexamethasone or dexamethasone acetate contraindicated individuals may be contraindicated with the Eversense CGM System as well.
- The administration of sorbitol or mannitol medications intravenously, or inform of an irrigation or peritoneal dialysis procedure, could increase the blood sorbitol or mannitol concentrations thereby giving inaccurate readings of the sensor glucose results (Fig. 6.4).

BlueStar diabetes application

Blue Star is a diabetes management application available on both Apple and Android devices created by a company named Welldoc. This Food and Drug administration approved application helps in decision-making, coaching, and supporting its customers in diabetes-related decisions. Blue Star's application gives out recommendations, from food to exercise to medications, and how these would affect blood sugar levels. Blue Star has free and

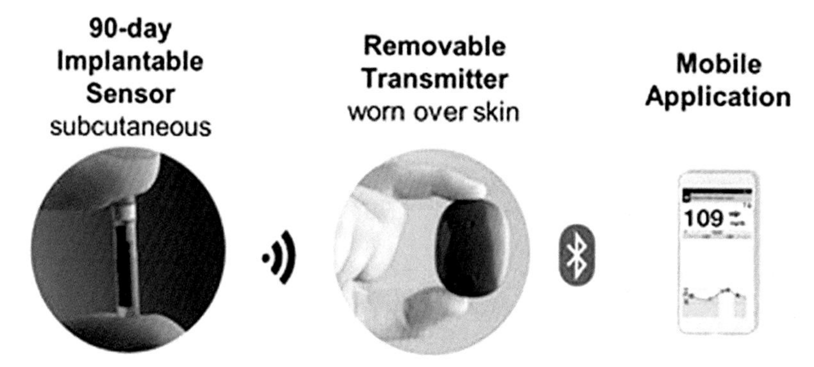

FIG. 6.4 Eversense CGM System.

"prescribed" versions, with employers or healthcare plans able to provide access codes for this application. It can connect to various other applications, such as fitness apps, pharmaceutical portals, and patient portals. Blue Star takes input about food, fitness, medications, and gives out artificial intelligence-driven feedback. There is also an option to share all of this information at the click of a button with a care team. In March of 2020, Welldoc announced a partnership with the company Dexcom (referenced above), to collaborate their two software programs. Patients using Dexcom will receive the recommendations of Blue Star to help manage diabetes (Welldoc, 2020).

Biosense breath ketone monitor and application

The Biosense breath ketone monitor measures the amount of acetone in the breath. Acetone correlates with the number of ketones being produced in the blood. Diabetic patients should get their ketone levels checked often especially when ill. Ketones can be produced when there is an insufficient supply of insulin available in the body. Due to an insufficient supply of insulin the body will produce ketones to help regulate sugar levels. The optimal ketone level is 1.0–3.0mmol/L. If the levels are high above 3.0mmol/L, that indicates too many ketones are produced and the patient may experience diabetic ketoacidosis. (Diabetes) Diabetic ketoacidosis is when the blood becomes too acidic which may cause sickness and/or damage to other organs. Ketone monitors are often expensive to buy individually, but insurance has been known to cover these costs especially with type 1 diabetic patients and other diabetic patients that are also being prescribed insulin. Physicians will also check ketone levels at appointments to help monitor insulin levels and create a plan to improve the health of a patient. The Biosense measures breath acetones, which is produced as a byproduct of ketogenesis (Biosense 1). This device allows you to see what level of ketosis your body is in. The application that corresponds to this device keeps track of your ACE levels and stores the data which can be given to your physician to help improve your health. Pictured below is the Biosense breath ketone monitor system (Fig. 6.5).

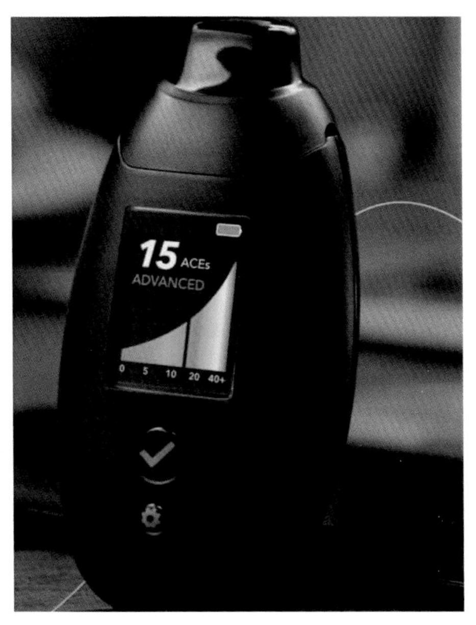

FIG. 6.5 Biosense Breath Ketone Monitor.

Dario blood glucose monitoring system

Dario Blood Glucose Monitoring System is an all-in-one smart pocket-sized device that helps users to monitor their blood sugar levels and trends, count their carbs, and help them build a healthy habit from the result they receive from the device. There is an in-built emergency hypo alert that incorporates users GPS location. Dario Blood Glucose Monitoring Device does not require a cable nor battery to function, and only needs a small blood drop of only 0.3 μL for glucose level measurement with results in less than 6 s. With Dario Blood Glucose Monitoring System, users can add other health metrics such as blood pressure, weight, thereby enabling them to have a full picture about their health vitals. Users of Dario Blood Glucose Monitoring System can also share their results with their healthcare providers and caregivers thereby keeping everyone in the know concerning their health conditions (dariohealth.com, diabetes management) (Fig. 6.6).

Table 6.1 provides a summary of the tool, related application, and the features available.

Discussion

The historical outcome of diabetes was undoubtedly death. As medical sciences progressed and the understanding of the human body increased, origins, and remedies for diseases were understood and created. For diabetes specifically, the discovery that blood

FIG. 6.6 Dario Blood Glucose Monitoring System.

TABLE 6.1 Digital health tools and devices summary.

Tool	Application
Diabeo System	Medical software and application system which patients can input vital information into. Providers can access real time information, and a dose calculator helps patients see recommended doses. A clinical decision support system is included to help providers
Dexcom Glucose Management System	Dexcom is a wearable medical software system which monitors blood sugar. A microsensor with a transmitter is inserted under the skin which transmits blood sugar to a smart device. It can transmit data to providers, and can interact with various diabetes management systems
Insulia	Medical application for smart devices for patients with type 2 diabetes. It has a web portal for providers and an application for patients. It provides insulin level recommendations, a doctor portal, and diabetes education
Eversense	Eversense is an application which measures blood glucose levels via a sensor implanted in the patient. It detects and transmits blood sugar levels, and attempts to predict high and low blood glucose episodes
Blue Star	Blue Star is a diabetes management application which records the input of patient vitals, provides recommendations and education, and can share this information with providers at the touch of a button
Biosense Breath Ketone Monitor	Devices used to measure acetone levels which correlate to the amount of ketones in the blood. Ketone monitors can help alert when there is excessive ketone in the body, and save patients from the dangers of these excessive ketones
Dario Health System	Dario system monitors blood glucose levels and carbs count using a pocket-sized smart device that is connected to any smartphone enabling users to easily track their key diabetes metrics for better management

sugar affected diabetic patients was a vital part of understanding the disease. The creation of insulin to offset the balance in the body saved hundreds of thousands of lives. While far less diabetics die from the disease now than just 100 years ago, this number has not yet fallen to zero. The Centers for Disease Control and Prevention estimates that 24.8 out of every 100,000 patients died from diabetes in 2016 (Huizen, 2019). While this number is surely less than 100 years ago, unfortunately it still has not fallen all the way to zero. The goal of this paper is to discuss some recent technologies which can help manage this disease and cause the diabetes related death rate to fall to zero.

The advancement of technology has greatly benefited the medical sciences. From the discovery of the cell in 1665 to the recent understanding of the parts that make up the cell, the progression of technology has exponentially increased the understanding of the human body (National Geographic Society, 2019). This same progression has also benefited diabetics throughout the world. In the current day and age when diabetes has been turned from a life-threatening disease to a chronic, treatable condition, there are a myriad of technologies and devices that can help diabetics. These devices include the basic blood insulin tests that were used in the past few decades to the implanted chips, applications, and smart, mobile devices that have risen in the past few years.

Wearable technology, implanted chips, and mobile health applications for use in diabetes management are used in a myriad of ways. They can help monitor blood glucose levels, send these records to doctors at the push of a button, provide education, and even provide recommended insulin dose levels per blood glucose. Using these devices and applications can help reduce the risk of developing hypoglycemia. Hypoglycemia is caused when too much insulin is in the bloodstream absorbing high amounts of glucose (Insulin Overdose). Glucose is used as a source of energy for the body so low levels of glucose cause fatigue and warn you that your body is going through an insulin overdose. Some symptoms of hypoglycemia are sweating, chills, dizziness, confusion, hunger, irritability, and blurred vision (Insulin Overdose). To reverse mild hypoglycemia, a person should eat food high in sugars to help regulate the glucose levels.

The systems and applications that have been covered in this paper are some of the most advanced technologies used to manage diabetes. Diabeo System, Dexcom Glucose Management System, Insulia, Eversense, Blue Star, Biosense Breath Ketone Monitor, and Dario Health Systems are some of the most popular and most advanced systems used today. This paper gave a short explanation on these systems, along with how some of these systems interact with each other.

Conclusion

Diabetes has ultimately turned from a life-threatening disease into a chronic, manageable condition. Advancements in digital technology have drastically improved the lifestyle and the disease outlook for diabetic patients. From our research, digital health consumer technologies ranging from implanted chips to mobile health apps and wearable technology, show some promising contributions in improving diabetes management. The technologies reviewed in this paper are alternative and supplemental tools diabetic patients can utilize to improve their diabetes control and make it a more easily manageable condition. Our review of multiple studies of the devices indicates that the use of digital health technologies to manage diabetes improved patient care clinically in certain cases and economically in some.

Overall, benefits realized by patients using these devices can help the overall healthcare system by lowering hospital readmissions, bad clinical outcomes, and reducing out of control healthcare costs. The applications, technologies, and devices we reviewed are among the most technologically advanced forms of diabetes management that exist today. These applications and devices allow diabetics to better manage their conditions, giving daily reminders and providing easy access to information. There are more devices and technologies in the pipeline that we believe will continue to improve care systems in the field of healthcare. Diabetes being one of the most common and costly diseases will ensure that it will continue to be the top disease that innovators will focus on for future devices and tools. We recommend more research and reviews of new technologies and devices that have been successful in managing other chronic conditions that can possibly be applied to diabetes care.

References

Centers for Disease Control and Prevention. (2020). *National Diabetes Statistics Report, 2020*. Atlanta, GA: Centers for Disease Control and Prevention, U.S. Dept of Health and Human Services. https://www.diabetes.org/resources/statistics/cost-diabetes.

Charpentier, G., Benhamou, P. Y., Dardari, D., Clergeot, A., Franc, S., Schaepelynck-Belicar, P., et al. (2011). The Diabeo software enabling individualized insulin dose adjustments combined with telemedicine support improves HbA1c in poorly controlled type 1 diabetic patients: A 6-month, randomized, open-label, parallel-group, multicenter trial (TeleDiab 1 Study). *Diabetes Care*, *34*(3), 533–539.

Dexcom Continuous Glucose Monitoring. (2021). https://www.dexcom.com/ (Accessed June, 2021).

Diabetes. (2020). www.diabetes.co.uk/diabetes_care/testing-for-ketones.html.

Eiland, L., et al. (2018). App-based insulin calculators: Current and future state. *Current Diabetes Reports*, *18*(11), 4. https://doi.org/10.1007/s11892-018-1097-y. Accessed 23 June 2021.

Franc, S., et al. (2020). DIABEO system combining a mobile app software with and without telemonitoring versus standard care: A randomized controlled trial in diabetes patients poorly controlled with a basal-bolus insulin regimen. *Diabetes Technology & Therapeutics*, *22*(12), 904–911. https://doi.org/10.1089/dia.2020.0021. Accessed 6 July 2021.

Heinemann, L., et al. (2018). Real-time continuous glucose monitoring in adults with type 1 diabetes and impaired Hypoglycemia Awareness or Severe Hypoglycemia Treated with Multiple Daily Insulin Injections (HypoDE): A Multicentre, Randomised Controlled Trial. *The Lancet*, *391*(10128), 1367–1377. https://doi.org/10.1016/s0140-6736(18)30297-6.

Higuera, V. (2020). *Diabetes: Past treatments, new discoveries*. Medical News Today. https://www.medicalnewstoday.com/articles/317484#modern-treatment.

Huizen, J. (2019). *Type 2 diabetes and life expectancy*. Medical News Today. https://www.medicalnewstoday.com/articles/317477.

National Geographic Society. (2019). *History of the Cell: Discovering the Cell*. https://www.nationalgeographic.org/article/history-cell-discovering-cell/.

Roze, S., et al. (2020). Long-term cost-effectiveness of Dexcom G6 real-time continuous glucose monitoring versus self-monitoring of blood glucose in patients with type 1 diabetes in the UK. *Diabetes Care*. https://doi.org/10.2337/dc19-2213, dc192213. Accessed 23 June 2021.

Welldoc Inc. (2020). *Welldoc and Dexcom Enhance BlueStar with the Dexcom G6® CGM System*. Welldoc | Chronic Care Platform. https://www.welldoc.com/news/welldoc-announces-collaboration-with-dexcom/.

Further reading

"Insulin Overdose: Signs and Risks". (2017). *Healthline*. www.healthline.com/health/diabetes/insulin-overdose#od-symptoms.

Akturk, H. K., et al. (2021). Real-world evidence and glycemic improvement using Dexcom G6 features. *Diabetes Technology & Therapeutics*, *23*(S1), S-21–S-26. https://doi.org/10.1089/dia.2020.0654. Accessed 23 June 2021.

Apple. (2013). *BlueStar diabetes*. App Store. https://apps.apple.com/us/app/bluestar-diabetes/id700329056.

Biosense. (2021). *BIOSENSE®—The Only Clinically Backed Breath Ketone Monitor*. Biosense®. 8 July 2021 mybiosense.com.

Comstock, J. (2017). *Eli Lilly gets FDA clearance for insulin dose calculator app*. https://www.mobihealthnews.com/content/eli-lilly-gets-fda-clearance-insulin-dose-calculator-app.

Dariohealth (2022) https://www.dariohealth.com/solutions/diabetes-management/.

Eversense Continuous Glucose Monitoring (CGM) System (2022) https://www.fda.gov/media/113491/download.

FDA (2022) https://www.fda.gov/medical-devices/recently-approved-devices/eversense-continuous-glucose-monitoring-system-p160048s006.

Hoyt, E. R., & Yoshihashi, A. (Eds.). (2014). *Health Informatics: Practical Guide for Healthcare and Information Technology Professionals*. Pensacola: Lulu Press.

Insulia App. (2021). https://insulia.com/ (Accessed June, 2021).

Rajpathak, S. (2016). *Prevalence of comorbidities high in type 2 diabetes*. Endocrine Today. https://www.healio.com/news/endocrinology/20160425/prevalence-of-comorbidities-high-in-type-2-diabetes.

The Cost of Diabetes. (2021). *ADA*. https://www.diabetes.org/resources/statistics/cost-diabetes. Accessed 23 June 2021.

Veazie, S., Winchell, K., Gilbert, J., Paynter, R., Ivlev, I., Eden, K. B., et al. (2018). Rapid evidence review of mobile applications for self-management of diabetes. *Journal of General Internal Medicine*, *33*(7), 1167–1176. https://doi.org/10.1007/s11606-018-4410-1.

CHAPTER

7

Recent advancements of pelvic inflammatory disease: A review on evidence-based medicine

Arshiya Sultana[a], Sumbul Mehdi[a], Khaleequr Rahman[b], M.J.A. Fazmiya[a], Md Belal Bin Heyat[c,d,e], Faijan Akhtar[f], and Atif Amin Baig[g]

[a]Department of Ilmul Qabalat wa Amraze Niswan (Gynecology and Obstetrics), National Institute of Unani Medicine, Ministry of AYUSH & Rajiv Gandhi University of Health Sciences, Bengaluru, Karnataka, India [b]Department of Ilmul Saidla, National Institute of Unani Medicine, Ministry of AYUSH & Rajiv Gandhi University of Health Sciences, Bengaluru, Karnataka, India [c]IoT Research Center, College of Computer Science and Software Engineering, Shenzhen University, Shenzhen, Guangdong, China [d]International Institute of Information Technology, Hyderabad, Telangana, India [e]Department of Science and Engineering, Novel Global Community Education Foundation, Hebersham, NSW, Australia [f]School of Computer Science and Engineering, University of Electronic Science and Technology of China, Chengdu, Sichuan, China [g]Faculty of Medicine, Universiti Sultan Zainal Abidin, Kuala, Terengganu, Malaysia

Introduction

Pelvic inflammatory disease (PID) is an inflammatory disorder and polymicrobial infection of the upper female genital tract, which encompass endometritis, parametritis, salpingitis, and oophoritis. It generally affects young, sexually active women (Shen et al., 2016). It poses a chief health problem for reproductive-age women in both developing and developed countries (Pandey, 2014). Subclinical PID/uncomplicated PID (uPID) is difficult to detect clinically as there are no signs or symptoms of acute PID, which is accountable for a more percentage of PID-related complications than a clinically recognized condition (Das, Ronda, & Trent, 2016; Sweet, 2011). Chronic PID refers to both late detected and untreated acute and sub-acute recurrence of a previous upper reproductive tract infection (Lamina & Hanif, 2008).

Computational Intelligence in Healthcare Applications
https://doi.org/10.1016/B978-0-323-99031-8.00016-8

PID is usually diagnosed clinically, however, a conclusive diagnosis can be made laparoscopically by viewing inflamed, purulent fallopian tubes directly. It can have serious concerns for the reproductive health of women, thus any woman who has abdominal or pelvic pain should be evaluated and treated (Mitchell & Prabhu, 2013). Inadequate treatment could result in complications such as infertility, chronic abdominal pain, recurrent PID (Haggerty et al., 2016) ectopic pregnancy, and internal pelvic scarring. Chronic pelvic pain, scarring and adhesions, infertility, ectopic pregnancy, and recurrent PID are long-term sequelae of PID. Following one bout of infection, the chances of infertility rise from 8% to 43% after three or more episodes of infection (Dayal, Singh, Chaturvedi, Krishna, & Gupta, 2016; Pandey, 2014; Ross, 2013; Sweet, 2011). The absence of Tubo-ovarian abscess (TOA) is termed as mild to moderate PID. The presence of severe systemic symptoms or TOA is referred to as severe or acute PID (Ross, 2013). Medical management for RTIs/PID includes antimicrobial/antibiotics.

We conducted a thorough literature search in UGC-CARE list journals, Web of Science, ScienceDirect, Springer, PubMed, EBSCO, Google Scholar, and Scopus electronic databases for an overview of pelvic inflammatory disease with recent advances and evidence-based medicine. Extensive exploration was conducted to comprehend the concept and practices of pelvic inflammatory in contemporary and complementary and alternative medicine for evidence-based studies. The search terms/keywords were "PID," "Historical background and PID," "pelvic inflammatory disease and risk factors," "management of PID," "overview on PID," "evidence-based studies for management of PID" "herbs and PID," "complementary and alternative medicine in PID," and Unani drugs useful in PID. All articles were meticulously evaluated without any language or time restriction. A data-charting in electronic form was jointly developed by the first author to determine which variables to extract. The first author (AS) and other authors continuously updated the data-charting form. A total of 100 articles and textbooks were retrieved to collect the information regarding the PID. Forty-nine articles were from PubMed, six textbooks, and six from other indexing sites. The articles that were included in the manuscripts encompassed research articles, review articles, and textbooks. This review was planned under the following headings: historical perspective, epidemiology, etiopathogenesis, clinical characteristics, diagnosis, differential diagnosis, investigations, complications, management and complementary and alternative medicine.

Historical perspectives

Ancient history (BC 3600-AD 500)

Since antiquity, RTIs/STDs have been discussed and gonorrhea and syphilis are the most commonly described sexually transmitted diseases in ancient times and dates back to the time of Hippocrates. Egyptian manuscripts, the *Papyrus Smith* and the *Papyrus Ebers* from 2000 B·C are the first sources briefly dealing with sexually transmitted diseases. The ancient Egyptians were well aware of the infection of genitalia (O'Dowd & Philipp, 2000). The description of gonorrhea is available in the Bible (Burg, 2012). *Kahun Papyrus* (1850 BC) contains a description of pruritus vulvae and treatment for sepsis of the uterus. *Hearst Papyrus* (1550 BC) also contends with the diseases of female genitalia. Treatment for pustular vulvar and vaginal eruptions, and to resolve the inflammation, is mentioned (O'Dowd & Philipp, 2000). The ancient Egyptian medicines declined with the occurrence of foreign invasion and Greek dominance.

Greek and roman medicine

Hippocrates (460–377 BC) defined "strangury" and stated that it is caused by sexual pleasure (Burg, 2012). He was aware of pyometra and pelvic abscess. Soranus of Ephesus (99–138 AD) described a rectal examination to differentiate between uterine and rectal inflammation, as well as treatment for inflammation of the vagina and uterus. Galen (131–121 AD) defined the term gonorrhea for the first time (O'Dowd & Philipp, 2000).

Postclassical era (500–1500)

Arabian Scholars preserved the ideas of Greek and Roman authors. The Arabic and other Unani physicians described *waram al-rahim* (uterine inflammation/reproductive tract infections). *The treatment of a pelvic abscess was documented by Aetius (600 AD). Gonorrhea, cystitis, and testicular inflammation were described by Rhazes (852 AD), Mesue (904 AD),* and *Ali Ibn Abbas Majusi (980 AD)* (Kabiruddin, 2006). Raban Tabari (838–870) in his classical source "*Firdaws al-Hikmat fi'l-Tibb*" had described etiology, clinical features, and management of PID (Tabri, 2010). Rhazes (865–925) gave a detailed description of PID as well as other gynecological diseases in his treatise "*Kitab al- Hawi fi'l Tibb*". Majusi (930–994) in his text "*Kāmil al-Sanā 'a al-Tibbiyya*" conversed about PID (Majoosi, 2010).

Avicenna (980–1030) in his treatise "*Al-Qānun Fi'l Tibb*" surmised PID in detail. He opined simple disease and compound disease are two kinds of diseases that occur in human beings (Sina, 2010). Lower abdomen discomfort is related to urinary symptoms such as dribbling, urine retention, and dysuria in anterior uterine wall inflammation, whereas constipation, dysentery, and rectum pain are associated with posterior uterine wall inflammation. Pain in the iliac area and thighs is caused by inflammation of the uterus's lateral side (ovaries and fallopian tubes). The cervix uteri inflammation is more severe and hard and can be identified on palpation because it is pain sensitive, but the fundus uteri inflammation is difficult to detect (Majoosi, 2010; Sina, 2010).

The associated symptoms are high-grade fever and black discoloration of the tongue. The patient also complained of headache (Qarshi, 2011). Pain in the perineum and umbilicus when a uterine anterior aspect is affected. Sometimes pain occurs in the back and low back/tail bone when a uterine posterior aspect is affected. Pain in the umbilicus and lower back radiates up to the thighs, and groin which causes distension and difficulty in getting up (Kabiruddin, 2006; Qarshi, 2011).

Epidemiology

Prevalence/incidence

Though PID's true prevalence is unknown because the majority of cases are asymptomatic, and its global incidence ranges from 0.28% to 1.67% (Nkwabong & Dingom, 2015). The Centre for Disease Control (CDC) in the United States estimates that over one million women have a

PID episode each year, including missing cases (Das et al., 2016). According to data from the United States, more than 10% of women have a history of PID. Over 1 million new cases of PID are reported each year (Ross, 2013). According to the largest UK study, *C. trachomatis* and *N. gonorrhoeae* are responsible for 39% and 14%of PID cases, respectively (Simms & Stephenson, 2000). PID or its complications claim the lives of almost 150 women each year (Pachori & Kulkarni, 2016). In a cross-sectional investigation, Wiesenfeld et al. used endometrial biopsies to detect subclinical PID. He discovered that *N. gonorrhoeae* and *C. trachomatis* are responsible for 26% and 27% of PID cases, respectively.

Subclinical PID does not show clinical signs and symptoms and hence must not be underestimated as it might destroy fallopian tubes as acute asymptomatic PID. Thus, asymptomatic STIs screening and early treatments are critical (Das et al., 2016).

Risk factors

The risk factors of PID are young age (<30 years) (Nkwabong & Dingom, 2015; Simms & Stephenson, 2000), minority race/ethnicity, lower socioeconomic status (Simms & Stephenson, 2000; Tamunomie, 2013) past history of pelvic inflammatory disease (Simms & Stephenson, 2000; Tamunomie, 2013) new or multiple sexual partners, substance abuse, past history of gonorrhea, and alcohol abuse (Jackson & Soper, 1999; Pandey, 2014; Ross, 2013; Tamunomie, 2013) douching (Jackson & Soper, 1999; Simms & Stephenson, 2000), bacterial vaginosis (BV) (Mitchell & Prabhu, 2013), uterine instrumentation (HSG, curettage), non-use of barrier contraceptive method, intrauterine contraceptive (IUCD) (Simms & Stephenson, 2000; Tamunomie, 2013) early coital, nulliparity, unsafe termination of pregnancies, previous stillbirths (Simms & Stephenson, 2000; Tamunomie, 2013), and manual removal of placenta (Fig. 7.1).

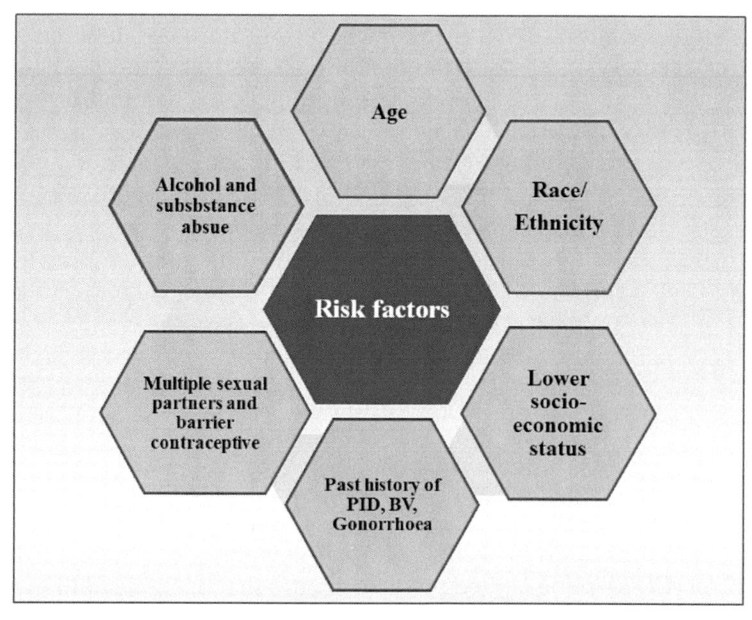

FIG. 7.1 Risk factors for pelvic inflammatory disease.

Age

The PEACH study reported recurrent PID was five times more common at age 19 and more likely to experience chronic pelvic pain (CPP) 7 years after being diagnosed with PID (Das et al., 2016). 16–24 years old's have the highest rate of bacterial STIs (Dhasmana, Hathom, McGrath, Tariq, & Ross, 2014), reflecting their increased risk in this age group. In the USA, 5.1% of women aged 15 to 44 years are to at least one episode of PID during their lifetime. In the United Kingdom, 1.7% of women aged 16 to 46 years are impacted each year (Dayal et al., 2016), and 15% of Swedish women in their life span (Qarshi, 2011), and 11 million women in the USA are treated every year (Haggerty & Ness, 2007).

Biological and behavioral factors contribute to the increased risk of PID in adolescents. Microorganisms can infect a larger surface area biologically. Adolescents are more likely to engage in unprotected sex, have several sex partners, and have short-term and frequent monogamous relationships (Crossman, 2006; Das et al., 2016; Pandey, 2014).

Although PID is the most frequent infection of the reproductive tract in sexually active women, it is rarely detected in postmenopausal women, who experience fever, nausea, lower abdominal pain, postmenopausal bleeding, and bowel irregularities (Dayal et al., 2016). In postmenopausal women, the prevalence is <2% (Jackson & Soper, 1999). Postmenopausal women are more likely to have TOA (Mitchell & Prabhu, 2013).

Habitat

It is responsible for 94% of all STIs morbidity. In the USA, more than a million women suffer from acute PID each year. In Latin America and the Caribbean, the prevalence is 36.8 per 100,000 (Savaris et al., 2007). In England, PID is found in 1.1% of young women seeking primary care (Duarte, Fuhrich, & Ross, 2015; Pachori & Kulkarni, 2016). It accounts for 17%–40% of hospital admissions in Sub-Saharan Africa, 15%–37% in Southeast Asia, and 3%–10% in India (Ross, 2013).

Race

According to Sutton et al., the incidence is two to three times higher in black women compared to white women (Das et al., 2016).

PID and bacterial vaginosis

BV has been linked to endometritis and salpingitis in human subjects. In animal models, BV-associated bacteria have been found to cause fallopian tube damage. In addition, BV-associated microbes may alter local cytokine responses, change mucosal immunity, and raise vaginal pH, increasing sensitivity to STIs and perhaps increasing the risk of bacterial ascent. As a result, BV may have a role in PID. *N. gonorrhoeae* and *C. trachomatis*, on the other hand, frequently co-infect patients with BV. As a result of the lower lactobacilli concentrations in BV, women are at increased risk for STI acquisition (Taylor, Darville, & Haggerty, 2013). The link between PID and BV is marked by the vaginal overgrowth of anaerobic bacterial species (Haggerty et al., 2016) that can all disrupt the vaginal flora, resulting in the loss of hydrogen peroxide-producing lactobacillus and an increase in anaerobic bacteria (Sweet, 2011; Taylor et al., 2013).

Cigarette smoking

It has also been linked to a higher incidence of PID. Cigarette smoking impairs the immunological response to infection or estrogen activity, and it also reflects poor health-seeking behavior in low socioeconomic groups (Pandey, 2014).

Aetiopathogenesis

The majority of instances of PID are caused by genital tract infections that pass through the endocervix and into the uterus and other uterine adjacent structures. Only 10–20% of infectious organisms from the lower reproductive tract make a way to the upper genital tract, where PID occurs. The immune system's response of the host and micro-organism infestation decides the clinical appearances of the disease manifestations that include cervicitis, endometritis, or abscess development, etc (Tamunomie, 2013). The spread of infection will be intra-abdominally, from the cervix to the endometrium and into the peritoneal cavity, or through the lymphatic system, as in the case of parametrium infection caused by an intrauterine device or hematogenous pathways, such as tuberculosis (Haggerty et al., 2016), but this is uncommon. Infection, generally bacterial, causes the inflammation seen in PID. The microorganisms responsible for it can be spread sexually (*Neisseria gonorrhoeae, Chlamydia trachomatis*) (Haggerty et al., 2016; Nkwabong & Dingom, 2015; Ross et al., 2006; Sweet, 2011; Taylor et al., 2013) or not (*Escherichia coli, Enterococcus faecalis, Klebsiella, Streptococcus sp., Staphylococcus sp.*) (Nkwabong & Dingom, 2015; Sweet, 2011; Taylor et al., 2013). It was assumed that PID occurs due to *N. gonorrhoeae* previously than the mid-1970s. The discovery of polymicrobial etiology came into existence with the advent of invasive techniques like culdocentesis, laparoscopy and/or endometrial aspirations (Sweet, 2011). Mycoplasmas of the vaginal tract, particularly *Mycoplasma genitalium* (Haggerty et al., 2016; Nkwabong & Dingom, 2015), have lately been discovered to cause PID. The other germs infiltrate the tissues when the most virulent germ has either begun to kill tissues or shifted vaginal flora to an aerobic condition, as in bacterial vaginosis (Nkwabong & Dingom, 2015). In some situations, cytomegalovirus and *Ureaplasma urealyticum* may be the causative agents (Ross et al., 2006). A study showed that the risk factors associated with PID 54% women were between the age group of 26- and 30-years age group and 74% women were from class whereas 24.7% were from the middle class (Vanamala, Pakyanadhan, Rachel, & Abraham, 2018).

Clinical characteristics

As the clinical manifestations of PID differ according to the responsible pathogens. *C. trachomatis* or *M. genitalium* organism usually present with clinical features of milder symptoms. Fever, adnexal tenderness, and mucopurulent cervicitis are more common in women with gonococcal infections. Less commonly there may be no symptoms at all or unusual symptoms, such as Fitz-Hugh-Curtis syndrome (Haggerty et al., 2016), which is characterized by right upper quadrant pain due to perihepatitis. Lower abdominal pain (Duarte et al., 2015; Jaiyeoba & Soper, 2011) or pelvic pain are common symptoms in women, however, they might be mild. Other clinical features include abnormal vaginal discharge, cramps, fever or chills, and dysuria (Haggerty et al., 2016; Jaiyeoba & Soper, 2011; Ross et al., 2017), deep dyspareunia (Duarte et al., 2015), and low back pain, abnormal or postcoital bleeding, nausea, vomiting (Haggerty et al., 2016; Jaiyeoba & Soper, 2011; Ross et al., 2017) secondary dysmenorrhea (Ross et al., 2017). The patient may also present with vulval itching and odor (Lamina & Hanif, 2008). Tenderness in the lower abdomen, cervical/uterine motion tenderness and adnexal tenderness (Duarte et al., 2015; Ross et al., 2017).

Diagnosis

Jacobson and Westrom developed the commonly used diagnostic protocol in 1969, that PID is defined as acute lower abdominal/pelvic pain with two or more symptoms/signs: fever, vomiting, abnormal vaginal bleeding, marked pelvic tenderness, proctitis, urinary symptoms, palpable abdominal mass or swelling with an erythrocytes sedimentation rate of >15 mm/h. The CDC modified Jacobson's diagnostic criteria to increase the specificity of clinical diagnosis criteria, establishing three minimal diagnostic criteria detected on pelvic examination as follows:

"*Minimum criteria: cervical motion tenderness; lower abdominal pain; uterine tenderness/adnexae tenderness are the minimum requirements.*

To improve the specificity of the minimum criteria and support a diagnosis of PID, one or more of the following additional criteria can be used:

Supportive criteria: abnormal cervical or vaginal mucopurulent discharge; Oral temperature > 101 °F (38.3 °C); elevated C-reactive protein; abundant WBC on saline microscopy of vaginal fluid; elevated erythrocyte sedimentation rate; laboratory documentation of cervical infection with N. gonorrhea or C. trachomatis" (CDC, 2015; Crossman, 2006; Das et al., 2016; Haggerty et al., 2016; Tamunomie, 2013).

The most common method for diagnosing uPID is clinical examination; however, it has a lower predictive value than laparoscopy (Dayal et al., 2016). The PEACH study reported an elevated leukocyte count ≥of 10,000 cells/mL had 41% and 76% sensitivity and specificity, respectively, for the presence of endometritis (Jaiyeoba & Soper, 2011). As per World Health Organization (WHO), Women who are sexually active with lower abdominal discomfort are recommended for abdominal and bimanual examination to test for PID manifestation.

In >90% of patients, bilateral lower abdominal pain and tenderness is found. The severe pain onset immediately after menses or during is highly susceptible for PID, as reported that in 75% of cases PID occurs within the first week of menses. Vaginal discharge as a new symptom is reported in 75% of patients however, it is not sensitive and specific for the diagnosis of PID. Abnormal uterine bleeding (AUB) affects 30% and 50% present with fever in PID. The pelvic examination is a very useful part of the physical examination for diagnosing PID to look for mucopurulent discharge bilateral adnexal tenderness, and cervical motion tenderness.

Investigations

Blood investigations

Hyperleucocytosis related to a high C-reactive protein (CRP) level suggests a complicated form of PID or a differential diagnosis such as acute appendicitis. The absence of hyperleukocytosis or a normal CRP concentration does not rule out a diagnosis of PID. When PID is suspected, complete blood count and a CRP assay are advised (grade C) (Brun, Castan, et al., 2020).

Erythrocyte sedimentation rate (ESR)

Mitchell and Oluwatosin found in their studies, an elevated ESR >15mm/h had 70% and 52% sensitivity and specificity, respectively, for endometritis or salpingitis (Jaiyeoba & Soper, 2011).

Wet mount test

Increased levels of vaginal neutrophils more than 3/hpf had a sensitivity of 78% and a specificity of 39% in two studies, whereas higher WBC had 57% and 88% sensitivity and specificity, respectively (Jaiyeoba & Soper, 2011). According to some investigators, the absence of vaginal white blood cells had a 95% negative predictive value. If there are no white blood cells in the wet mount/vaginal smear, it is suggested that an alternative diagnosis to PID be considered (Crossman, 2006; Jaiyeoba & Soper, 2011).

Urinalysis

Urinalysis is recommended to exclude a urinary tract infection.

Urine pregnancy test

A pregnancy test should be performed on all patients with suspected PID to rule out ectopic pregnancy and septic abortion.

Swab culture

Endocervical cultures for *C. trachomatis* and *N. gonorrhoeae* should be recommended in suspected patients (Brun et al., 2020).

Transvaginal ultrasonography (TVS)

TVS reveals fluid in the cul-de-sac, tubal wall thickness >5mm, incomplete septae inside the tube, inclusion cyst, and on cross-sectional view a "cogwheel" appearance of the tube as noted (Brun et al., 2020; Crossman, 2006; Haggerty et al., 2016; Jaiyeoba & Soper, 2011).

Doppler

Doppler investigations show pelvic infection, such as tubal hyperemia, according to Doppler studies. The vascularity and pulsatility indices are measured using Color Doppler flow or power Doppler. A study reported that the power Doppler accurately recognizes laparoscopically confirmed instances of acute PID, making it 100% sensitive for this diagnosis (Crossman, 2006).

Magnetic resonance imaging technique (MRI)

The magnetic resonance imaging (MRI) approach is extremely expensive and unreachable to women who seek outpatient assessment for PID. Thickened, fluid-filled tubes, pelvic free fluid, pyosalpinx, and TOAs are among the conditions that it can detect and is very sensitive and specific (Haggerty et al., 2016).

Endometrial biopsy (EB)

In asymptomatic women, EB is the gold standard investigation to diagnose PID. Endometritis is histopathologic evidence and the only sign of PID (CDC, 2015).

Laparoscopy

Laparoscopy was compared to other types of diagnostic procedures. Clinical diagnosis has a sensitivity rate of 87%, while laparoscopy has a sensitivity rate of 81% and a specificity rate of 100% (CDC, 2015). PID can be diagnosed via laparoscopy if there is distinct tubal surface congestion, tubal wall oedema, and exudate on the tube's surface and fimbriae endings. These criteria can also be used to diagnose chronic PID sequelae such as pyosalpinx (Taylor et al., 2013) (Fig. 7.2).

Differential diagnosis

- Inflammatory bowel disease (Haggerty et al., 2016).
- Appendicitis: Younger people are more likely to have appendicitis. Ultrasonography and CT scans can confirm the diagnosis. A non-compressible tubular structure with thickened muscular walls can be seen in transabdominal ultrasonography (TAS) of the right lower quadrant (Jackson & Soper, 1999).
- Urinary tract infections, pyelonephritis, and renal calculi are examples of genitourinary disorders.
- Obstetric and gynecologic disorders, such as ovarian tumors, cysts, and torsion, ectopic pregnancy, and functional pelvic pain (Haggerty et al., 2016).
- In postmenopausal women, diverticulitis is the most important differential diagnosis. The patient presents with prominent bowel symptoms, unlike PID. For the evaluation of suspected diverticulitis, computed tomography (CT) is the diagnostic modality of choice (Jackson & Soper, 1999).

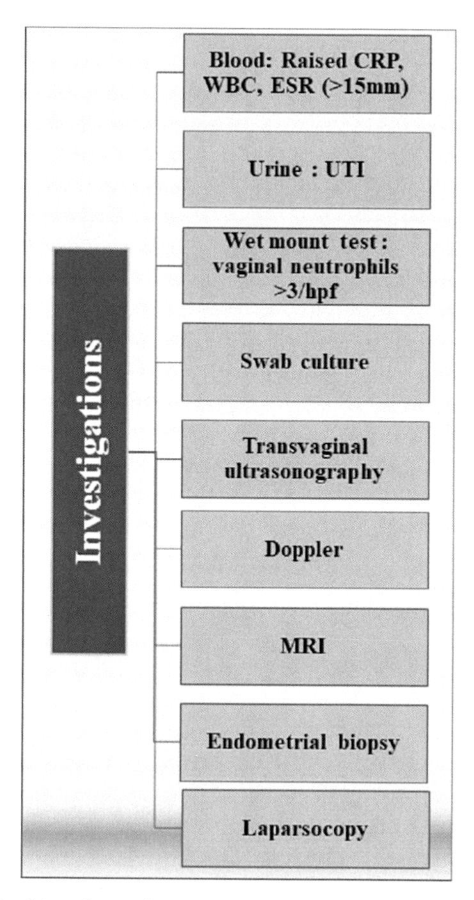

FIG. 7.2 Investigations required in pelvic inflammatory diseases.

Complications

The complications are divided into immediate and late.

Immediate

Pelvic peritonitis or even generalized peritonitis, septicemia, Fitz-Hugh-Curtis syndrome and TOA (Brun et al., 2020).

Late

Dyspareunia, without adequate treatment, if not treated properly, can lead to pelvic adhesions and scarring, and other aforementioned long sequelae (Williamson & Aldeen, 2010). Following three or more episodes of infection, the chances of infertility increase from 8% to 43% (Pandey, 2014; Ross, 2013; Sweet, 2011). These complications have been linked to an increased incidence of hysterectomy (Pandey, 2014). In the PEACH trial, women were at the highest risk who had two or more episodes of PID with 36% reporting chronic pelvic pain. The infertility rate was 16% in a Swedish cohort study of women with laparoscopically confirmed salpingitis who were followed for an average of 94 months, with 67% of it attributed to tubal factor infertility, compared to 2.7% in women without salpingitis. 9% of women with salpingitis who became pregnant had an ectopic pregnancy in comparison to 1.9% of control women. The severity of salpingitis and the frequency of PID episodes were associated with infertility. Repeat Chlamydial cervicitis infections increase the likelihood of ectopic pregnancy and women who had three or more episodes had a 4.5-fold higher risk of PID (Mitchell & Prabhu, 2013).

Tubo-ovarian abscess (TOA)

TOA is a polymicrobial infection that causes abdominal pain to worsen, as well as unilateral adnexal tenderness and fever (Curry et al., 2019; Jennings & Krywko, 2020).

Fitz-Hugh-Curtis syndrome

Fitz-Hugh-Curtis syndrome is a kind of perihepatitis caused by PID spreading into the peritoneal cavity. Pleuritic right upper quadrant pain is caused by inflammation and swelling of the liver capsule.

Chronic pelvic pain

Infraumbilical discomfort that lasts at least 6 months and is severe enough to induce functional disability is referred to as chronic pelvic pain. It is one of the factors that contribute to long-term morbidity following a PID episode (Bartlett, Levison, & Munday, 2013; Jennings & Krywko, 2020) (Fig. 7.3).

Management

The major goal of management is to limit the risk of long-term consequences such as infertility, recurrent infection, ectopic pregnancy and pelvic pain, in addition to alleviating the acute inflammatory condition, clinical cure, and microbiological cure. As a result, an early and accurate diagnosis is of greater importance for avoiding PID-related complications and morbidity (Dayal et al., 2016).

Prevention

PID prevention is divided into three categories:

(1) Preventing the initial episode of PID and
(2) Recurrent disease prevention.

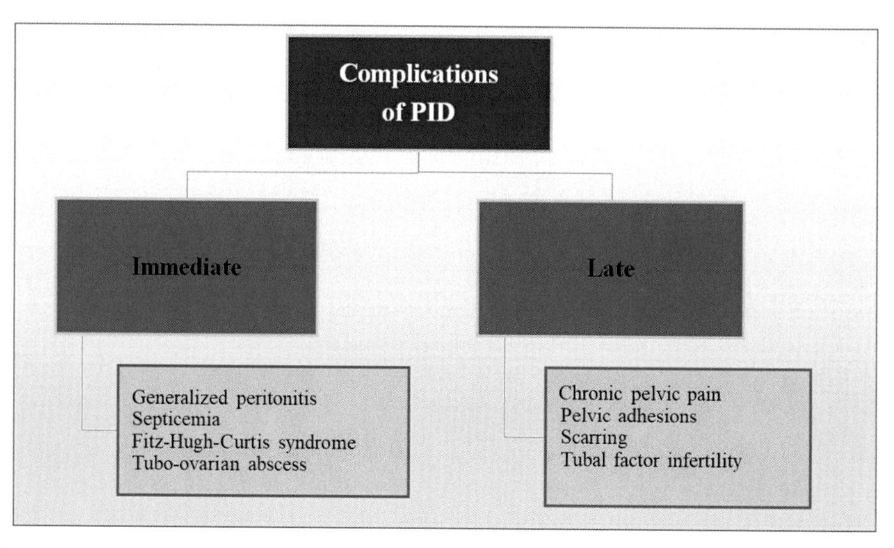

FIG. 7.3 Complications of PID.

Recurrent PID prevention is also a public health concern. Recurrent disease is typical among women with PID. According to the PEACH trial, 14.5% of individuals had recurrent PID 3 years after their initial diagnosis, and >21% experienced repeat PID 7 years later. According to these findings, additional efforts to implement clinical interventions aiming at adequate treatment and prevention of recurrent disease are necessary (Crossman, 2006; Das et al., 2016; Haggerty et al., 2016).

Intervention

Oral or intramuscular (IM) (CDC)

"Ceftriaxone 250 mg IM one dose or cefoxitin 2 g IM one dose with probenecid 1 g orally or other parenteral 3rd generation cephalosporin and doxycycline 100 mg orally twice daily for 14 days with or without metronidazole 500 mg twice daily for 14 days with or without metronidazole 500 mg orally twice daily for 14 days. 3 g of ampicillin/sulbactam IV every 6 h, plus 100 mg of doxycycline orally (or IV) every 12 h is recommended as an alternative regimen."

Indications for the hospitalization

Pregnancy, unresponsiveness to oral antibiotics; probable surgical emergency (e.g., appendicitis), inability to adhere or endure outpatient therapy; severe illness (e.g., high fever, vomiting, chronic pain) and TOA (CDC, 2015; Das et al., 2016; Mitchell & Prabhu, 2013; Williamson & Aldeen, 2010).

Moxifloxacin vs CDC-recommended regimen

Several parenteral antibiotic regimens show high clinical and microbiological efficacy in the short term. For 14 days, oral doxycycline 100 mg twice a day should be given. In cases of TOA, prolonged therapy with metronidazole 500 mg orally or clindamycin 450 mg orally QID is more active exposure against anaerobic bacteria. Gentamicin single dose is a parenteral regimen B to treat other pelvic and abdominal infections. While the CDC suggests taking doxycycline 100 mg twice a day or clindamycin 450 mg four times a day for a total of 14 days of treatment, the author prefers Clindamycin oral treatment since it gives more anaerobic coverage. In the case of severe PID, particularly tuboovarian abscess, the CDC recommends continuous clindamycin treatment. Alternative regimens include IV ofloxacin 400 mg twice daily with IV metronidazole 500 mg three times daily for 14 days, or IV ciprofloxacin 200 mg twice daily plus IV or oral doxycycline 100 mg twice daily plus IV metronidazole 500 mg three times daily, according to the European guideline. The CDC no longer recommends quinolone-containing regimens in acute PID because of quinolone-resistant *N. gonorrhoeae*.

Moxifloxacin has more effective in-vitro activity against *Mycoplasma genitalium*, *C. trachomatis*, and anaerobic bacteria than other fluoroquinolones (Mitchell & Prabhu, 2013; Williamson & Aldeen, 2010).

Azithromycin once a day or doxycycline once a day for 14 days

Many individuals find it challenging to stick to the 14-day antibiotic prescription. One randomized, double-blind trial assessed the efficacy of a single IM ceftriaxone dose followed by either twice-daily doxycycline or once-daily azithromycin for 14 days. The cure rate was 90% with once-daily azithromycin vs 72% for twice-daily doxycycline (Brun et al., 2020; Williamson & Aldeen, 2010).

Treatment based on several guidelines

In both the United States and Europe guidelines (2014) for the treatment of PID with antibiotics have been developed representing the most recent evidence-based guidance currently available. The European and CDC recommendations differ for intramuscular ceftriaxone dose for outpatient regimens and a higher dose is advised in *N. gonorrhoeae* infections to reduce the risk of resistance. Further, they recommended oral ofloxacin or levofloxacin plus metronidazole as an outpatient antibiotic regimen as quinolone resistance gonorrhea infection are relatively high in many regions of the world. The CDC does not recommend oral moxifloxacin; however, the European guideline suggests it as an alternative if gonococcal PID is deemed improbable (Duarte et al., 2015).

According to the CDC, if a patient had sexual contact with a male partner 60 days before the onset of symptoms, the male partner should also be treated with anti-N, gonorrhoeae, and anti-C trachomatis regimens. Sexual intercourse should be avoided by women with acute PID until both partners are asymptomatic and their therapy is completed (CDC, 2015; Duarte et al., 2015; Haggerty et al., 2016; Sweet, 2011).

PID has a significant morbidity rate; approximately 20% of women who are affected become infertile, 40% experience chronic pelvic pain, and 1% of those who conceive have an ectopic pregnancy. Clinical symptoms and indications appear to resolve in a considerable proportion of untreated women, according to uncontrolled observations (Ross, 2013).

Follow up

It is critical to check up on the patient to see how they are reacting to the outpatient treatment. If clinical symptoms improve within 72h, the treatment is working, and if they don't, more testing is needed. Additional tests will be performed to rule out other possibilities, such as a TOA, which may be required in some patients, as well as assessment for additional antibiotic medication, parenteral antimicrobials, and hospitalization (Haggerty et al., 2016).

Complementary and alternative medicine (CAM)

Shivagutika was given twice daily with honey after each meal for 60 days to treat PID and the majority of the patients showed a gradual improvement in symptoms over time, demonstrating the formulation's efficacy in PID. In conclusion, Shivagutika effectively reduced PID symptoms and clinically controlled infection (Vishwesh & Bhat, 2014). The study details are summarized in Table 7.1. The main ingredient in Shivagutika is Bitumen (Shilajith) and other herbal medicine properties are summarized in Table 7.2. Few single-blind, randomized controlled trials have proven their efficacy in cervicitis (Zahid, 2016), endocervicitis (Habib et al., 2011), and pelvic inflammatory diseases (Qayyum & Arshiya, 2018; Sayed et al., 2016; Vishwesh & Bhat, 2014). The aforementioned studies were phase II, single-blind with a small sample size. Hence, further phase III and IV, double-blind, randomized controlled trials in larger samples are recommended to confirm the Unani and other traditional medicine are effective in RTIs.

In another study, Pachori and co-workers assessed medicinal plants as an alternative treatment for PID. Twelve plants were selected for antimicrobial activity against test pathogens. The results on antimicrobial activity revealed that six plants, viz. *Ocimum sanctum*, *Tribulus terrestris*, *Curculigo orchioides*, *Phyllanthus niruri*, *Ficus racemosa*, and *Solanum xanthocarpum* showed prominent antimicrobial activity against test pathogens, viz. *S. aureus*, *Streptococci species*, *Klebsiella species*, *Salmonella species* and *Candida albicans*. The antimicrobial activity of composite herbal formulation was found to be at par with commercial antibiotic of choice against *Bacteroides species*. The active components showing antimicrobial activity against test pathogens were found to be saponins, flavonoids, alkaloids, terpenoids, phenols and fixed oil (Pachori, Kulkarni, & Bodhankar, 2015). Anti-inflammatory phytochemicals e.g., flavonoids, and dicaffeoylquinic acid affect the synthesis and liberation of inflammatory mediators, thus abolishing inflammation and associated tenderness (Chaturvedi et al., 2012).

Balogun and coworkers reported management of chronic PID with shortwave diathermy. After nine shortwave diathermy (SWD) treatments using a modified crossfire technique, the patient was completely relieved of her abdominal and back pains (Balogun & Okonofua, 1988).

In one study, Yudan Liang discussed acupuncture for chronic PID. Out of 15 participants, nine patients reported favorable body changes and relief from lower abdominal pain, seven patients reported recovering their menstrual cycle, and two patients reported feeling worse at times (Liang & Gong, 2014) (Fig. 7.4).

TABLE 7.1 Studies in complementary and alternative medicine.

First author	Patient age (Y)	Diagnosis method	Type of intervention	Control group	Duration of treatment	Outcome	Adverse event	Refer
Vishwesh and Bhat	Child bearing age	Clinical features of PID and USG finding	Shivaguitaka (500 mg) BID with honey ($n=15$)	–	60 days	Reduction in symptoms and changes in USG findings	Not reported	Vishwesh and Bhat (2014)
Zahid et al.	18–40	Clinical features of cervicitis	Sufoofe sailan (*Woodfordia fructicosa Kurz.*, *Areca catechu L*, *Salamalia malabarica Scholts &* *Mimusops* elangi L. 5 g BID Locally pessary of *Plantago ovata* Forsk. and *Linum usitatissimum* L. ($n=30$)	Placebo ($n=30$)	Oral-90 days and local 10 days in each cycle after menses for 3 cycles	Reduction in clinical symptoms assessed by Visual analog scale for clinical features	No adverse effects	Zahid (2016)
Habib et al.	18–40	Clinical features of endocervicitis	*Majoon Ushba* 10 g BID and per vaginum *Nakuna* ointment for 15 days ($n=30$)	Orally Tab doxcycline 100 g BID and vaginal pessary of clindamycin and clotrimazole ($n=15$)	Test: Oral–Majoon and per vaginum 15 days for 3 consecutive cycles Control: 7 days	Resolution of clinical features	No adverse effect	Habib et al. (2011)
Sayed et al.	20–40	VAS and McCormack pain scale core for PID	Arq Brinjasif (Ingredent: *Achillea millefolium*, *Foeniculum vulgare*, *Cichorium intybus*, *Tamarix gallica*, *Mako khushk* *Solanum nigrum* and *Artemisia absinthium* 60 mL BID ($n=20$)	Ofloxacin 400 mg and ornidazole 500 mg	14 days	Resolution of clinical features	No adverse effect	Sayed, Shameem, and Mubeen (2016)
Qayyum and Sultana	18–45	VAS and McCormack pain scale core, WBC >10/hpf in vaginal wet mount test for uPID	Oral placebo capsules and pessary of *P. ovata* Forsk. and L. *usitatissimum* L. ($n=33$)	Tab doxycline 100 mg BID and Tab metrogyl 400 mg TID ($n=33$) and per vaginum placebo	14 days	Primary outcome: Resolution of clinical features,Secondary outcome Change in WBC cells and improvement in Quality of life (SF12 QoL)	No adverse effect	Qayyum and Arshiya (2018)

TABLE 7.2 Antimicrobial activity and chemical constituents of Herbal medicine.

Name	Botanical name	Relevant activity	Active phyto/chemical constituents	Refer
Shilajat	*Bitumen*	Anti-inflammatory, analgesic, immunomodulatory, antiviral and antioxidant activity. Antimicrobial activities against s. aureus, *E. coli* and *Candida albicans*	Fulvic acid, dibenzo-α-pyrones, selenium, tirucallane triterpenes, phenolic lipids, small tannins	Carrasco-Gallardo, Guzmán, and Maccioni (2012); Trivedi, Mazumdar, Bhatt, and Hemavathi (2004); Vishwesh and Bhat (2014)
Gul-e-Dhawa	*Woodfordia fructicosa* Kurz	Anti-inflammatory, analgesic, antimicrobial	Methanol, flavonoids, tannins, sterols, anthroquinones, saponins	Baravalia, Vaghasiya, and Chanda (2012); Ghante, Bhusari, Duragkar, and Ghiware (2014); Parekh and Chanda (2007)
Gul-e-fufal	*Areca catechu* L.	Methanolic fruit extract of Areca Nut showed an antibacterial effect on both Gram-positive and Gram-negative bacteria	four alkaloids, including Arecaidine, Arecoline, Guvacine, and Guvacoline	Jam, Hajimohammadi, Gharbani, and Mehrizad (2021)
Moochras	*Salamalia malabarica* Scholts	Anti-inflammatory, antioxidant, antimicrobial and hepatoprotective activity	Arabinose, galactose, galacturonic acid, rhamnose	Karole, Gautam, and Gupta (2017)
Alsi	*Linum usitatissimum* L.	Anti-inflammatory, analgesic, antipyretic Active *against S. aureus, B. cereus, K. pneumonia*	Phenolic acids, cinnamic acids, flavonoids and lignins	Hanaa, Ismail, Mahmoud, and Ibrahim (2017); Kaithwas, Mukherjee, Chaurasia, and Majumdar (2011)
Gond-e-molasri	Mimusops elangi L.	Antinociceptive, diuretic effects, gastroprotective, antibacterial, antifungal, anticariogenic	β-Sitosterol and β-sitosterol β-D-glucoside	Gami, Pathak, and Parabia (2012)
Isapgol	*Plantago ovata* Forsk	Anti-inflammatory, antimicrobial activity against *S. aureus, E. coli*	Alkaloids, caffeic acid derivatives flavonoids iridoid glycosides, vitamins, Plantamajoside	Reddy, Vandana, and Prakash (2018)
Barg Jhau	*Tamarix gallica*	Anti-inflammatory, analgesic (Chaturvedi)	Flavonoids	Chaturvedi, Drabu, and Sharma (2012)
BIranjasif	*Achillea millefolium* L.	Antimicrobial, antioxidant	Essential oil, dicaffeoylquinic acids and flavonoids proazulenes and other sesquiterpene lactones	Hemmati et al. (2011)
Afsanteen	*Artemisia absinthium* L.	Anti-inflammatory, analgesic	Nerolido, Santolina triene, αpinene, trans-β-Farnesene	Hadi, Hossein, Shirin, Najmeh, and Abolfazi (2014)

TABLE 7.2 Antimicrobial activity and chemical constituents of Herbal medicine—cont'd

Name	Botanical name	Relevant activity	Active phyto/chemical constituents	Refer
Zanjabeel	*Zingiber officinalis* L.	Antioxidant, analgesic, antibacterial, anti-inflammatory, antipyretic	Monoterpenes, diterpenes, curcumene, gingerols, phenolic and ketone derivatives, diaryl heptenones, gingesulphonic acid and monoacyldigalactosyl glycerols	Ahmed, Hwang, Choi, and Han (2017); Viljoen, Visser, Koen, and Musekiwa (2014)

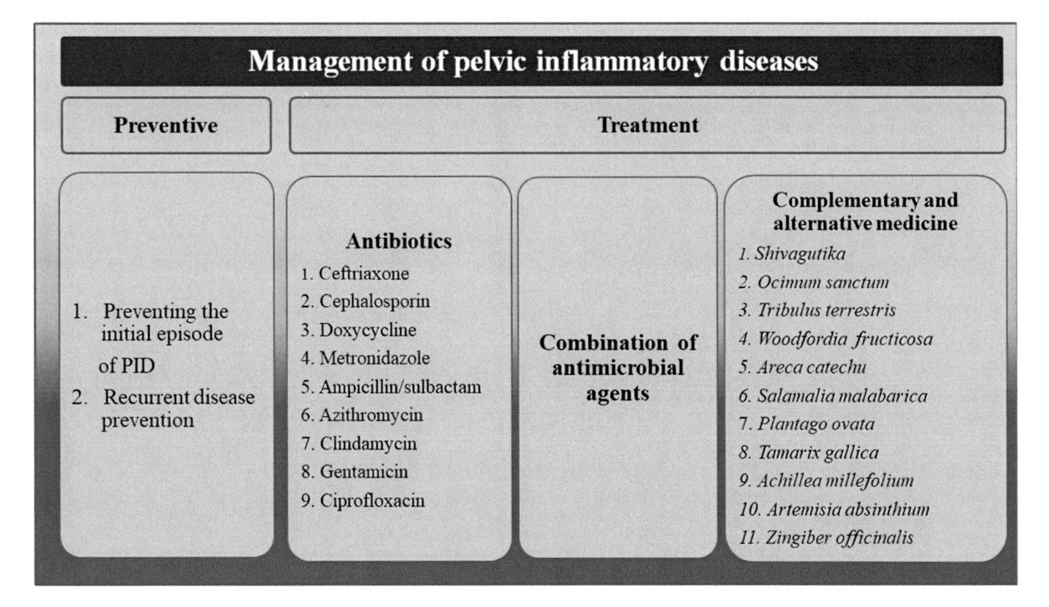

FIG. 7.4 Management of pelvic inflammatory disease.

Conclusion

PID is an inflammatory disorder and polymicrobial infection of the upper female genital tract poses a chief health problem of reproductive age women in both developing and developed countries. The major goal of management is to limit the risk of long-term consequences such as infertility, recurrent infection, ectopic pregnancy and pelvic pain, in addition to alleviating the acute inflammatory condition, clinical cure, and microbiological cure. Hence, an early and accurate diagnosis is of greater importance for avoiding PID-related complications and morbidity. In both the United States and Europe guidelines (2014) for the treatment of PID with antibiotics have been developed to represent the most recent evidence-based guidance currently available. Few complementary and alternative medicine studies have proven the efficiency of herbal medicine. Further, studies are warranted in large sample sizes to generalize PID treatment in mass. Hence, there has been a re-emergence and renaissance of awareness in medicine, which is regarded as cost-effective, readily available, safe, and easily affordable, with minimal or no adverse effects.

Acknowledgments

We are thankful to Prof. Ansari, Prof. Mehrotra, Prof. Lai, Dr. Adhikari, Dr. Guragai, Dr. Abla, Dr. Ijaz, Dr. Nasir, and Dr. Tripathi for the useful discussion of this study.

References

Ahmed, M., Hwang, J. H., Choi, S., & Han, D. (2017). Safety classification of herbal medicines used among pregnant women in Asian countries: A systematic review. *BMC Complementary and Alternative Medicine*, *17*, 489. https://doi.org/10.1186/s12906-017-1995-6.

Balogun, J. A., & Okonofua, F. E. (1988). Management of chronic pelvic inflammatory disease with shortwave diathermy: A case report. *Physical Therapy*, *68*(10), 1541–1545.

Baravalia, S., Vaghasiya, Y. K., & Chanda, S. (2012). Brine shrimp, cytotoxicity, antiinflammatory & analgesic properties of *Woodfordia fruticosa* Kurz flowers. *Iranian Journal of Pharmaceutical Research*, *11*, 851–861.

Bartlett, E., Levison, W., & Munday, P. (2013). Pelvic inflammatory disease. *BMJ*, *346*, 1–3.

Brun, J. L., Castan, B., et al. (2020). Pelvic inflammatory diseases: Updated French guidelines. *Journal of Gynecology Obstetrics and Human Reproduction*, *49*, 101714.

Burg, G. (2012). The history of sexually transmitted diseases. *Giornale Italiano di Dermatologia e Venereologia*, *147*(4), 329–340.

Carrasco-Gallardo, C., Guzmán, L., & Maccioni, R. B. (2012). Shilajit: A natural phytocomplex with potential procognitive activity. *International Journal of Alzheimer's Disease*, *2012*. https://doi.org/10.1155/2012/674142, 674142.

CDC. (2015). *Sexually Transmitted Diseases Treatment Guidelines*. Available from: https://www.cdc.gov/std/tg2015/pid.htm. [Accessed on 20-1-16].

Chaturvedi, S., Drabu, S., & Sharma, M. (2012). Anti-inflammatory and analgesic effect of *Tamarix gallica*. *International Journal of Pharmacy and Pharmaceutical Sciences*, *4*(3), 653–658.

Crossman, S. H. (2006). The challenge of pelvic inflammatory disease. *American Family Physician*, *73*(5), 859–864.

Curry, A., et al. (2019). Pelvic inflammatory disease: Diagnosis, management and prevention. *American Family Physician*, *100*(6), 357–364.

Das, B. B., Ronda, J., & Trent, J. (2016). Pelvic inflammatory disease: Improving awareness, prevention, and treatment. *Infection and Drug Resistance*, *9*, 191–197.

Dayal, S., Singh, A., Chaturvedi, V., Krishna, M., & Gupta, V. (2016). Pattern of pelvic inflammatory disease in women who attended the tertiary care hospital among the rural population of North India. *Muller Journal of Medical Sciences and Research*, *7*(2), 100–104.

Dhasmana, D., Hathom, E., McGrath, R., Tariq, A., & Ross, J. D. C. (2014). The effectiveness of nonsteroidal anti-inflammatory agents in the treatment of pelvic inflammatory disease: A systematic review. *Systemic Reviews*, *3*, 79. Available from: http://www.systematicreviewsjournal.com/content/3/1/79 [Accessed on 23-2-16].

Duarte, R., Fuhrich, D., & Ross, J. D. (2015). A review of antibiotic therapy for pelvic inflammatory disease. *International Journal of Antimicrobial Agents*, *46*(3), 272–277.

Gami, B., Pathak, S., & Parabia, M. (2012). Ethnobotanical, phytochemical and pharmacological review of *Mimusops elengi* Linn. *Asian Pacific Journal of Tropical Biomedicine*, *2*(9), 743–748. https://doi.org/10.1016/S2221-1691(12)60221-4.

Ghante, M. H., Bhusari, K. P., Duragkar, N. J., & Ghiware, N. B. (2014). Pharmacological evaluation for anti-asthmatic and anti-inflammatory potential of *Woodfordia fruticosa* flower extracts. *Pharmaceutical Biology*, *52*(7), 804–813. https://doi.org/10.3109/13880209.2013.869232.

Habib, S., Begum, W., Shameem, I., Sofi, G., Lone, A. H., & Bilal, A. (2011). Clinical efficacy of a Unani herbal formulation of *Hemidesmus indicus* and ointment of *Astragalus Hamosus* in endocervicitis—A randomized single blind standard controlled trial. *International Journal of Current Research and Review., 3*(11), 170–176.

Hadi, A., Hossein, N., Shirin, P., Najmeh, N., & Abolfazi, M. (2014). Anti-inflammatory and analgesic activities of *Artemisia absinthium* and chemical composition of its essential oil. *International Journal of Pharmaceutical Sciences Review and Research, 24*(2), 237–244.

Haggerty, C., & Ness, R. (2007). Newest approaches to treatment of pelvic inflammatory disease: A review of recent randomized clinical trials. *Clinical Infectious Diseases, 44*, 953–960.

Haggerty, C., Totten, P., Tang, G., Astete, S., Ferris, M., Morori, J., et al. (2016). Identification of novel microbes associated with pelvic inflammatory disease and infertility. *Sexually Transmitted Infections, 92*(6), 441–446.

Hanaa, M. H., Ismail, H. A., Mahmoud, M. E., & Ibrahim, H. M. (2017). Antioxidant activity and phytochemical analysis of flaxseeds (*Linum usitatisimum* L). *Minia Journal of Agricultural Research and Development, 37*(1), 129–140.

Hemmati, A. A., Arzi, A., Adinehvand, A., Mostofi, N. E., Mozaffari, A. R., & Jalali, A. (2011). Yarrow *Achillea millefolium* L. extract impairs the fibrogenic effect of bleomycin in rat lung. *Journal of Medicinal Plant Research: Planta Medica, 5*(10), 1843–1849.

Jackson, S., & Soper, D. (1999). Pelvic inflammatory disease in the postmenopousal women. *Infectious Diseases in Obstetrics and Gynecology, 7*, 248–252.

Jaiyeoba, O., & Soper, D. E. (2011). A practical approach to the diagnosis of pelvic inflammatory disease. *IDOG, 2011*, 753037. 1–7.

Jam, N., Hajimohammadi, R., Gharbani, P., & Mehrizad, A. (2021). Evaluation of antibacterial activity of aqueous, ethanolic and methanolic extracts of Areca nut fruit on selected Bacteria. *BioMed Research International, 2021*, 6663399. https://doi.org/10.1155/2021/6663399.

Jennings, L. K., & Krywko, D. M. (2020). *Pelvic Inflammatory Disease*. Treasure Island (FL): Stat Pearls.

Kabiruddin, M. (2006). *Kulliyyat-i-Qanun* (pp. 56–58). New Delhi: I'jaz Publishing House. 97–100, 198–9, 243–5, 356–3.

Kaithwas, G., Mukherjee, A., Chaurasia, A. K., & Majumdar, D. K. (2011). Antiinflammatory analgesic & antipyretic activities of *Linum usitatissimum* L. (flaxseed/linseed) fixed oil. *Indian Journal of Experimental Biology, 49*, 932–938.

Karole, S., Gautam, G., & Gupta, S. (2017). Pharmacognostic and pharmacological profile of *Bombax ceiba*. *Asian Journal of Pharmaceutical Education and Research, 6*, 16–27.

Lamina, S., & Hanif, S. (2008). Shortwave diathermy in the management of chronic pelvic inflammatory disease pain: Case reports. *Journal of the Nigeria Society of Physiotherapy, 16*(1), 31–36.

Liang, Y., & Gong, D. (2014). Acupuncture for chronic pelvic inflammatory disease: A qualitative study of patient's insistence on treatment. *BMC Complementary and Alternative Medicine, 14*(345), 1–6.

Majoosi, A. B. A. (2010). *Kamil al-Sana'a (Urdu trans. By Kantoori GH). Vol. I* (pp. 42–46). New Delhi: Idara Kitabus Shifa. 422–435, 536, Vol II. p. 492.

Mitchell, C., & Prabhu, M. (2013). Pelvic inflammatory disease: Current concepts in pathogenesis, diagnosis and treatment. *Infectious Disease Clinics of North America, 27*(4), 1–21.

Nkwabong, E., & Dingom, M. A. N. (2015). Acute pelvic inflammatory disease in Cameroon: A cross sectional descriptive study. *African Journal of Reproductive Health, 19*(4), 87–91.

O'Dowd, M. J., & Philipp, E. E. (2000). *The history of obstetrics & gynecology* (pp. 225–227). New York: The Parthenon Publishing Group.

Pachori, R., & Kulkarni, N. (2016). Studies on the incidence of pelvic inflammatory disease and association clinical consequences in reproductive women. *Journal of Nurse Midwifery and Maternal Health, 5*(3), 1329–1337.

Pachori, R., Kulkarni, N., & Bodhankar, M. (2015). Medicinal plants as an alternative for treating pelvic inflammatory disease. *International Journal of Biochemistry and Molecular Biology, 3*, 23–32.

Pandey, B. (2014). Epidemiology and risk factors of pelvic inflammatory disease. *MJSBH, 13*(1), 4–8.

Parekh, J., & Chanda, S. (2007). In vitro antibacterial activity of the crude methanol extract of *Woodfordia fruticosa* Kurz. Flower (Lythracea). *Brazilian Journal of Microbiology, 38*, 204–207.

Qarshi, M. H. (2011). *Jam-ul-Hikmat* (p. 1108). New Delhi: Idara Kitabus Shifa.

Qayyum, S., & Arshiya, S. (2018). *Efficacy of tukhme katan and isapghol in waram al-rahim—A Randomized Standard Controlled Study*. [dissertation], Karnataka: Rajiv Gandhi University of Health Sciences.

Reddy, P. R., Vandana, K. V., & Prakash, S. (2018). Antibacterial and anti-inflammatory properties of *Plantago ovata* Forssk. Leaves and seeds against periodontal pathogens: An in vitro study. *AYU, 39*, 226–229.

Ross, J. (2013). Pelvic inflammatory disease: Clinical evidence. *BMJ, 12*, 1606.

Ross, J., Cole, M., Evans, C., Lyons, D., Dean, G., & Cousins, D. (2017). *United Kingdom National Guidelines for the Management of Pelvic Inflammatory Disease*. Available from: https://www.bashhguidelines.org/media/1144/pid-guidelines-2017-for-consultation.pdf. [Accessed on 14-10-17].

Ross, J., Cronje, H., Paszkowski, T., Rakoczi, I., VilDaite, D., Kureishi, A., et al. (2006). Moxifloxacin versus ofloxacin plus metronidazole in uncomplicated pelvic inflammatory disease: Results of a multicentre, double blind, randomized trial. *Sexually Transmitted Infections*, *82*(6), 446–451.

Savaris, R. F., Teixeira, L. M., Torres, T. G., Edelwiss, M. I. A., Moncada, J., & Schachter, J. (2007). Comparing ceftriaxone plus azithromycin or doxycycline for pelvic inflammatory disease: A randomized controlled trial. *The Obstetrician and Gynaecologist*, *110*, 53.

Sayed, A., Shameem, I., & Mubeen, U. (2016). Effect of Arq Brinjasif in mild pelvic inflammatory disease—A randomised controlled trial. *International Journal of Medicine and Research*, *1*(2), 91–96.

Shen, C. C., Hu, L. Y., Yang, A. C., Chiang, Y. Y., Hung, J. H., & Tsai, S. J. (2016). Risk of uterine, ovarian and breast cancer following pelvic inflammatory disease: A nationwide population- based retrospective cohort study. *BMC Cancer*, *16*, 839.

Simms, I., & Stephenson, J. (2000). Pelvic inflammatory disease epidemiology: What do we know and what do we need to know? *Sexually Transmitted Infections*, *76*(2), 80–87.

Sina, I. (2010). *Al-Qānun Fi'l Tibb* (pp. 1101–1104). New Delhi: Idara Kitabus Shifa.

Sweet, R. (2011). Treatment of acute pelvic inflammatory disease: Review article. *Infectious Diseases in Obstetrics and Gynecology*, 561909. https://doi.org/10.1155/2011/561909.

Tabri, A. H. (2010). *Firdaws al-Hikmat fi'l Tibb (Urdu trans: Hkm. Mohd Awwal.S.S)* (pp. 256–258). Delhi: Idara Kitabus Shifa.

Tamunomie, N. K. (2013). Pelvic inflammatory disease, correlation between clinical and laparoscopic diagnosis—A review. *IOSR-JDMS*, *7*(5), 16–20.

Taylor, B., Darville, T., & Haggerty, C. (2013). Does bacterial vaginosis cause pelvic inflammatory disease? *Sexually Transmitted Diseases*, *40*(2), 117–122.

Trivedi, N. A., Mazumdar, B., Bhatt, J. D., & Hemavathi, K. G. (2004). Effect of shilajit on blood glucose and lipid profile in alloxan-induced diabetic rats. *Indian Journal of Pharmacology*, *36*, 373–376.

Vanamala, V. G., Pakyanadhan, S., Rachel, A., & Abraham, S. (2018). Pelvic inflammatory disease and the risk factors. *International Journal of Reproduction, Contraception, Obstetrics and Gynecology*, *7*(9), 3572–3575.

Viljoen, E., Visser, J., Koen, N., & Musekiwa, A. (2014). A systematic review and meta-analysis of the effect and safety of ginger in the treatment of pregnancy-associated nausea and vomiting. *Nutrition Journal*, *13*, 20. https://doi.org/10.1186/1475-2891-13-20.

Vishwesh, B. N., & Bhat, S. (2014). A clinical study to evaluate the role of Shivagutika in pelvic inflammatory disease. *Journal of Ayurveda and Holistic Medicine*, *2*(1), 27–32.

Williamson, B. K., & Aldeen, A. Z. (2010). *Focus on: Emergent Evaluation and Management of Pelvic Inflammatory Disease*. Available from: https://www.acep.org/Clinical—Practice-Management/Focus-On–Emergent-Evaluation-and-Management-of-Pelvic-Inflammatory-Disease. [Accessed on 14-10–17].

Zahid, S. (2016). Effect of Unani formulation in cervicitis (Warme unqur-rahm): A single-blind randomized placebo-controlled trial. *Alternative and Integrative Medicine*, *5*, 213. https://doi.org/10.4172/2327-5162.1000213.

CHAPTER

8

A review of amenorrhea toward Unani to modern system with emerging technology: Current advancements, research gap, and future direction

Sumbul Mehdi[a], *Arshiya Sultana*[a,*], *Md Belal Bin Heyat*[b,c,d,*], *Channabasava Chola*[e], *Faijan Akhtar*[f,*], *Hirpesa Kebede Gutema*[g], *Dawood M.R. Al-qadasi*[h], *and Atif Amin Baig*[i]

[a]Department of Ilmul Qabalat wa Amraze Niswan (Gynecology and Obstetrics), National Institute of Unani Medicine, Ministry of AYUSH & Rajiv Gandhi University of Health Sciences, Bengaluru, Karnataka, India [b]IoT Research Center, College of Computer Science and Software Engineering, Shenzhen University, Shenzhen, Guangdong, China [c]International Institute of Information Technology, Hyderabad, Telangana, India [d]Department of Science and Engineering, Novel Global Community Education Foundation, Hebersham, NSW, Australia [e]Department of Computer Science and Engineering, Indian Institute of Information Technology, Kottayam, Kerala, India [f]School of Computer Science and Engineering, University of Electronic Science and Technology of China, Chengdu, Sichuan, China [g]School of Information and Software Engineering, University of Electronic Science and Technology of China, Chengdu, China [h]School of Material Science, Tongji University, Shanghai, China [i]Faculty of Medicine, Universiti Sultan Zainal Abidin, Kuala, Terengganu, Malaysia

Introduction

Nowadays, amenorrhea is the commonest gynecologic problem among reproductive age women. Amenorrhea affects about 3%–4% of women. It is defined as a change in regular menstrual cycles, including the lack of menstruation, regardless of the diagnosis. The absence

*Authors have equal distribution.

https://doi.org/10.1016/B978-0-323-99031-8.00013-2

of menses in a reproductive age woman is known as amenorrhea (Golden & Carlson, 2008; Nichols, Bieber, & Gell, 2010). *Ihtibas al-Tamth* (amenorrhoea) is defined in the Unani system of medicine as the absence of monthly bleeding for more than 2 months or a decrease in the quantity of menstrual blood. The lack or total cessation of menstrual blood is referred to as amenorrhea. Primary amenorrhea is woman who had never menstruated whereas secondary is woman who had attained menarche, but had no periods for three consecutive months. Primary amenorrhoea is rare whereas secondary amenorrhea is common. It affects roughly 3%–4% of women. In the United States, about 5%–7% of menstrual women endure three months of secondary amenorrhea each year (Izzaty, Imandiri, & Suciati, 2017; Mokaberinejad et al., 2012; Pereira & Brown, 2017). Secondary amenorrhea in reproductive age could be a sign of an undiscovered chronic disease, and proper treatment is contingent on accurate diagnosis of the underlying cause (Klein, Paradise, & Reeder, 2019; Klein & Poth, 2013; Shrivastava, Shrivastava, Himanshu, & Jha, 2013; Sultana & Rahman, 2021).

Amenorrhoea is a symptom that indicates anatomical, genetic, and neuroendocrine problems, and it is accurately understood in terms of the menstrual cycle pathophysiology. Menstruation is triggered by a combination of circumstances that culminate in the visible discharge of menstrual blood. The release of gonadotrophin-releasing hormone (GnRH) in the hypothalamus initiates the cascade. Follicle-stimulating hormone (FSH) and luteinizing hormone (LH) are two gonadotrophins secreted by the pituitary gland. The ovary secretes estrogen and progesterone at varying rates as a result of FSH and LH stimulation. Until monthly shedding, a responsive endometrium grows and changes its shape in response to ovarian hormone fluctuations. Menstrual bleeding can occur at that time if the outflow tract is normal (Golden & Carlson, 2008; Morrison, Fleming, & Levy, 2021). There are two types of causes for amenorrhoea: (a) anatomical anomalies of the vaginal organs and (b) endocrine disorders. Amenorrhea is defined as a failure of the hypothalamic–pituitary-gonadal axis to cause cyclic changes in the endometrium that ordinarily result in menses. It can also be caused by a lack of end organs or a blockage of the outflow route. The causes of amenorrhoea are related to expulsive faculty of body (*Quwwat Dafi al-Badan*), morbid matter (*Madda*) and Menstrual blood (*Johar khun-i-Hayd*), or menstrual flow organs (*Ala Makhraj-i-Hayd*). Cold (*Barid*), dry (*Yabis*), warm (*Harr*) or warm-dry (*Harr wa Yabis*) altered temperament (*Su-i-Mizaj*) or simple altered temperament (*Su-i-Mizaj Sada*) or with humoral altered temperament (*Su-i-Mizaj Maddi*) can lead to weakness in expulsive faculty (*Du'f al-Quwwat-i-Dafia*). Derangement in quality and quantity of morbid matter leads humoral altered temperament can also lead to amenorrhoea.

Traditional Unani Medicine (USM)/Traditional Persian Medicine (TPM) is a comprehensive system of medicine based on concept of temperament (*Mizaj*) and humors (*Akhlat*) and it is one of the most well-known branches of complementary and alternative medicine (CAM) and has been practiced for thousands of years. *Mizaj* (Temperament) is formed by the activity and response of four key elements (*Arkan Arba'a*) (fire (*Ag*), Air (*Hawa'*), water (*Pani*)and *Mitti*), and it gives life its many features (Moini Jazani et al., 2018; Parvizi, Nimrouzi, Bagheri Lankarani, Emami Alorizi, & Hajimonfarednejad, 2017; Shojaee-Mend, Ayatollahi, & Abdolahadi, 2020). Temperament is divided into four categories according to USM: Cold (*Barid*), dry (*Yabis*), wet (*Ratb*), and warm (*Harr*). Diseases can be caused by any altered temperament of organs (also known as mal-temperaments/dystemperament). TPM/USM classifies amenorrhoea as cessation of menses, infrequent menstruation >35 days to 3 months (oligomenorrhea), and scanty menstrual flow (hypomenorrhea). Anatomical and functional diseases (dystemperament) in the female genital system (uterus and ovaries) and involvement of other organs (liver, brain) are the main causes of oligo-menorrhea and amenorrhea (Moini Jazani et al., 2018; Parvizi et al., 2017; Shojaee-Mend et al., 2020).

Cold altered temperament (*Su'-i-Mizaj barid*) produces obstruction (*Sudda*) development in the uterine blood vessels as a result of excessive fluid intake, resulting in amenorrhoea and infertility.

Viscous humors (*Akhlat-i-Ghaliz*) particularly phelgm (*Balgham*), increases blood viscosity. After consuming *Ghaliz* and heavy diet (*Sakhil Ghiza*), this *Ghaliz Madda* accumulates in the blood and generates obstruction which blocks the uterine capillaries and causes amenorrhea.

Amenorrhoea is caused by weakness of liver (*Duf-i-Jigar*) for three reasons:

- Blood flow to distant organs is disrupted because the liver is unable to distinguish blood from other body fluids.
- Improper defective hemopoesis (*Tawlid-i-Khun*).
- *Sudda* formation in the liver produces a blockage in blood supply to the uterus.
- Amenorrhoea is caused by obesity (*Siman Mufrit*) in three ways:
- Excessive fat deposition on the uterus compresses the uterine blood vessels, resulting in amenorrhoea.
- *Sudda* development as a result of an excess of phelgm in the uterine vessels.
- Chronic anovulation is caused by a change in ovarian function caused by the dominance of wetness (*Rutubat*) and coldness (*Burudat*) in the body, resulting in menstrual irregularities and infertility.

Etiology of the amenorrhea

Amenorrhea can be caused by problems with the uterus, brain, or circulation, according to USM/TPM. It was also thought that any restriction in the uterus and related organs or tissues, whether physical or functional, may cause the body to develop this illness. According to current medical theory, any restriction in the genitalia or birth canal and endocrine dysfunctions might result in developing this disorder.

The major causes of amenorrhea are:

- *Physiologic Amenorrhea*: Pregnancy, contraception, breastfeeding, exogenous androgens, menopause Pregnancy
- *Hypothalamic*: Eating disorders, immaturity of the HPO axis, exercise-induced amenorrhea (overall energy deficit), medication-induced amenorrhea, chronic illness, stress-induced amenorrhea, Kallmann syndrome (Gonadotrophin deficiency), infections (e.g., tuberculosis meningitis, syphilis), traumatic brain injury
- *Pituitary*: Hyperprolactinemia, empty sella syndrome, prolactinoma, Sheehan syndrome, craniopharyngioma, isolated gonadotropin deficiency, autoimmune disease, infiltrative disease (e.g., sarcoidosis), medications (Antidepressants Antihistamines Antihypertensive Antipsychotics Opiates), other pituitary, or central nervous system tumor.
- *Thyroid*: Hypothyroidism and hyperthyroidism

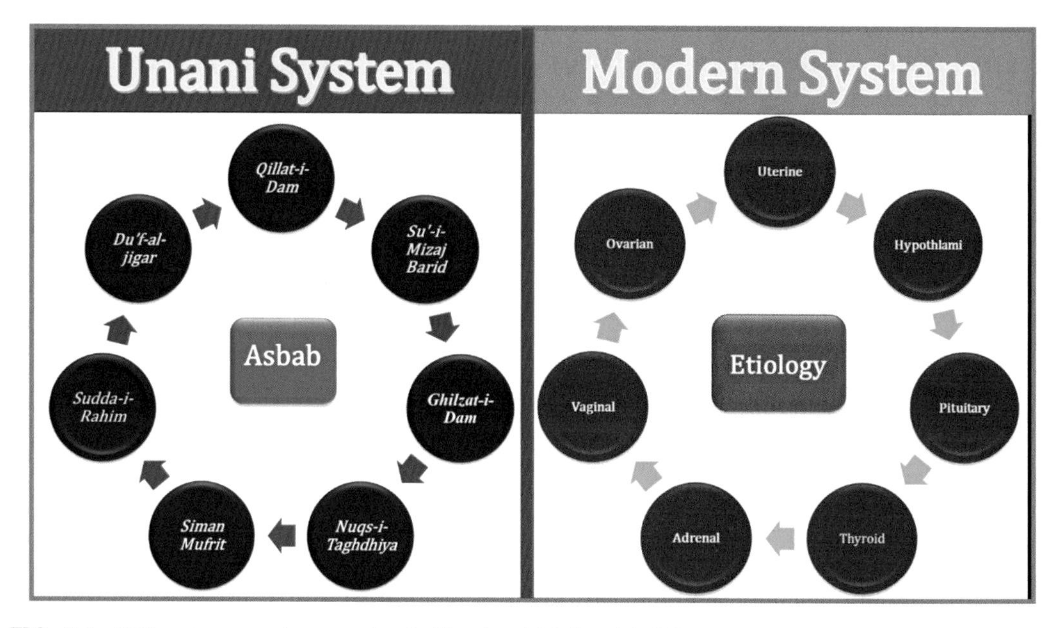

FIG. 8.1 Different causes of amenorrhea in Unani and Modern Medicine.

- *Adrenal*: Congenital adrenal hyperplasia, cushing syndrome.
- *Ovarian*: congenital, polycystic ovary syndrome, gonadal dysgenesis (Turner syndrome), premature ovarian failure, ovarian tumor, acquired primary ovarian insufficiency (chemotherapy or irradiation),
- *Uterine*: androgen insensitivity, uterine adhesions (Asher man syndrome), Mullerian agenesis, and cervical agenesis.
- *Vaginal*: Imperforate hymen, transverse vaginal septum, vaginal agenesis (Klein & Poth, 2013) (see Fig. 8.1).

Unani concept of amenorrhoea (Ihtibas Al-Tamth)

According to Unani physicians, the disease is caused by Phelgm and black bile (Khilt -i-Balgham *and* Sauda). *According to* Avicenna (*Al-Qanun*), Jurjani (*Dhakhira Khawarizm Shahi*), *Abbas Majusi* (*Kamil al-Sana'a*), *Rabban Tabari*, (*Firdaws al-Hikma fi'l Tibb*), Akbar Arzani (*Tibb-i-Akbar*) *the major causes of amenorhhoea are cold altered temperament* (*Su-i-Mizaj* Barid), *Morbid humor* (Akhlat-i-Ghaliz), *and weakness of liver* (Sharma, 2020). *It proves the theories mentioned by eminent Unani physicians in pathophysiology of amenorrhoea, as it is caused by abnormal production of phelgm, which in turn causes weakness in liver resulting in* amenorrhoea. According to *Ghina Muna*, *cold altered temperament* obstructs uterine and ovarian blood vessels, preventing blood flow and resulting in *amenorrhoea*. According to traditional Unani literature, obesity is linked to amenorrhoea and infertility. Unani physicians have found a link between visceral obesity (fat deposition on the omentum), amenorrhoea, anovulation, and infertility, all of which are linked to polycystic ovarian disease (Islam, 2018; Parveen et al., 2020; Sharma, 2020). In obese infertile women, a qualitative and quantitative disruption of the humors (*Akhlat's*) equilibrium leads to increased phlegm production, leading in *weakness of liver* and weakness on reproductive faculty (*Duf-i- Quwwat-i-Tawlid-i- Mani*) and chronic anovulation. Weakness in retentive faculty of uterus (*Duf-i- Quwwat-i- Masika Rahim*) inhibits implantation and raises the risk of miscarriage. It causes amenorrhoea by forming *Sudda* in uterine blood vessels. *It usually occurs in women having phlegm temperament and fair complexion and such women generally suffer from dysmenorrhoea* (Usr-i-Tamth) *as heavy menstrual bleeding may occur after a long period of amenorrhea* (Moini Jazani et al., 2018; Sultana & Rahman, 2021). Symptoms usually associated with amenorrhoea are headache, nausea, backache and lower abdominal pain, tiredness, and some respiratory problems (Sultana and Rahman, 2021).

Diagnosis

If thyroid-stimulating hormone (TSH) and prolactin levels are normal, a progestogen challenge test (PCT) will help discover endogenous estrogen that is persuading the endometrium and evaluate a patent genital outflow tract. After the PCT challenge test, 2 to 7 days later, a withdrawal bleed typically happens. A failed PCT indicates an insufficient estrogenization or an anomaly in the genital outflow tract (Moini Jazani et al., 2018). The two diagnoses can be distinguished using an estrogen/PCT. A genital outflow tract obstruction is usually indicated by a failed estrogen/progestogen challenge test. A positive test suggests that the ovaries or HPO axis are abnormal. Subclinical hypothyroidism has a reduced risk of amenorrhea than overt hypothyroidism. TSH levels should be measured since the consequences of subclinical hypothyroidism on menstruation and fertility are unknown, and aberrant TSH levels may impact prolactin levels. Gonadotropin levels can also be used to figure out what is causing the problem FSH or LH levels that are abnormally high indicate an ovarian problem (hyper gonadotropic hypogonadism) (Kitasaka, Tokoro, Kojima, Fukunaga, & Asada, 2021; Morrison et al., 2021; Schmidt, 2005). FSH or LH levels that are normal or low indicate a hypothalamic or pituitary problem (hypogonadotropic hypogonadism). A pituitary tumor can be ruled out by Magnetic Resonance Imaging (MRI) of the sella turcica. The presence of a hypothalamic etiology of amenorrhea is indicated by a normal MRI. Complete Androgen Insensitivity Syndrome, defined by scarce or missing pubic and axillary hair, normal breast growth, and a blind vaginal pouch; and 5-alpha reductase insufficiency, characterized by partially virtualized genitalia, are two rare causes of amenorrhea with normal serum testosterone levels as those reported in males of the same age group in similar circumstances and the karyotype will be 46, XY. To avoid malignant transformation, testicular tissue should be excised. *Asherman syndrome, that is, uterine adhesion, produced by gynecologic or obstetric uterine instrumentation procedures and can be assessed and treated by hysteroscopy, is a structural cause of secondary amenorrhea.*

Treatment

The mainstay of treatment for these disorders is hormone therapy based on estrogen and progesterone chemicals. In modern medicine, the sole therapy for secondary amenorrhea is hormone supplements. Though successful, have side effects such as weight gain, headaches or migraines, mood swings, abdominal distention, and so on (Arentz, Abbott, Smith, & Bensoussan, 2014). Patients with thromboembolism, stroke, hypertension, myocardial infarction, or liver disease should avoid them. As a result, an alternative therapy that is safe, effective, easy to obtain, and has long-term effects is required. The Unani system of medicine treats this illness by modifying one's lifestyle, using appropriate herbal remedies, and by regimental therapy. The use of complementary and alternative medicine is becoming more popular these days (CAM). Therapeutic cupping is an effective method of therapy that is natural, holistic, preventive, and curative. It has been used for thousands of years and has been found to be good for everyone with any ailment. Lifestyle changes (particularly dietary habits, physical activity, and sleep), medicine, and nonmedical treatments such as wet and dry cupping, as well as surgery, are all used to treat oligo-menorrhea and amenorrhea. Herbal drugs are one of the most important therapy tenets since they can change organ temperaments. The treatment focuses on changing one's temperament.

Dietotherapy (Ilaj bi'l Ghiza)

- Use light diet (*Mulattif aghzia*) like lukewarm water or vinegar (*Sirka/Kanji*) in empty stomach. The diet advised for amenorrhea patient is to add spices such as pepper (*Filfil*), *raai*, Cumin (*Zeera*), garlic (*lehsan*) to the vegetables. Use plain soup and vegetables with dry chapatti.
- Avoid cold water, milk, butter, mutton, fish, oily, and fried food.
- Use a light and nutritious diet (*Qalil al-Taghziya wa Kasir al-Kamiya't Ghiza*) like vegetables and fruits, etc., filling the stomach.
- In the case of obesity, dietary control (*Taqlil-i-ghiza*) is mainly required.

Regimenal therapy ('Ilaj bi'l Tadbir)

Some regimenal therapies are also beneficial in amenorrhoea, such as venesection of saphaneous vein (*fasd-i-Safin*), sitz bath (*Hammam*) and cupping therapy (*Hijama*) on calf area. The different regimenal therapy useful in amenorrhea are.

Abzan (sitz bath): *Joshanda* of *mulattif* drugs such as *Astragalus hamosus* Linn (*Aqleelul Malik*), Valeriana wallichi (*Asaroon*), *Matricaria chamomilla* (*Babuna*), *Lawsonia inermis* (*Heena*), *Nigella sativa* (*Kalonji*), *Marzanjosh*, Mint leaves (*Pudina*), *Shibbat*, *Sudab*, *Sa'atar*, *Tagar*, *Qardmana*.

Dalk (*massage*): *Dalk* with *zift balut* or *Natrun* followed by *hammam* for to reduce fat accumulation and weight reduction.

Hijama bit Shart (*cupping*) *and Fasd* (*venesection*): Application of wet cupping for bloodletting over the calf muscles produces menstruation in amenorrhea by diverting blood flow toward the uterus and facilitating detoxifying or cleansing of body (*Tanqiya Badan*) by removing harmful particles in the form of menstruation. Wet cupping on ventral part of the leg is advisable if amenorrhoea caused by obesity or increase in viscosity of blood or cold temperament. The procedure described by *Ismail Jurjani* for wet cupping in *Dhakhira Khawarizm shahi* is to apply cups laterally on the calf approximately 4 in. from the *Shataaling*. Prior to application of cups, to increase the blood flow toward the leg, the patient is directed for *hammam* (bath), pour warm water on legs, and walk. After the procedure mentioned above, cups are applied for three times for dry cupping followed by wet cupping on one leg and Venesection of saphenous vein on other leg and viva versa on next day. At the time of wet cupping, the patient is advised to keep the legs firm on the floor, when blood start to flow. This method of dry and wet cupping is useful in amenorrhea as well as other conditions such as epilepsy, skin diseases, melancholia, and sciatica. Usually, this procedure is advised 2 days before the menstrual period so that both types of eliminating methods (*Istfragh*) will not be mixed. To induce menstruation, *Lateef* and hot (*Garam*) drugs are given near expected date of menstruation. Ibn Sina advised wet cupping on calf muscle, especially in obese women. *Mishktaramasee* (*Mentha arvensis* L.) and *Abhal* (*Juniperus communis* L.) are very strong emmenagogue drugs. *Fasd* of *rag-i-safin* and *rag-i-mabiz* as it diverts the flow of blood toward the uterus to induce menstruation.

Firzaja (*Vaginal Suppositories*): *Ro'ghan Sosan*, *Ma'al-asal*, and *Murmakki*.

Hamul (*Suppositories*): *Asal Musaffa*, *Ro'ghan Sosan*, *Mur*, *Muqil*, *Samagh Kankaz*, *Shehad*, *Sakbeenaj*.

Hammam-i-Yabis (*Dry Bath*): In obese women, before meals, *Hammam-i-Yabis* is suggested followed by nap for few hours and less diet to reduce the weight.

Huqna: *Boriq,Namak*, *Ro'ghan zaitoon*, *Sheham Hanzal Huqna* is useful in amenorrhea.

Riyazat (*Exercises*): *Riyazat-i-Qawi* for weight reduction in obese women.

Takmid (*Fomentation*): *Takmid* with *Harr Advia* (*Advia-i-Muhammira*) *over* lower abdomen to stimulate blood flow toward the uterus.

Zimad (*Paste*): *Joshanda* of *Harr* and *Mulattif Advia* over lower abdomen.

Unani pharmacotherapy (Ilaj bi'l Dawa)

The main principle of treatment to treat amenorrhea includes blood producing *Tawleed-i-Dam*, *Tanqiya-i-Akhlat Ghaleeza*, *Talteef-i-khilt*, *Tafteeh-i-UruqRahim*, and *Tahzeel*. According to USM/TPM's pharmacotherapy, medications used to treat amenorrhea should have emmenagogue and diuretic effects. *Mudirr-i-Hayd* medications (emmenagogues) usually have warm (*Harr*) and light (*Latif*) temperament and are employed in the treatment of amenorrheoa in the USM. These drugs correct the functional defect in the uterus and liquefy blood, and eliminate obstruction increasing blood circulation in uterine arteries, aggravating surrounding organs, stimulating uterine muscles and nerves, and generally raising the blood formation in order to act as emmenagogue clearing the way for free menstrual blood flow. Further, these drugs correct faulty *Balgham* and metabolize it to blood. One of the causes of amenorrhea is liver dysfunction (*Duf-i-Jigar*) caused by inflammation and obstruction. This change can be attributed to the properties of *Habb* mudir's liver tonic (*Muqawwi-i-Jigar*), resolvent or anti-inflammatory for liver inflammation (*Muhalil-i-Warm-i-Jigar*), and deobstrucent for liver obstruction (*Mufattit Sudad-i-Jigar*), which induce menstruation by improving liver functions, as most hormones are metabolized in the liver. *Mentha longifolia* is a therapeutic Unani herb that probably impacts menstrual periods, according to Iranian traditional medicine (ITM) books, particularly the "Canon of medicine" and "Al-Hawi" (Mokaberinejad et al., 2012). The single Unani drugs beneficial in amenorrhea are *Abhal, Afsanteen, Anisoon, Ashnan, Asarun, Ayarij Feeqra, Bakhur Maryam, Darchini, Javitri, Junbedastar, Jausheer, Kalonji, Kardmana, Kibr, Karafs, Irsa, Izkhar, Majeeth, Mushkatramashi, Tukhm Marzanjosh, Pudina Nehri, Pudina Kohi, Soanf, Sudab, Qust, Habb-Ul-Ghar, O'od Balsan, Ushq, Murmakki, Indrain, Farfiyoon, Sakbeenaj, Turmus, Lobiya, Zarawand*, etc. Some compound Unani drugs beneficial in amenorrhoea are Ayarij Feeqra, Qurs Abhal, Dhamarsa, Ma'jun Abhal, and *Sharbat Afsanteen. Sharbat Kasoos, Sharbat Ja'ada, Sharbat Buzuri, Sharbat Biranjasif, Sharbat Saleekha, Naqu'buzur, Loghazia, Aqras mur, Habb mudir, Safoof baboona, Kushta Faulad*, and *Safoof Muhazzil* (Ahmed, Nizami, & Aslam, 2005). Depending upon the cause, the treatment is as follows:

- *Qillat-i-Dam*: Half boiled egg, chicken, goat's milk and meat, advice to take sweet items for instant energy, take proper rest, and *Hammam Ratb.*
- *Ghilzat-i-Khilt*: *Majun* of *Ayarij, Tukhm Karafs, Anisoon, Pudina, Badiyaan*, mixed with honey, or sugar. It will reduce the viscosity of blood.
- *Sudda al-Rahim*: *Farzaja* of *Sheer-i-Khisht, Simaq, Nabat* or *Maghz-i-Tukhm Kaddu, Khubbazi*, or *Badiyan* with honey.
- *Siman-i-mufrit*: Venesection of saphenous vein, use of diuretics, walking before meal, and advice to take a bath before food. *Itrifal Sagheer, Kamooni, Gulqand*, and *Badiyaan Romi* are effective in reducing weight.

Ethno-medicinal plants used in Amenorrhoea and abnormal menstruation

- PIYAAZ (*Allium Cepa* L.) (Liliaeae)

  To begin menstruation, young bulbs are consumed in large quantities.
  The use of plant decoction to regulate the menstrual cycle and heavy bleeding is highly prevalent among bhils.

- GAAJAR (*Daucus carota* L.) (Apiaceae).

  Decoction of seeds is given to regularize menstruation as it has emmenagogue property

- ABHAL (*Juniperus communis* L) (Cupressaceae)

  The plant is helpful in amenorrhea, leucorrhea, chronic cystitis (Khare)

- PUDINA KOHI (*Mentha spicata* L.) (Lamiaceae)

  The plant is useful in amenorrhea

- PUDINA NEHRI (*Mentha aquatica* Linn)

  The Plant is useful in amenorrhea

- AFSANTEEN VILAYATI (*Artemisia absinthium* L.) (Asteraceae)

  It has emmenagogue property.

- AFSANTEEN-*E*-HINDI (*Artemisia vulgaris* Linn) (Asteraceae)

Leaf—emmenagogue, menstrual regulator, nervine, stomachic (in anorexia and dyspepsia), anhelminthic, choleretic, diaphoretic

- MURMAKKI (Commiphora molmol) (Burseraceae)

The oleo-gum resin is used for irregular menstruation painful periods.

Evidence-based recent studies

In a randomized placebo-controlled experiment, only *Foeniculum vulgare* showed therapeutic efficacy for depot medroxyprogesterone acetate-induced amenorrhea. *Sesamum indicum* L. (sesame) is one of the medicinal herbs that can produce menstrual bleeding with negligible adverse effects, according to traditional medicine books such as the "Canon of Medicine," "Al-Hawi," and "Makhzan al Advia." In the Unani system of medicine, the principle of treatment is *'Ilaj bi'l Zid*; the temperament of the disease is *Barid Ratb* and drugs with *HarrYabisMizaj*; properties like *Mudirr-i-Hayd, Mulattif balgham*, and *Mufattih* are used in the management of *Ihtibas al-*Tamth *such* drugs stimulate the flow of blood toward the uterus and its blood vessels and dilate them; such drugs stimulate the flow of blood, corrects the functional defect of the uterus, liquefies blood, and eliminates *Sudda*.

Modern technology

Disturbance of the release in hormone secretion of hypothalamic gonadotropins leading to Functional Hypothalamic Amenorrhea (FHA) as well as hypoestrogenism, which acts in reduced Bone Mineral density (BMD) and increased bone Marrow Adipose Tissue (MAT). It also observed increased BMD in young adults and adolescents due to the management of transdermal estradiol at physiological dosages. We hypothesized that the physiological replacement estrogens would cause a decrease in MAT, coupled with an increase in BMD. The experiment carried out with a size 15 includes adolescent and young women between 14 and 25 years with FHA. The whole population has induced with a17β-estradiol transdermal patch at a dose of 0.1 mg/day applied weekly twice for 12 months and received cyclic progestin for 10–12 days every month. Single proton (1H)-magnetic resonance spectroscopy were used to quantify the MAT(lipid/water ratio) of the fourth lumbar (L4) vertebral body and femoral diaphysis, as well as compartmental volumetric BMD of the distal radius and tibia utilizing high-resolution peripheral quantitative computed tomography. They found that following transdermal estrogenic therapy, MAT declines in adolescent and young adult women with FHA and that these changes are related to higher cortical vBMD (Singhal et al., 2021). Disorders of sex development (DSD) are addressed with the challenging study of sex chromosome mosaicism. The study's objective was to address the DSD patients with the help of urothelial cells to uncover complex and hidden cell lines. In this study, females with primary amenorrhea of 19-year-old, short stature without ambiguous exterior genitals. The patient had the presence of 45, X/46, XY karyotype in leukocytes. Interphase FISH found hidden 45, X/47, XYY/47, XXY/46, XY/46, XX mosaicism in urothelial cells and leukocytes. They concluded that to concentrate on the importance of investigating the mosaicism of sex chromosomes in other tissues. Urine epithelium cells are absorbing in the case of DSD, which can reflect the cell composition of the urogenital crest. The analysis of these cells should be considered in the clinical evaluation of DSD patients (Sevilla-Montoya et al., 2021). Choriocarcinoma is a rare malignancy originating from trophoblastic cells which is known to raise from the placenta. In this study, 28-year-old female was consulted for amenorrhea and increased βhCG, imitating pregnancy of an unfamiliar location, which has turned out to be primary lung choriocarcinoma (Ben Abdallah et al., 2021). Puberty is often referred to as a life-changing period among youths, which changes physiological, physical, and social factors. In clinical terms, primary Amenorrhea is a lack of menstrual cycle or absence of menstruation. This condition is categorized into primary and secondary. The article's focus is primary amenorrhea; this condition resulted in anxiety in patients also can be seen by family members and often reported to a general pediatrician. History of patients and appropriate usage of investigations are essential for in-time diagnosis and management (Varughese, Ryan, & Makaya, 2021). Women with HA are enhanced for RSVs in genes leads IHH, suggesting that gene variation is relative to the release of gonadotropin hormone neuronal ontogeny and function can be a major determinant of individual sensitivity to HA development with to exercise, stress, and diet (Delaney et al., 2021). The population of size 5734 qualified patients considered for the analysis. Serum AMH levels and antral follicular counts were notably lower in women with short cycles and higher in women with Amenorrhea than in women with a normal menstrual cycle in women without Polycystic Ovarian Syndrome (PCOS). In contrast, scanty periods and amenorrhea were associated with greater antral follicle count and serum AMH in women with PCOS than women with a normal menstrual cycle. In women with and without PCOS, the 75–100th percentile group of AMH levels had a substantially higher risk of oligo/amenorrhea than the 0–25th percentile group. Poor ovarian response was more prevalent in women with short periods and less common in those with amenorrhea than in women with normal cycles in women without PCOS. Short cycles and amenorrhea were linked to a higher likelihood of poor response when AMH levels were considered. Women with short periods and amenorrhea were more likely to have a low ovarian response in women without PCOS and low AMH levels (Hu et al., 2021). Chemotherapy-Related Amenorrhea (CRA) at primary endpoint as 18 months less likely in recipients of adjuvant T-DM1 than TH. It can be seen an increased in CRA over time in T-DM1 due to intensified endocrine therapy. T-DM1 has to be preferred over TH. Assessment of amenorrhea rates for T-DM1 is administered after standard chemotherapy led by HER2 (Ruddy et al., 2021). Women aged less than or equal to 40 age has been considered and median as 36 age among the population enrolled from 2006 to 2016 are effected from breast cancer, young age is considered as prone to disease recurrence. This population is experienced unique treatment and survivorship issues associated with Treatment-Related Amenorrhea (TRA), which consists of fertility conservation and ovarian function management as endocrine therapy. The same population is shown with lower BMI. It is noted as an independent predictor of TRA. Results of menstruation were self-reported through continuous surveys. Factors related to TRA were evaluated with logistic regression. The present study substantially impacts previous interpreting studies, future research, and patient care in the growing obesity population. Furthermore, the existing data express the relation of TRA associated to usage of docetaxel/cyclophosphamide, which drastically used instead of diets consisting of anthracycline. Together, this information can help predict the use of fertility-preserving strategies in women who require treatment and the potential need for ovarian due to modern chemotherapy for the young population of women with estrogen receptor-positive breast cancer (Poorvu et al., 2021).

Research gaps and future directions

We found a lack of data and appropriate models for predictive analysis of certain medical conditions in the current literature. There exist few statistical methods to quantify the results proposed with the help of linear regression and some static methods. There were several studies with the use of machine learning-based approaches, good prediction models were proposed for kidney, liver, brain, lung, and brain. We should consider multimodal systems by which we can understand the subject or disease and diagnose it. In terms of analysis of disease, we can provide questionnaire's to the patients that will be able to do analyses, and results produced just-in-time by the help of artificially intelligent catboats trained on task-specific. Remote monitoring of patients and providing remote assistance could be done as future work. Considering the patient's historic medical diagnosis, which consists of various types of questionnaires, X-ray, MRI, blood sample, gene, etc., data collection and analysis will give more appropriate decision-making ability. So, in the combination of Artificial intelligence (AI), methods would produce and could be aid for existing medical diagnostics systems. Image processing and deep learning methods have outperformed the object's classification, localization, and detection for any given image. This could help enhance the performance of decision-making methods for various image-based diagnoses of cells and tissues. Overall, we would like to conclude that AI would add benefits to the biological research field.

According to Ibn Sina, when blood flows toward the uterus (which is a natural passage for the excretion of menstrual blood) and fails to find a way out, it returns to the body, resulting in complications such as *Ikhtinaq al-Rahim* (hysteria), *Sayalan al-Rahim* (leucorrhoea), *Waram al-Sulb Sawdawi wa Saqirus* of *Rahim* (uterine tumors and malignancy), *Uqr* (infertility), *Waram al Jjigar* (hepatitis) *Istisqa* (ascites) *Awram-i-Ahsha* (visceral inflammation) *Malankhuliya*, generalized anasarca, etc.

Amenorrhea early detection based on machine learning models on Image or signal data. The diagnostic process consists of various types of data such as Electro encephalogram (EEG) signal, Electrocardiogram (ECG) signal, Electromyogram (EMG) signal, and MRI, functional magnetic resonance imaging (fMRI), and X-ray images [(Alshorman et al., 2021; Mehdi et al., 2016), (Bin Heyat, Akhtar, Ammar, Hayat, & Azad, 2016; Bin Heyat, Akhtar, Singh, & Siddiqui, 2017; Bin Heyat, Lai, et al., 2020; Bin Heyat et al., 2020a, 2020b), (Bin Heyat, 2016a; Bin Heyat, Akhtar, & Azad, 2016, 2017; Bin Heyat et al., 2020a; Bin Heyat, Lai, Khan, & Zhang, 2019; Bin Heyat & Siddiqui, 2015; Heyat, Lai, Akhtar, Hayat, & Azad, 2020; Lai, Bin Heyat, Khan, & Zhang, 2019)]. In collaboration with machine learning, deep learning, and computer vision techniques with the help of the above-discussed data types could result in automated detection of Amenorrhea. To design and develop a drug based on telemedicine for the treatment of amenorrhea. Use AI for accurate detection and treatment of amenorrhea.

In contrast, we can study the amenorrhea-related gene and chromosome activities in bioinformatics, considering chemotherapy outcomes that could cause adverse effects like gene modifications. We should consider histopathological images to analyze the blood cell samples into different classes for various tissues in the larger picture. Furthermore, genomic deoxyribonucleic acid (DNA) information extraction and exosome sequences are taken into account for understanding genotype, phenotype details, and their behavior for understanding different outcomes for various diseases. Also, analysis of the relationships among protein, chromosome, and genes. There is scope for different statistical methods, most likely to be machine learning and pattern analysis. We can use artificial neural networks, convolution neural networks, and machine learning (Akhtar et al., 2020; AlShorman et al., 2020; Bin Heyat, 2016b; Chola et al., 2021; Guragai, Alshorman, Masadeh, & Bin Heyat, 2020; Lai, Zhang, Zhang, & Bin Heyat, 2019; Lai, Zhang, Zhang, Su, & Bin Heyat, 2019; Pal et al., 2020). When we combine AI and biology, the outcome would be more robust and relevant concerning the simulation of medicine effectiveness over certain diseases.

Conclusion

We conclude that this chapter is divided into introduction, pathophysiology, etiology, diagnosis, management, complications, and emergency technology in amenorrhoea. In addition, amenorrhoea is a condition of lack or total cessation of menstrual blood. Disturbance of the release in hormone secretion of hypothalamic gonadotropins leading to FHA as well as hypoestrogenism, anatomical, and functional diseases. In Unani concept, altered temperament in the female genital system (uterus and ovaries) and other organs (liver, brain) are the leading causes of oligo-menorrhea and amenorrhea. Various single drugs and compound formulations for amenorrhea have been mentioned in USM. Hence, it is the need of the hour for clinical trials in this field is a mandate to validate Unani scholars' claims and evaluate the efficacy of Unani herbs in confirmatory therapeutic and post-marketing surveillance trials.

In collaboration with machine learning, deep learning, and computer vision techniques automated detection of amenorrhea evaluation is a new scope for development in bioscience and biotechnology for a robust approach toward amenorrhea patients. To design and develop a drug based on telemedicine for the treatment of amenorrhea. Use AI for accurate detection and treatment of amenorrhea. We can use artificial neural networks, convolution neural networks, and machine learning. When we combine AI and biology, the outcome would be more robust and relevant concerning the simulation of medicine effectiveness over certain diseases.

Acknowledgments

We are thankful to Prof. Ansari, Prof. Rajat, Prof. Lai, Prof. Naseem, and Dr. Ijaz Gul for the valuable discussion. We also acknowledge the National Institute of Unani Medicine, Bangalore, provided the research laboratory for this study.

References

Ahmed, F., Nizami, Q., & Aslam, M. (2005). *Classification of Unani drugs with English and scientific names* (pp. 237–239). Delhi: Maktaba Eshaatul Quran.

Akhtar, F., Bin Heyat, M. B., Li, J. P., Patel, P. K., Pal, R., & Guragai, B. (2020). *Role of machine learning in human stress: A review*. https://doi.org/10.1109/ICCWAMTIP51612.2020.9317396.

AlShorman, O., Masadeh, M., Alzyoud, A., Heyat, M. B. B., Akhtar, F., & Pal, R. (2020). The effects of emotional stress on learning and memory cognitive functions: An EEG review study in education. In *2020 Sixth International Conference on e-Learning (econf)* (pp. 177–182). https://doi.org/10.1109/econf51404.2020.9385468.

Alshorman, O., et al. (2021). Frontal lobe real-time EEG analysis using machine learning techniques for mental stress detection. *Journal of Integrative Neuroscience, 21*, 1–11.

Arentz, S., Abbott, J. A., Smith, C. A., & Bensoussan, A. (2014). Herbal medicine for the management of polycystic ovary syndrome (PCOS) and associated oligo/amenorrhoea and hyperandrogenism; a review of the laboratory evidence for effects with corroborative clinical findings. *BMC Complementary and Alternative Medicine*. https://doi.org/10.1186/1472-6882-14-511.

Ben Abdallah, I., et al. (2021). Amenorrhea and elevated βhuman chorionic gonadotropin of unknown origin: An unexpected location of choriocarcinoma. *Gynecologic Oncology Reports*. https://doi.org/10.1016/j.gore.2021.100746.

Bin Heyat, M. B. (2016a). *Insomnia: Medical sleep disorder & diagnosis* (1st ed.). Hamburg, Germany: Anchor Academic Publishing.

Bin Heyat, B. (2016b). *A review on neurological disorder epilepsy affected in the human body*. no. 3 (pp. 1–4).

Bin Heyat, M. B., Akhtar, F., Ammar, M., Hayat, B., & Azad, S. (2016). Power spectral density are used in the investigation of insomnia neurological disorder. In *XL- pre congress symposium* (pp. 45–50).

Bin Heyat, M. B., Akhtar, F., & Azad, S. (2016). Comparative analysis of original wave and filtered wave of EEG signal used in the prognostic of bruxism medical sleep syndrome. *International Journal of Trend in Scientific Research and Development, 1*(1), 7–9. https://doi.org/10.31142/ijtsrd53.

Bin Heyat, M. B., Akhtar, F., & Azad, S. (2017). A review on use of sunlight in human life. *International Journal of Trend in Scientific Research and Development*. https://doi.org/10.31142/ijtsrd70.

Bin Heyat, B., Akhtar, F., Singh, S. K., & Siddiqui, M. M. (2017). Hamming window are used in the prognostic of insomnia. In *International seminar present scenario future prospectives Res. Eng. Sci. (ISPSFPRES)* (pp. 65–71).

Bin Heyat, M. B., Lai, D., Akhtar, F., Ansari, M. A., Khan, A., & Alkahtani, F. (2020). Progress in detection of insomnia sleep disorder: A comprehensive review. *Current Drug Targets*. https://doi.org/10.2174/1389450121666201027125828.

Bin Heyat, M. B., Lai, D., Khan, F. I., & Zhang, Y. (2019). Sleep bruxism detection using decision tree method by the combination of C4-P4 and C4-A1 channels of scalp EEG. *IEEE Access, 7*, 102542–102553. https://doi.org/10.1109/access.2019.2928020.

Bin Heyat, M., & Siddiqui, M. M. (2015). Recording of EEG, ECG, EMG signal. *International Journal of Advanced Research in Computer Science and Software Engineering, 5*(10), 813–815. [Online]. Available www.ivline.org.

Bin Heyat, M. B., et al. (2020a). Bruxism detection using single-channel C4-A1 on human sleep S2 stage recording. In D. Gupta, et al. (Eds.), *Intelligent data analysis: From data gathering to data comprehension* (pp. 347–367). USA: John Wiley & Sons Ltd.

Bin Heyat, M. B., et al. (2020b). Detection, treatment planning, and genetic predisposition of bruxism: A systematic mapping process and network visualization technique. *CNS & Neurological Disorders: Drug Targets*. https://doi.org/10.2174/1871527319666201110124954.

Chola, C., et al. (2021). *IoT based intelligent computer-aided diagnosis and decision making system for health care*. https://doi.org/10.1109/icit52682.2021.9491707.

Delaney, A., et al. (2021). Increased burden of rare sequence variants in GnRH-associated genes in women with hypothalamic amenorrhea. *The Journal of Clinical Endocrinology and Metabolism*. https://doi.org/10.1210/clinem/dgaa609.

Golden, N. H., & Carlson, J. L. (2008). The pathophysiology of amenorrhea in the adolescent. *Annals of the New York Academy of Sciences, 1135*, 163–178. https://doi.org/10.1196/annals.1429.014.

Guragai, B., Alshorman, O., Masadeh, M., & Bin Heyat, M. B. (2020). *A survey on deep learning classification algorithms for motor imagery*. https://doi.org/10.1109/ICM50269.2020.9331503.

Heyat, M. B. B., Lai, D., Akhtar, F., Hayat, M. A. B., & Azad, S. (2020). Short time frequency analysis of theta activity for the diagnosis of bruxism on EEG sleep. In D. Gupta, & A. Hassanien (Eds.), *Advanced computational intelligence techniques for virtual reality in healthcare. studies in computational intelligence* (pp. 63–83). Springer.

Hu, K.-L., et al. (2021). Oligo/amenorrhea is an independent risk factor associated with low ovarian response. *Frontiers in Endocrinology, 12*. https://doi.org/10.3389/fendo.2021.612042.

Islam, A. (2018). Unani methods of cure in the Indian subcontinent: An analytical study. *International Journal of Engineering and Technology, 7*(2.29), 480. https://doi.org/10.14419/ijet.v7i2.29.13802.

Izzaty, N. R., Imandiri, A., & Suciati, S. (2017). Secondary amenorrhea therapy with accupuncture and turmeric—Fenugreek herbal. *Journal of Vocational Health Studies, 1*(1), 27. https://doi.org/10.20473/jvhs.V1.I1.2017.27-31.

Kitasaka, H., Tokoro, M., Kojima, M., Fukunaga, N., & Asada, Y. (2021). Gonadotropin levels at the start of ovarian stimulation predict normal fertilization after hCG re-trigger in GnRH antagonist cycles. *Reproductive Medicine and Biology*. https://doi.org/10.1002/rmb2.12359.

Klein, D. A., Paradise, S. L., & Reeder, R. M. (2019). Amenorrhea: A systematic approach to diagnosis and management. *American Family Physician*. https://doi.org/10.2310/obg.19117.

Klein, D. A., & Poth, M. A. (2013). Amenorrhea: An approach to diagnosis and management. *American Family Physician, 87*, 781–788.

Lai, D., Bin Heyat, M. B., Khan, F. I., & Zhang, Y. (2019). Prognosis of sleep bruxism using power spectral density approach applied on EEG signal of both EMG1-EMG2 and ECG1-ECG2 channels. *IEEE Access, 7*, 82553–82562. https://doi.org/10.1109/ACCESS.2019.2924181.

Lai, D., Zhang, X., Zhang, Y., & Bin Heyat, M. B. (2019). Convolutional neural network based detection of atrial fibrillation combing R-R intervals and F-wave frequency spectrum. In *2019 41st annual international conference of the IEEE engineering in medicine and biology society (EMBC)* (pp. 4897–4900). https://doi.org/10.1109/EMBC.2019.8856342.

Lai, D., Zhang, Y., Zhang, X., Su, Y., & Bin Heyat, M. B. (2019). An automated strategy for early risk identification of sudden cardiac death by using machine learning approach on measurable arrhythmic risk markers. *IEEE Access*. https://doi.org/10.1109/access.2019.2925847.

Mehdi, S., Bin Heyat, B., Akhtar, F., Ammar, M., Heyat, B., & Gupta, T. (2016). Cure of epilepsy by different system of medicine. *International Journal of Technical Research & Science, 1*(8), 244–247.

Moini Jazani, A., et al. (2018). Herbal medicine for oligomenorrhea and amenorrhea: A systematic review of ancient and conventional medicine. *BioMed Research International, 2018*, 1–22. https://doi.org/10.1155/2018/3052768.

Mokaberinejad, R., et al. (2012). Mentha longifolia syrup in secondary amenorrhea: A double-blind, placebo-controlled, randomized trials, DARU. *Journal of Pharmaceutical Sciences*. https://doi.org/10.1186/2008-2231-20-97.

Morrison, A. E., Fleming, S., & Levy, M. J. (2021). A review of the pathophysiology of functional hypothalamic amenorrhoea in women subject to psychological stress, disordered eating, excessive exercise or a combination of these factors. *Clinical Endocrinology, 95*(2), 229–238. https://doi.org/10.1111/cen.14399.

Nichols, J. L., Bieber, E. J., & Gell, J. S. (2010). Secondary amenorrhea attributed to occlusion of microperforate transverse vaginal septum. *Fertility and Sterility*. https://doi.org/10.1016/j.fertnstert.2009.12.052.

Pal, R., et al. (2020). *Effect of Maha Mrityunjaya HYMN recitation on human brain for the analysis of single EEG channel C4-A1 using machine learning classifiers on yoga practitioner*. https://doi.org/10.1109/ICCWAMTIP51612.2020.9317384.

Parveen, A., Parveen, R., Akhatar, A., Parveen, B., Siddiqui, K. M., & Iqbal, M. (2020). Concepts and quality considerations in Unani system of medicine. *Journal of AOAC International, 103*(3), 609–633. https://doi.org/10.5740/jaoacint.19-0284.

Parvizi, M. M., Nimrouzi, M., Bagheri Lankarani, K., Emami Alorizi, S. M., & Hajimonfarednejad, M. (2017). Health recommendations for the elderly in the viewpoint of traditional Persian medicine. *Shiraz E Medical Journal*. https://doi.org/10.5812/semj.14201. vol. In Press, no. In Press.

Pereira, K., & Brown, A. J. (2017). Secondary amenorrhea. *The Nurse Practitioner, 42*(9), 34–41. https://doi.org/10.1097/01.NPR.0000520832.14406.76.

Poorvu, P. D., et al. (2021). Treatment-related amenorrhea in a modern, prospective cohort study of young women with breast cancer. *Npj Breast Cancer, 7*(1), 99. https://doi.org/10.1038/s41523-021-00307-8.

Ruddy, K. J., et al. (2021). Chemotherapy-related amenorrhea (CRA) after adjuvant ado-trastuzumab emtansine (T-DM1) compared to paclitaxel in combination with trastuzumab (TH) (TBCRC033: ATEMPT trial). *Breast Cancer Research and Treatment*. https://doi.org/10.1007/s10549-021-06267-8.

Schmidt, P. J. (2005). Depression, the perimenopause, and estrogen therapy. *Annals of the New York Academy of Sciences, 1052*, 27–40. https://doi.org/10.1196/annals.1347.003.

Sevilla-Montoya, R., et al. (2021). Unravelling complex mosaicism of sex chromosomes in a patient with primary amenorrhea through cytogenetic analysis on urothelial cells. *Taiwanese Journal of Obstetrics & Gynecology*. https://doi.org/10.1016/j.tjog.2021.03.025.

Sharma, P. (2020). From medical pluralism to medical marginality: Changing dynamics within Unani system of medicine. *Sociological Bulletin*. https://doi.org/10.1177/0038022920923208.

Shojaee-Mend, H., Ayatollahi, H., & Abdolahadi, A. (2020). Developing a mobile-based disease ontology for traditional Persian medicine. *Informatics in Medicine Unlocked*. https://doi.org/10.1016/j.imu.2020.100353.

Shrivastava, P., Shrivastava, R. K., Himanshu, H., & Jha, R. K. (2013). Use of algorithmic approach to evaluate the cause of secondary amenorrhea. *Journal of Evolution of Medical and Dental Sciences, 2*(25), 4620–4629. https://doi.org/10.14260/jemds/888.

Singhal, V., et al. (2021). Changes in marrow adipose tissue in relation to changes in bone parameters following estradiol replacement in adolescent and young adult females with functional hypothalamic amenorrhea. *Bone, 145*. https://doi.org/10.1016/j.bone.2021.115841, 115841.

Sultana, A., & Rahman, K. (2021). Evaluation of general body temperament and uterine dystemperament in amenorrhoea: A cross-sectional analytical study. *Journal of Complementary and Integrative Medicine*. https://doi.org/10.1515/jcim-2020-0334.

Varughese, R., Ryan, F., & Makaya, T. (2021). Fifteen-minute consultation: A structured approach to the child with primary amenorrhea. *Archives of Disease in Childhood. Education and Practice Edition*. https://doi.org/10.1136/archdischild-2019-317999.

CHAPTER

9

Wearable EEG technology for the brain-computer interface

Meenakshi Bisla and R.S. Anand

Department of Electrical Engineering, Indian Institute of Technology–Roorkee, Roorkee, Uttarakhand, India

Introduction

Overview

Wireless electroencephalography (EEG) system design is a research field that is experiencing large development, we can expect substantial progress in designing an EEG acquisition system that would afford superior signal quality required for precise temporal and spatial resolution. EEG is one of the most important noninvasive technique to monitor the electrical pattern of the brain. Among noninvasive brain activity monitoring modalities, EEG is the only modality that monitors brain activity using sensors and mounting competences such that it can be worn during free locomotion. As compared to functional near infrared spectroscopy (fNIRS), EEG has temporal resolution and fNIRS can only be used areas not covered with hairs like forehead and detects the blood oxygenation level (similar to fMRI), which cannot capture rapid physiological changes in brain waves. Other modalities, such as fMRI and positron emission tomography and magnetoencephalography (MEG) have low temporal resolution than EEG and are unfeasible to be modified as wearable devices to monitor brain activity (Toga & Thompson, 2001).

Noninvasive EEG stalks from neuronal pattern below the scalp area so the electrical potentials of neurons have to travel through the surrounding tissue to the scalp surface. There are millions and trillions of neurons present within different brain regions; hence, neuronal activity of individual neuron cannot be recorded. Activity recorded by individual electrode represents the summation of electrical activity of thousands to millions of spatially aligned neurons. As compared to fMRI and fNIRS, EEG provides poor spatial resolution, but EEG

Computational Intelligence in Healthcare Applications
https://doi.org/10.1016/B978-0-323-99031-8.00005-3

provides high temporal resolution that matches the speed of cognition. As MEG is bulky and expensive, hence not suitable for daily life monitoring; EEG is the ideal choice for studying dynamic processes of brain and coupling of different regions in the brain during challenges imposed in our daily life. fNIRS can only be used as an add-on to EEG as it is difficult to apply through hairy regions, records activity from nonhairy regions like forehead and also offers low temporal resolution of the signal (Kaiser et al., 2014). Using it in conjunction with EEG device would provide high spatial resolution at the frontal brain regions.

The monitoring of brain activity in real-life EEG applications would become possible with further understanding on how synchronous firing of neurons results in EEG patterns and how cortical fields are generated (Matthews, McDonald, Hervieux, Turner, & Steindorf, 2007). Different inputs to our senses and internal inputs correspond to different brain activity. Advancement in research and technology transformed EEG systems from wired cumbersome systems used mostly in clinical environment to intelligent wearable and wireless systems that are convenient and comfortable lifestyle solutions. Smart wireless wearables must be equipped with application-driven design, end-user driven development, standardization and sharing of EEG data, and development of sophisticated approaches to handle EEG artifacts. An intelligent wearable EEG system should acquire superior signal quality providing meaningful information when used in different real-life conditions while providing comfort and convenience to the users.

With a daily life brain activity monitoring device, changes in mental state and cognitive processes can be measured. For example, EEG signals with eyes closed result in a completely different pattern all over the brain compared to the same person having eyes open. Current EEG monitoring systems and the existing data formats are not capable of implementing such information sources in daily life (Casson & Rodriguez-Villegas, 2009). EEG monitoring is mostly done in a skillful lab environment with detailed specification of standard recording protocols. However, as soon as the EEG equipment moves into an uncontrolled daily life environment, monitoring the contextual information and especially the sensory input would become an important design requirement. Similarly, capturing and storage of personal traits, this also calls for standardizing contextual information for the purpose of proper documentation and exchange.

More sophisticated technologies such as electrocardiography (ECG) and EEG have been addressed in the research community in recent years to provide health status of our heart and brain, respectively. The need of today is to move this device to a completely uncontrolled environment without expert surveillance and specially designed labs (Casson, Smith, Duncan, & Rodriguez-Villegas, 2008). EEG monitoring in hospitals requires an unwieldy and time-consuming preparation procedure, which involves skin preparation and gel-electrode application, mounting of wired sensors, and connecting electrodes to the main acquisition unit and a PC. This is far from being user-friendly, comfortable, and convenient, also results in headache and unpleasant experience to users, and limits the use to clinical or ambulatory setup. We need extensive progress in developing technology for a convenient wearable EEG monitoring system and in developing applications that can utilize the technology for the making of a database of knowledge in different areas of brain research.

Need of a wearable EEG technology-based brain-computer interface (BCI)

Having technology like wearable devices empowered users to take care of their own health and that will assist healthcare providers in understanding and improving patient's conditions. The first noninvasive and portable EEG monitoring system was designed for epilepsy patients by performing ambulatory EEG monitoring (Gilliam, Kuzniecky, & Faught, 1999). Since then, not much has changed in the systems used to trace user's brain activity. With innovation in technology, EEG recording units got smaller over time. g.Nautilus (g.Tec), Smarting(mBrainTrain), Neuroelectrics Enobio, Cognionics, and so on are some truly wearable brain monitoring systems appeared in the market recently. The key factors toward this change are development in the fields of compact electronics components and chip design.

BCI using wearable EEG devices is having a bright future to be widely used in personal healthcare and home diagnostics, such as wellness and health monitoring, home rehabilitation, and the early detection of brain disorders (Gu et al., 2020). A wireless wearable EEG system has inordinate potential to understand the functioning of human brain and helps in improving the quality of life by monitoring real-life interaction of humans with their environment. In recent years with development in technology wearable, EEG can be a dominant modality for studying brain dynamics and performance. Technology has to be advanced to a level that it can be used in daily life activities as an unremarkable option to monitor and improve health.

Electrical activity of the brain is affected by maturing and aging of the brain and is largely obstructed by one's personal traits (Duffy, Albert, McAnulty, & Garvey, 1984). Using wearable EEG device, we can monitor development and cognitive evolution of our brain activity, which is not possible with cumbersome and unfeasible procedure of traditional EEG monitoring. With the help of evolution in technology, EEG can be a part of routine health monitoring and included in systematic research on impact of age, gender, and race/ethnicity on brain activity.

There is lack of large publicly available databases containing EEG activity of the normal healthy population. Neither is there a consensus on how to store the EEG information and the context in which it is recorded nor on what features to extract from EEG signals for characterizing mental traits and states of a person. With EEG systems that facilitate daily life monitoring of brain activity would simplify monitoring larger populations with diverse behavioral traits enabling generation of large databases (Debener, Minow, Emkes, Gandras, & de Vos, 2012). Useful information can be extract from the recordings stored into the database. Therefore with the availability of vast dataset, researches can be done for understanding of EEG features and topologies across different traits of different users, enabling standardization in terms of recording and exchange formats, as well as integrating EEG information with information from other sensors for increasing degree of freedom.

Challenges for wearable EEG technology

It is challenging to moving the technology used in hospitals to users' daily routine without expert aid. Most in-hospital techniques for monitoring health are too bulky and complex to be used outside hospitals. To have a more holistic view of health, we need to be aware of more aspects other than body weight, blood pressure, and activity. Despite all the research efforts, there are still some challenges that are needed to be considered while designing an efficient wearable EEG recording system (Casson, 2019).

1. Data produced by the recordings of current outpatient systems have limited use to health care physician, doctor, or for research purpose. Experts are less confident about the quality of recordings and the sources of physiological changes that are recorded by such outpatient devices.
2. The EEG signals measured from the scalp are very weak. It can be easily corrupted by external noise. Anything that has electricity can produce a small electric field that could be picked up by the EEG signal if it is close enough or if it is strong enough. Before launching such device for routine monitoring, it must be ensured, whether people can really on them at home where they freely move around without compromising on the information.
3. Recorded data should be accurately related to the cognitive state you are trying to interpret.
4. State-of-the-art bioamplifier architectures and low-power circuit design techniques intended for wearable EEG acquisition should take factors like high medical grade signal quality, electrode offset tolerance, common-mode rejection ratio, input impedance, power dissipation, and so on into consideration.
5. The brain information is still known very little. If the regular consumer were to use this wearable EEG device, interpretations of any data might be inaccurate, misleading, or even incorrect.
6. EEG is used as a very important brain monitoring modality used in clinical environment; it is considered to be too prone to noise to be used in daily life wearable applications. The main difficulty in utilization of wearable EEG device monitoring is the insufficiency of solutions used to acquire and process electric potentials of brain signals and less effective feedback information.
7. User lacks sufficient knowledge to ensure proper usage of wearable devices and usually cannot interpret the extracted information in the correct way. Significant research and development efforts are required to bridge this gap.
8. Lifestyle wearable EEG technology of the future have to be compatible for integration with sensors that can capture other physiological signals (e.g., ECG, GSR) or contextual information (cameras, GPS locators). Formats used for storing EEG recordings also have to be extended somehow or have the capability to be linked to the data coming from these secondary sensors.

There are clearly lots of factors to consider. It may seem like that there are so many doubts and so few answers at this point. The same thing was said for the heart rate monitors or pedometers years ago. These challenges do not mean that consumer-ready wearable EEG devices would not be extensive and operative soon. There is positively scope for wearable EEG to be useful for regular consumers without experts. This absolutely requires good scientists, engineers, and software developers to build a wearable EEG device reliable for everyday recordings. In this chapter, we discussed advancements in wireless and wearable EEG systems and a number of aspects where this field need rapid research progress and provided guidelines on how these aspects can be achieved for the development of intelligent wireless wearable EEG solutions.

Wireless EEG data acquisition

In a typical ambulatory EEG recording system, electrodes are glued with conductive gel at the prescribed location in a controlled lab environment by a technician and the signal quality is tested. Electrodes are supported with a head cap to keep them in place. Long wires are used to transmit the recorded brain activity from electrodes to the main electronic unit that user wears around the waist. This unit amplifies and records the signal with a suitable recording medium.

The progression of ambulatory EEG to wearable EEG must be unobtrusive, light weighted, discrete, and reliable (Gilliam et al., 1999). A large ambulatory EEG acquisition system and wires linking it to the electrodes are substituted by microchips embedded with the required circuitry like amplifiers, quantizers, and wireless transmitter, which are mounted on top of the electrodes themselves. EEG data are then wirelessly sent via bluetooth to a mobile phone, laptop, or any other device that user usually keep within a short distance of themselves.

Acquisition

A block diagram for EEG data acquisition is shown in Fig. 9.1. Stimulus can be internal or external trigger; changes in neural activity are measured in response to stimulus. Electrodes convert the ionic potentials generated by the brain into electric potentials, which are then amplified and stored. Differential architecture is used to confiscate common-mode interference signals, and recording from a pair of sensors, thus, forms an EEG channel. The signals vary

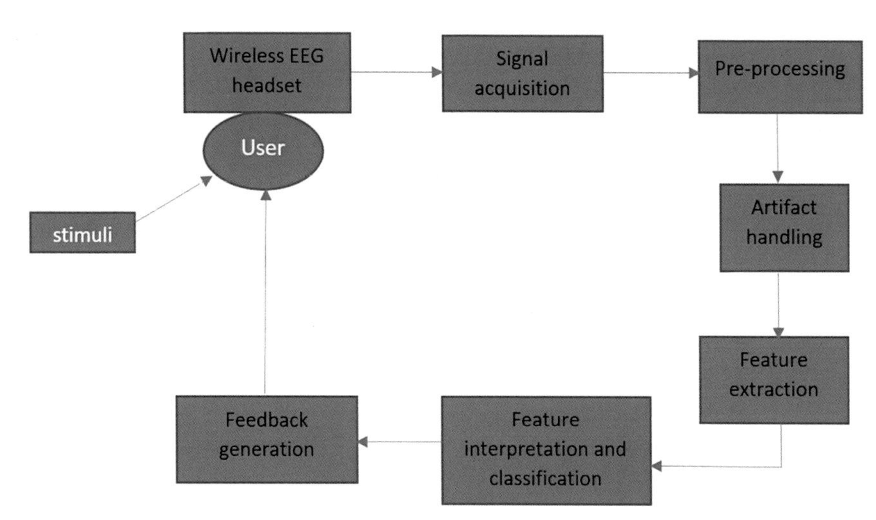

FIG. 9.1 Block diagram of EEG data acquisition process. *No Permission Required.*

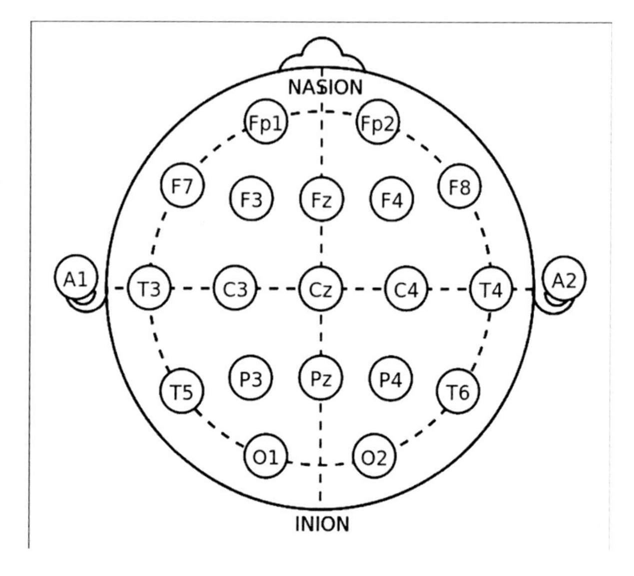

FIG. 9.2 10–20 international standard for electrode placement. *From Rojas, G., Alvarez, C., Montoya Moya, C., de la Iglesia Vaya, M., Cisternas, J., & Gálvez, M. (2018). Study of resting-state functional connectivity networks using EEG electrodes position as seed.* Frontiers in Neuroscience, 12. *https://doi.org/10.3389/fnins.2018.00235 10-20.*

both temporally and spatially so multiple channels are used with electrode positions usually determined using the international 10–20 standard, as shown in Fig. 9.2, to record both spatial and temporal information in response to the stimulus presented as shown in. After amplification and bandwidth limiting, the signals are stored digitally in a suitable location.

EEG signal preprocessing

It aims to remove noise and artifacts that entered during data acquisition by apply signal processing steps mentioned in the following:

- Electrode rereferencing: common average reference is commonly a used technique. Referencing methods might not always improve signal quality; proper referencing methods need to be carefully chosen for a specific application.
- Low pass/band pass filtering: depending on frequency range of interest. An improper filtering technique can result in removing an important portion of EEG activity.
- Altering the sampling rate: resampling can reduce noise, which can be useful for saving transmission power in the case of real-time system design.
- Epoch extraction: EEG analysis is typically performed per epoch; depending on the type of analysis and application, overlap is introduced between epochs. Different epoch duration and overlap percentage result in different performance.
- Selecting the suitable epochs: the most commonly used approach is visual inspection and manual removal of contaminated EEG segments.

Artifact handling

An electrical signal captured by the EEG device but produced by sources other than physiological activity of brain is term as artifact. EEG is a highly sensitive recording device; it can sense nearby electrical interferences as signals that resemble cerebral activity. It is easy to do artifact free recording in an isolated lab environment, but while recording in a real-time daily life environment, artifact handling is a very challenging and compulsory task for the successful implementation of wearable wireless EEG devices in daily life monitoring. Ceasing head and body movements in order to avoid artifacts would oppose the very purpose of real-time daily life EEG monitoring. Developing a reliable technological solution for identification of physiological and motion artifacts and reducing the impact of these artifacts is a great challenge for the design of smart wireless wearable EEG systems (Mihajlovic, Grundlehner, Vullers, & Penders, 2015).

These are the sources of artifacts that can condense the quality of wireless wearable EEG recordings during:

- instrumental
- measurement
- environmental
- physiological
- movement artifact

Instrumental artifacts

These artifacts originate from circuit components while recording the EEG signal. This kind of artifacts typically results in thermal noise, spiky signal, cross-talk between different channels, and so on. Proper system design and evaluation is primarily needed to eliminate such artifact components.

Measurement artifact

This kind of artifact arises due to improper usage of the EEG system. They usually do not exist in the laboratory environment, but in real-time monitoring by untrained individuals, they are quite common during preparation or setup in the early phases of monitoring due to improper positioning of electrodes or a headset. These kinds of artifacts can be avoided by providing a user interface for fixing, correcting, and trying the setup.

Environmental artifacts

Such artifacts are caused due to power lines or electromagnetic interference, and so on. Environmental artifacts are usually not a big issue in wireless EEG headsets because of less cumbersome design.

Physiological artifacts

These artifacts are produced by eye movement and eye blinks (EOG), muscle tension (EMG), and cardiac activities (ECG). EOG artifact has larger impact in frontal areas of head and can have amplitude up to thousands of millivolts . EMG artifacts are created due to jaw, face, and neck muscle movement. These kinds of artifacts are most prominent over the temporal lobes. Movement of head muscle is most common in daily life activities so a reliable solution to such kind of artifact is must for efficient wearable EEG system design. Similarly, ECG artifacts can also contaminate the EEG signal.

Movement artifacts

These artifacts are dominantly caused due to change in relative position between the skin and electrode due to any kind of movement caused. They are relatively larger in dry electrode monitoring systems as compared to traditional conductive gel electrodes. Handling motion artifact is one among major concerns of designing an efficient, artifact free daily life monitoring EEG device, where use is free to move and perform usual activities.

Feature extraction

EEG feature extraction and interpretation algorithms and techniques are very miscellaneous and involve different linear or nonlinear spatiotemporal signal processing tools and techniques (Boonyakitanont, Lek-uthai, Chomtho, & Songsiri, 2020). In terms of algorithm complexity, wireless wearable EEG monitoring devices have to consider more rigorous boundaries for real-time processing of signals with reduced number of electrodes. EEGLAB, Open Vibe, and Field trip are the most frequently used toolboxes for EEG feature extraction and interpretation. Feature extraction and interpretation algorithms usually depend on the aim of the application. For successful implementation of a real-time wearable wireless EEG device, time and memory efficient implementation of signal processing algorithms that support real analysis with instantaneous information processing and fusion is a very important concern.

Brief overview of available wireless EEG headsets and headbands

In current years, increasing number of products and concepts target on wearable EEG monitoring using dry electrodes. The mechanical designs may vary from product to product that keeps the electrodes in place since gel electrodes are not glued in such devices. High-speed application and uncompromised signal quality are very important constraints for wireless devices that target medical diagnostics or BCI applications, where the design needs to support in acquiring the highest possible signal quality. For serious applications based on BCI approaches, acquisition design should be user friendly, easy to use, robust, and "foolproof" (Wolpaw, Birbaumer, McFarland, Pfurtscheller, & Vaughan, 2002).

Numerous research organizations have recognized the need for wearable EEG solutions and involved in developing outpatient wireless EEG systems or technology to be used in such systems. There are some commercially available EEG systems: MindWave, MindSet, and Necomimi by Neurosky, Emotiv Epoc headset, Muse by InteraXon, g.tec gSahara, Quasar DSI series, Neuroelectrics Enobio, and Cognionics 16–64 channel headsets. Also, within Holst Centre/imec, an EEG monitoring headset with four active sensor chips is available. The different wireless EEG systems available are useful in different application areas. The number of existing devices is too many to be conversed here exhaustively; we have discussed selected number of devices that are commercially available and that vary greatly in design and application.

Neurosky

Neurosky's product range provides a single-channel measurement platform with dry electrodes (Neurosky, n.d.). The channel made up of stainless alloy is typically situated at the forehead for frontal recordings ground, where ground and reference are located at the ear clip. Data is wirelessly transmitted using bluetooth to a PC or a smartphone. This device can potentially target the low-end consumer market as it has the lowest price tag of all wireless wearable EEG devices available.

Emotive

The Epoc (Emotive) is 14 channel wireless wearable EEG headset that provides a flexible and versatile research platform and being low cost (EPOC, n.d.). It is most commonly used low-cost devices available in market. Data is wirelessly transmitted through a registered radio link and allows for 12-h constant transmission. However, a considerably expensive license is required to access the raw EEG data to use Emotive for research platforms.

Imec

Imec's headset has four channels wireless acquisition platform with a flexible headset design, as shown in Fig. 9.3. It uses dry Ag/AgCl-coated electrodes with Biopac EL120 pins to make contact with skin through the hair. Electrodes are mounted within spring-loaded holders to facilitate sufficient force at each electrode and enhance user comfort. A patented protocol is used for transmitting data wirelessly with battery that allows for 20 h of continuous data transmission.

g.Nautilus

G.tec's g.Nautilus platform is a recent invention consists of a cap with active dry electrodes fabricated with gold-plated pins embedded according to the 10–20 International System. EEG data can be transmitted continuously for 8 h wirelessly through a radio link without charging. They are used in research test centers around the world for BCI investigations or for gaming applications built using BCI technology.

FIG. 9.3 Four channel Imec's EEG headset. *From Imec. (n.d.). Retrieved August 9, 2021, from https://www.imec-int.com/en/eeg.*

Muse

Muse headset is designed by InteraXon, a Toronto-based innovation in the field of brain-sensing technology to measure alertness status of brain and guide the user in clearing it. With the headset on user can make comparisons with different sessions, durations of a particular day or days of the week, and hit daily targeted goals.

SSE cap

A team of researchers at the Georgia Institute of Technology combined wireless soft scalp electronics and virtual reality in a brain-machine interface (BMI) system in which a wheelchair with robotic arm is controlled wirelessly by imagining an action (Mahmood et al., 2021). This system is soft, comfortable, and wireless as compared to other existing wearable EEG monitoring devices. BMI is a kind rehabilitation technology that turns intention into actions by monitoring neural activity of an individual and interprets it into commands. The most common traditional noninvasive EEG device requires a cumbersome and tangled web of wires.

This new wireless wearable BMI system can be of great importance to people with motor dysfunction, paralysis, or locked-in syndrome patients. This portable wireless EEG system design integrates imperceptible microneedle electrodes with soft wireless circuits offering better-quality signal acquisition.

Smart EEG headband by BrainBit

A wearable EEG monitoring system by BrainBit has four dry electrodes, one reference electrode, and one common electrode. They claim highest spatial resolution than any other available device in the consumer market. Recent technology advancements allow us to use EEG monitoring in many more ways that just for clinical use. Individual brain activity monitoring to our routine daily activities is now possible with the invention of wearable EEG technology.

Electrodes mounted inside the BrainBit headband follows 10–20 International Standard for electrode placement, forming a direct contact at the T3 and T4 temporal lobe regions as well as at the O1 and O2 occipital lobe regions. Analog EEG data acquired from the electrodes is converted into digital data and transmitted wirelessly to a smartphone, a tablet, or a computer via bluetooth. BrainBit headband records the electrical neural activity as raw EEG data, which can be administered for user legibility. Important design specifications of the headset are given in the following.

- 250 Hz sampling rate
- noise cancellation
- 12-h active battery
- 4.0 LE bluetooth connectivity
- Wearable and lightweight design allows the user to wear BrainBit headband in the daytime and during sleep.
- The headband features the highest quality dry electrodes and components.

- EEG data is collected in exclusive BrainBit cloud for data mining and machine learning analysis.

It is the only available device in consumer market to recognize different states of brain wave to detect different stages of consciousness and connection between brain waves and mental state. BrainBit headset helps in improving client's understanding with brain visions and understanding what is going inside the brain. This system can be employed in BCI applications for turning intuitions into actions.

Nio wear

Nio wear is a medical grade brain wearable invented by Evercot AI that allows acquisition of EEG data outside the laboratory environment just by wearing an EEG device and performing usual activities. Acquired EEG data can be wirelessly sent to the physician or to the cloud automatically. Device can be used for collecting neural data for research and development purposes or for BCI applications. It can be used in areas like mental health and depression monitoring, epilepsy seizure analysis while performing daily routine activity, and so on. This wearable headset can used in combination with the virtual reality system to capture the neural patterns, responses, and reactions of the patient during BCIs to interpret human intensions and perceptions.

EEG data acquisition when an individual is performing day-to-day activities usually involve gestures leading to movement artifacts in EEG data recording. Nio Wear headset is equipped with advance signal processing techniques to deal with these unwanted artifacts and ensures high-quality mobile EEG data recoding. Evercot AI is also concerned about data-sharing partnerships with medical health centers and research teams who aim to utilize machine learning and Big Data analysis to improve early disease diagnostics or in BCI applications to increase the efficiency of health care workflows.

Deep learning and some reinforcement learning are also incorporated to the build analytical model for BCI applications. Time series data mining techniques and a broader group of supervised and unsupervised machine learning model are used for efficient decoding and classification of neural responses. Recorded EEG data is divided into training and test dataset, and the predictive model is trained on the training set and the performance of the model is tested on the test dataset. The model can efficiently provide a prediction result when an unseen user EEG data is uploaded, like weather; the mental state is relaxed or stressed or the patient is having a certain disease or not and can also help in classification of BCI commands.

DSI series dry EEG headsets

Dry sensing interface series by neuracle's wearable sensing uses QUASAR's high-fidelity dry electrode technology for high-density EEG monitoring (DSI series, n.d.). Being compatible with wide range of research tools for integration and analysis, they facilitate data acquisition for research and development purposes. They offer fast and easy to use high signal quality wearable headsets for real-world applications. The DSI series headsets enable fast and easy setup for large-scale transmissions and also provide coziness during long or repeated monitoring. They are resistant to electrical and motion artifacts that are likely to occur during ambulatory recordings outside the lab. We will discuss in detail about DSI-24 out of all other DSI series EEG systems as it is most commonly used wearable headset out of other DSIs for BCI and research applications.

DSI series headsets are ideally suited for BCI and neurogaming applications as they offer easy setup and are comfortable for hours of continuous monitoring. Dry electrode technology that does not necessitate skin scratching and does not dry up facilitates long-term real-life monitoring without negotiating on signal quality, which is a blessing for rehabilitation, communication, and BCI applications.

Quasar's DSI 10/20

It is a 21-channel EEG headset designed with aim of an ambulatory EEG recording, and mechanical and electrical mechanisms to reduce motion artifacts are also included in system design. Recoded data is transmitted wirelessly through a registered system with the help of a USB dongle. It allows continuous EEG transmission for 24h. This EEG recorded device is high on price but provides the highest signal quality possible.

The DSI-24

It is a research-grade wireless dry electrode smart EEG headset. It is integrated with 21 sensors as per the standards of 10–20 International System. Wearable Sensing is selected as a Technology Partner of the Mental Work human-machine interaction installation at the Ecole Polytechnique Federale de Lausanne, Switzerland. A DSI-24 EEG monitoring system can be used for recording EEG data of factory workers who are involved in mental workload tasks to interact with the installation.

Its design specifications include:

- Easy to set up to monitor EEG in under 5min.
- Comfortable to wearing at least up to 8h of continuous monitoring.
- Portable design suitable for real-time daily life monitoring.
- 21 dry electrodes are placed according to the 10–20 International System
- Continuous impedance check.
- Wireless transmission via bluetooth with a sampling rate of 300Hz.
- Hot-swappable Li-ion batteries that allow continuous monitoring up to 8h.
- 8-bit digital trigger input.
- Adaptable to fit a wide range of head sizes (52–62cm in circumference).

The DSI-24's level-headedness, compactness, and high signal quality take numerous applications that were earlier limited to lab environments to real-life monitoring. Such applications include:

- neurofeedback
- neuroergonomics
- neuroeconomics/neuromarketing
- augmented cognition
- BCIs
- biometrics
- cognitive stress or workload monitoring

- psychological research
- sports peak performance training

Irrespective of all the research and development efforts made toward designing a wearable wireless EEG technology, there are still complaints like headset not fitting all head shapes well, electrodes not recording from the claimed locations on the scalp, or the recording device being too cumbersome and uncomfortable to wear for long-term monitoring. There is still a need to put more emphasis on designing the optimal headset that can be conveniently used for daily life EEG monitoring.

Electrode-skin impedance

It is an important factor of system performance and design analysis. Electrode is the first component of the signal acquisition chain. Primary concerns for electrodes design are fabrication material, polarization voltage, electrode–tissue impedance (ETI) and user comfort. Electrode design for a wearable EEG monitoring system can be grouped into the following categories: metal-plate electrodes (long-term), disposable foam-pad electrodes (low cost), metallic suction electrodes (no strap), floating electrodes (minimize motion artifacts), flexible electrode (comfortable), and internal needle electrode (subdermal).

Elastic and flexible metal/polymer electrodes with pins sliding through hair ensure high-quality scalp contact for a wearable scalp EEG system. Gold (Au), platinum, and Ag/AgCl are commonly used electrode fabrication materials. Gold and platinum electrodes are expensive but are robust and easy to maintain. As polarizable electrodes (capacitive), gold electrodes provide a good signal quality at frequencies above 0.1 Hz (not suitable for dc recordings). Ag/AgCl electrodes can be used for dc recordings. Electrodes made of tin or stainless steel usually suffer from varied degrees of polarization, baseline drift, low-frequency noise, and high resistance so they are less preferred for high-quality EEG recording. Noninvasive body surface electrodes can be divided into three categories: wet, dry contact, and dry noncontact. Wet electrodes confirm an easy and low-impedance conversion between ionic current and electron current. A dry contact electrode does not need to be glued with gel, but it results in high contact impedance. Dry noncontact electrodes lead to even higher electrode-skin impedance and are more prone to motion artifact as they isolate the electrode and skin by capacitive coupling (Kam et al., 2019).

Electrodes used for EEG recordings in clinical environment are typically made of silver/silver-chloride (Ag/AgCl) that are glued with the scalp through electrolytic gel. The electrolyte bounds the ionic current flow from the scalp and the electron flow in the electrode and also increases grip of the electrode to the scalp. The scalp is frequently cleaned and skin on the scalp is abraded to facilitate good ETI.

A traditional EEG model offers a typical impedance of order of 10 kΩ at frequency of interest below 100 Hz provided that the skin is prepared properly and electrodes are glued well. The contact potentials of the interface remain steady up until the electrolyte starts to dry out. The abrasion process and the use of conductive gel is a lengthy process that requires expert assistance and makes daily life EEG monitoring inconvenient and messy. Recent research and development in the field of wearable EEG systems are focusing on sensors that do not require conductive gel and skin preparation.

Quasar, imec, gTech, Mindo, and Cognionics have provided some solutions to the design of a wearable wireless EEG system. Electrodes used are either Ag/AgCl or gold plated. Electrodes with spring-loaded pins are sometimes used to provide comfort to the user along with a good electrode-skin contact.

Such dry EEG sensors can be reused by the time their top conductive layer does not wear off. Despite offering short setup time and improved user comfort, they somewhat compromise on signal quality as compared to ambulatory EEG monitoring devices. The decrease in signal quality is caused by the fragile and complex ETI caused due to using dry electrodes in the absence of the electrolyte that complicates the transition of ionic tissue currents to electrode electron currents much more complex. In real-life monitoring, where a person is free to perform daily life activities while wearing an EEG monitoring system, the artifacts curtailing from movements can be significantly amplified. Monitoring contact properties in terms of ETI can be of great significance in approximating the EEG signal quality. To efficiently replace gel electrodes by dry ones in the future, we need to work more on ETI improvement research methods.

Various electrode technologies used in wearable EEG devices

Predominantly, new electrode technology with lower-power electronics is a major imminent research challenge for the design of such devices. Typical old school EEG devices have their dependence on gel electrodes that can enhance reliability at the cost of inconvenience and discomfort. As the gel eventually dries out resulting in degraded recording quality. This need for electrode replacement prevents long-term and continuous EEG monitoring using traditional EEG systems, especially when number of electrodes used for analysis is large.

Dry electrodes

Dry electrodes employed with rigid conductive metal pins can penetrate the hair (as shown in Fig. 9.4) and facilitate long-term EEG monitoring but at the cost of discomfort and pain. Silver-coated polymer bristles, dry foam electrodes, polymer electrodes can provide soft contact to the skin while still providing low electrode impedance are some capable and promising substitutes for wearable EEG systems.

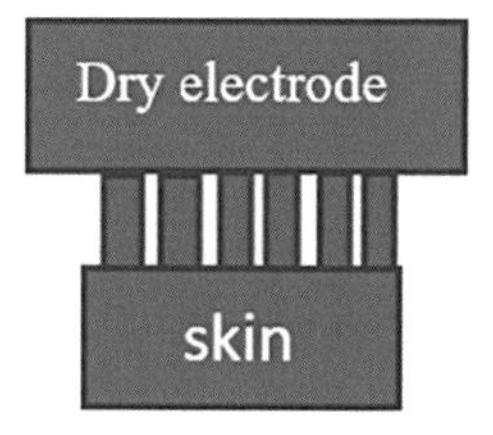

FIG. 9.4 Dry electrode with microneedles to make contact with skin. *No Permission Required.*

Clinical practice for EEG acquisition is still considered as gold standards leading to prevention in widespread use of wearable EEG devices. Dry electrode technology is the latest revolution in wearable brain monitoring that offers increased user comfort, short setup time without expert assistant, and has the potential to change the current scenario of exploiting brain activity monitoring devices for real-time daily life monitoring and medical applications (Fonseca et al., 2007). Dry electrode sensors come with many shortcomings in terms of increased susceptibility to noise and interference and delicateness of the skin-electrode interface, but this technology enables EEG acquisition outside controlled laboratory environments at one's own pace. Soon after the research models signifying the potential of dry EEG monitoring, the first commercial devices appeared on the market are NeuroSky and Quasar. Invention of such devices cemented the way for many new dry electrode solutions entered the market recently like in Fig. 9.5.

Dry electrodes enable a faster setup time and greater user comfort by solving this problem as the eliminating the need for gel but at the expense of increased electrode-skin impedance. Dry-electrode impedance falls into a range from a few hundreds of kilohms to a few tens of megaohms leading to a significant increase in environmental noise and interference. Flat electrodes are used for forehead regions that are not covered with hairs, and a metal pin structure is used for hairy areas. The Spring-like mechanism is introduced within pins or on the top of electrodes to facilitate flexibility and comfort. Flat electrodes with conductive polymer coating would be better option for areas not covered with hairs.

None of the existing dry electrode solutions sustained long-term use assessments. Users have raised complain regarding discomfort after wearing the latest imec's wireless EEG headset with the best suited headset size, flexible polymer electrodes, and the spring electrode support mechanism. Complaints are often raised after wearing the headset for about an hour. New fabrication constituents used in the manufacturing of electrodes are usually not tested for toxicity and infection on the skin. This is particularly the case with new polymer electrodes containing carbon material that coating layers on the tips of the pins may get in contact with damaged skin or tissue.

Risk minimization and analysis procedure is required before deploying dry electrodes in wearable products and especially before any medical use. Toxicity analysis must be done before the use of any of the newly developed electrodes on human.

FIG. 9.5 g.SAHARA's dry electrode. *From Dry electrode. (n.d.). Retrieved August 14, 2021, from https://www.gtec.at/product/g-sahara-hybrid-eeg-electrodes/.*

Quasidry electrodes

In current technology, a new electrode concept called quasidry electrodes is also proposed to reduce the electrode impedance to a few tens of kilohms. Notion amid "wet" and "dry" electrodes hydrates the local scalp area by releasing a small amount of moisturizing solution from the electrode reservoir (Mota et al., 2013).

Active electrodes

They are progressively being employed in wearable healthcare and lifestyle applications. Active electrodes have inbuilt readout circuitry that amplifies and buffers EEG signals before driving any cabling (Fonseca et al., 2007); hence, they are robust to environmental interference and low output impedance of an AE alleviates cable motion artifacts, thus enabling the use of high-impedance dry electrodes for greater user comfort. Latest innovations in wearable device technologies, integrated circuits, sensors, and data analysis techniques have fast tracked the development of wearable EEG technology for brain signal monitoring applications. These innovations integrate low-power and small-size electrodes into a wearable accessory that detects, process, and send physiological information of an individual while performing daily life activities.

Active sensors perform on chip signal amplification immediately after electrode skin-contact minimizing the impact of environmental noise on dry electrode output signals. Pre-amplifier is cointegrated with the electrode itself with minimal wire routing in between. They offer high input impedance and low output impedance, ensuring that the high impedance due to dry electrodes do not affect the EEG signal quality and also minimizing the artifacts due to cable motion and power line interference. Active sensors provide good signal quality over the passive electrode but at the expense of a more complex and cumbersome sensor unit and system design. Hence, contraction and wiring optimization are a fundamental part of an active sensor system design.

Printed electronics

EEG electrode caps are somewhat difficult to wear, that is, why they are suited for laboratory use only. Devices made of printed electronics are more comfortable and easier to wear and allow users to continue their daily activities (Inzelberg & Hanein, 2019). The electrodes are printed on a flexible material using silver ink that makes long-term EEG monitoring more convenient and comfortable. The idea behind printed electrodes EEG is that long-term EEG recordings are uneasy for patients. The electrodes in any conventional EEG device are either pasted to the scalp or held in place with an elastic fixing strap or a cap and also conductive gel is applied to confirm low contact impedance between the scalp and electrodes. Longer recording sessions with such an uneasy setup can also cause tension headaches. A printed EEG monitoring system can be worn on the skin quite comfortably and for a longer time as printed electronics is a very lightweight and flexible technology.

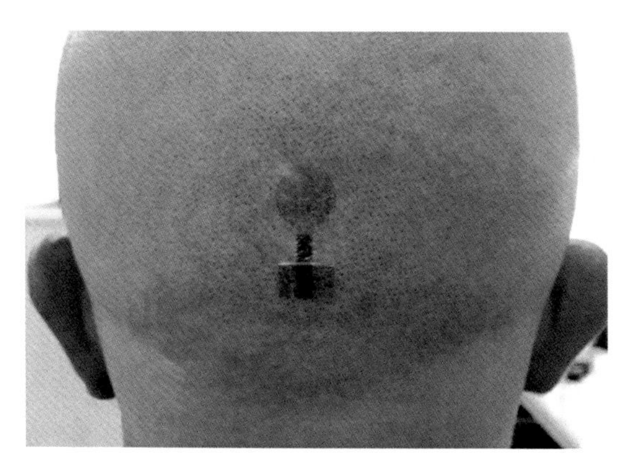

FIG. 9.6 Inkjet printed tattoo electrode printed on skull. *From Ferrari, L. M., Ismailov, U., Badier, J. M., Greco, F., & Ismailova, E. (2020). Conducting polymer tattoo electrodes in clinical electro- and magneto-encephalography.* Npj Flexible Electronics, *4(1). https://doi.org/10.1038/s41528-020-0067-z.*

Inkjet-printed tattoo electrodes

Researcher Francesco Greco has developed temporary ultralight tattoo electrodes that are barely visible on the skin and make long-term measurements of brain activity via EEG inexpensive and easier (Ferrari, Ismailov, Badier, Greco, & Ismailova, 2020). Innovative EEG technology that measures brain potentials as accurate as traditional electrodes using temporary electrodes tattooed on the scalp, as shown in Fig. 9.6. The technology is economical and can be formed using an inkjet printer. They facilitate stable and accurate skin contact for long period of time. Due to inkjet printing and the commercially available substrates this technology is significantly less expensive than current EEG electrodes and also offer more advantages in terms of wearing comfort during long-term monitoring. The tattoo electrodes are made up of conductive polymers that can be printed through a standard inkjet printer. According to the reports of scientific tests, this new technology is found to be as efficient as conventional electrodes, which entail time-consuming applications by professionals. A wet conductive gel required to paste the conventional electrodes to scalp also dries in a few hours, and hence, constant monitoring over a long period of time seems unreliable.

Elegant wearable EEG technology with extraordinary performance eliminates the need to handle multiple electrodes and wires. Gels and adhesive pastes sometimes cause skin irritation, and short circuits between adjacent electrodes and signal quality also worsen as the gel dries. These electrodes provide convenience to the user and achieve high-resolution recording. The material of the ink has to be durable in bending, stretching, flexibility, and can sustain extreme folding without breaking. The electrical conductivity of the ink is important in EEG applications when long traces' resistance values get comparable with the skin-electrode impedance.

Summary

The emerging need for long-term continuous and comfortable brain signal monitoring have encouraged the development of wearable EEG technology for both medical and BCI applications like deep sleep monitoring, seizure detection, mental state analysis, gaming, sports, military use, and so on. In a typical EEG system, a lot of time is spent in placing the electrodes, marking their position, and confirming good conductivity. Depending on the number of measuring electrodes, this may take from a few minutes to an hour.

In this chapter, we have discussed how EEG recording went from being implementable in a medical setting to being accessible in the form of streamlined headbands and hairbands. With the research and advances in wearable technology, EEG that was once only available in a research lab has made its way into our living rooms. First, we have discussed the need of wearable EEG devices and then link these to the applications of portable devices in the field of BCI. We have investigated recent progress in this field and performance validation of novel technology, showing that an inclusive range of research paths are present.

References

Boonyakitanont, P., Lek-uthai, A., Chomtho, K., & Songsiri, J. (2020). A review of feature extraction and performance evaluation in epileptic seizure detection using EEG. *Biomedical Signal Processing and Control*, *57*, 101702. https://doi.org/10.1016/j.bspc.2019.101702.

Casson, A. J. (2019). Wearable EEG and beyond. *Biomedical Engineering Letters*, *9*(1), 53–71. https://doi.org/10.1007/s13534-018-00093-6.

Casson, A. J., & Rodriguez-Villegas, E. (2009). Toward online data reduction for portable electroencephalography systems in epilepsy. *IEEE Transactions on Biomedical Engineering*, *56*(12), 2816–2825. https://doi.org/10.1109/TBME.2009.2027607.

Casson, A. J., Smith, S., Duncan, J. S., & Rodriguez-Villegas, E. (2008). Wearable EEG: What is it, why is it needed and what does it entail? In *Proceedings of the 30th annual international conference of the IEEE engineering in medicine and biology society, EMBS'08—"Personalized healthcare through technology"* (pp. 5867–5870). IEEE Computer Society. https://doi.org/10.1109/iembs.2008.4650549.

Debener, S., Minow, F., Emkes, R., Gandras, K., & de Vos, M. (2012). How about taking a low-cost, small, and wireless EEG for a walk? *Psychophysiology*, *49*(11), 1617–1621. https://doi.org/10.1111/j.1469-8986.2012.01471.x.

DSI series. (n.d.). Retrieved August 15, 2021, from https://wearablesensing.com/.

Duffy, F. H., Albert, M. S., McAnulty, G., & Garvey, A. J. (1984). Age-related differences in brain electrical activity of healthy subjects. *Annals of Neurology*, *16*(4), 430–438. https://doi.org/10.1002/ana.410160403.

EPOC. (n.d.). Retrieved August 15, 2021, from https://www.emotiv.com/.

Ferrari, L. M., Ismailov, U., Badier, J. M., Greco, F., & Ismailova, E. (2020). Conducting polymer tattoo electrodes in clinical electro- and magneto-encephalography. *Npj Flexible Electronics*, *4*(1). https://doi.org/10.1038/s41528-020-0067-z.

Fonseca, C., Cunha, J. P. S., Martins, R. E., Ferreira, V. M., de Sa, J. P. M., Barbosa, M. A., et al. (2007). A novel dry active electrode for EEG recording. *IEEE Transactions on Biomedical Engineering*, *54*(1), 162–165. https://doi.org/10.1109/TBME.2006.884649.

Gilliam, F., Kuzniecky, R., & Faught, E. (1999). Ambulatory EEG monitoring. *Journal of Clinical Neurophysiology*, *16*(2), 111–115. https://doi.org/10.1097/00004691-199903000-00003.

Gu, X., Cao, Z., Jolfaei, A., Xu, P., Wu, D., Jung, T., et al. (2020). EEG-based brain-computer interfaces (BCIs): A survey of recent studies on signal sensing technologies and computational intelligence approaches and their applications. *IEEE/ACM Transactions on Computational Biology and Bioinformatics*, *18*(5), 1645–1666.

Inzelberg, L., & Hanein, Y. (2019). Electrophysiology meets printed electronics: The beginning of a beautiful friendship. *Frontiers in Neuroscience*, *12*, 992. https://www.frontiersin.org/article/10.3389/fnins.2018.00992.

Kaiser, V., Bauernfeind, G., Kreilinger, A., Kaufmann, T., Kübler, A., Neuper, C., et al. (2014). Cortical effects of user training in a motor imagery based brain-computer interface measured by fNIRS and EEG. *NeuroImage*, *85*, 432–444. https://doi.org/10.1016/j.neuroimage.2013.04.097.

Kam, J. W. Y., Griffin, S., Shen, A., Patel, S., Hinrichs, H., Heinze, H. J., et al. (2019). Systematic comparison between a wireless EEG system with dry electrodes and a wired EEG system with wet electrodes. *NeuroImage*, *184*, 119–129. https://doi.org/10.1016/j.neuroimage.2018.09.012.

Mahmood, M., Kwon, S., Kim, H., Kim, Y. S., Siriaraya, P., Choi, J., et al. (2021). Wireless soft scalp electronics and virtual reality system for motor imagery-based brain–machine interfaces. *Advanced Science*. https://doi.org/10.1002/advs.202101129.

Matthews, R., McDonald, N. J., Hervieux, P., Turner, P. J., & Steindorf, M. A. (2007). A wearable physiological sensor suite for unobtrusive monitoring of physiological and cognitive state. In *Annual international conference of the IEEE engineering in medicine and biology—Proceedings* (pp. 5276–5281). https://doi.org/10.1109/IEMBS.2007.4353532.

Mihajlovic, V., Grundlehner, B., Vullers, R., & Penders, J. (2015). Wearable, wireless EEG solutions in daily life applications: What are we missing? *IEEE Journal of Biomedical and Health Informatics, 19*(1), 6–21. https://doi.org/10.1109/JBHI.2014.2328317.

Mota, A. R., Duarte, L., Rodrigues, D., Martins, A. C., Machado, A. V., Vaz, F., et al. (2013). Development of a quasi-dry electrode for EEG recording. *Sensors and Actuators, A: Physical, 199*, 310–317. https://doi.org/10.1016/j.sna.2013.06.013.

Neurosky. (n.d.). Retrieved August 19, 2021, from http://neurosky.com.

Toga, A. W., & Thompson, P. M. (2001). Maps of the brain. *The Anatomical Record, 265*, 37–53. https://doi.org/10.1002/ar.1057.

Wolpaw, J. R., Birbaumer, N., McFarland, D. J., Pfurtscheller, G., & Vaughan, T. M. (2002). Brain–computer interfaces for communication and control. *Clinical Neurophysiology*, 767–791. https://doi.org/10.1016/s1388-2457(02)00057-3.

CHAPTER

10

Automatic epileptic seizure detection based on the discrete wavelet transform approach using an artificial neural network classifier on the scalp electroencephalogram signal

Pragati Tripathi[a], M.A. *Ansari*[j,†], *Faijan Akhtar*[b,†], *Md Belal Bin Heyat*[c,d,e,†], *Rajat Mehrotra*[f], *Akhter Hussain Yatoo*[g], *Bibi Nushrina Teelhawod*[h], *Ashamo Betelihem Asfaw*[h], *and Atif Amin Baig*[i]

[a]Department of Electrical Engineering, Gautam Buddha University, Greater Noida, UP, India [b]School of Computer Science and Engineering, University of Electronic Science and Technology of China, Chengdu, Sichuan, China [c]IoT Research Center, College of Computer Science and Software Engineering, Shenzhen University, Shenzhen, Guangdong, China [d]International Institute of Information Technology, Hyderabad, Telangana, India [e]Department of Science and Engineering, Novel Global Community Educational Foundation, Hebersham, NSW, Australia [f]Department of Electrical & Electronics Engineering, GL Bajaj Institute of Technology & Management Greater Noida, UP, India [g]School of Mathematical Sciences, School of Electronics Science and Engineering, University of Electronic Science and Technology of China, Chengdu, Sichuan, China [h]School of Information and Software Engineering, University of Electronic Science and Technology of China, Chengdu, Sichuan, China [i]Faculty of Medicine, Universiti Sultan Zainal Abidin, Kuala, Terengganu, Malaysia [j]Department of Electrical Engineering, School of Engineering, Gautam Buddha University, Greater Noida, UP, India

[†]Authors have equal distribution.

https://doi.org/10.1016/B978-0-323-99031-8.00012-0

Introduction

An epileptic seizure is a common neurological disorder represented comprehensively by recurrent and unexpected disruptions in the brain. Epilepsy is one of the most general neural disorders that affect 0.60.8% of the world's population (Guo, Rivero, Dorado, Rabuñal, & Pazos, 2010). Human suffering from epilepsy is two or three times more likely to die at an early age than a healthy human. Hence detecting and predicting epileptic seizures is precisely seen as crucial to bring more efficient precaution and cure for the patients. Electroencephalogram (EEG) is a vital dimension of the brain's electrical function obtained by the cerebral cortex's nerve cells; it has been a crucial clinical tool for epilepsy estimation and cure. As a result of the complicated interdependence of millions of neurons, the recorded EEG signals are complex, nonlinear, nonstationary, and casual in behavior (Mormann, Andrzejak, Elger, & Lehnertz, 2007). Recently, many signals analyzing and processing approaches have been examined for epilepsy detection. In general words, the mechanism of these techniques can be described as three tracks, namely, preprocessing, feature extraction, and classification. In the preprocessing phase, the artifact detection and removal are processed (Hasib, Nayak, & Huang, 2018). Feature extraction aims to acquire the sensible aspects concealed in EEG signals, which precisely command the concluding classification precision. Therefore feature extraction acts significantly in pattern recognition. Various approaches have been applicable in EEG feature extraction, namely, spectral analysis, wavelet transform (WT), nonlinear dynamics analysis (Alhagry, Aly, & El-Khoribi, 2017; Tripathi, Acharya, Sharma, Mittal, & Bhattacharya, 2017). By the diversity of methods, WT is most frequently used in epilepsy detection because of its diversified potential in characterizing by multiresolution and elaborated arrangement.

WT gives the aspects of time and frequency, making it desirable and accurate to grab and restrict temporal features in the data set like epileptic spears (Direito, Ventura, Teixeira, & Dourado, 2011). Presently, scientists endeavor to examine the potential of Discrete Wavelet Transform (DWT) analysis in the field of EEG signals processing. Yet, DWT finds to be a very efficient tool and has been successfully utilized in seizure detection (Donos, Maliia, Dümpelmann, & Schulze-Bonhage, 2018; Li, Zhou, Yuan, & Liu, 2013). Scientists proclaimed that the combined form of WT and FFT for EEG classifications is rarely experienced. In this paper, WT combination with FFT is received as the feature extractor. Classification is also another aspect of signal analysis. The classifier artificial neural networks (ANNs) are broadly applicable in epileptic seizure classification. KNN is one of the most well-known techniques due to its peculiar characteristics of self-learning, flexibility, durability (Gadhoumi, Lina, & Gotman, 2012; Moghim & Corne, 2011). An array of ANN-based techniques has been investigated in the literature survey on epilepsy detection. Tzallas, Tsipouras, and Fotiadis (2007) have utilized time-frequency approaches and ANN in applicative seizure identification. Ghosh-Dastidar, Adeli, and Dadmehr (2007) elaborated an approach for the classification of epileptic seizures utilizing mixed band feature space and back propagation neural network. *Lai* and *Siddiquigroups* (Bin Heyat, 2016; Bin Heyat, Akhtar, Ammar, Hayat, & Azad, 2016; Bin Heyat, Akhtar, & Azad, 2016; ; Bin Heyat, Akhtar, Singh, & Siddiqui, 2017; Bin Heyat et al., 2020a, 2020b, 2020c, 2020d; Bin Heyat, Lai, Khan, & Zhang, 2019Hasan, Bin Heyat, Siddiqui, Azad, & Akhtar, 2015; Lai, Bin Heyat, Khan, & Zhang, 2019; Lai, Zhang, Zhang, & Bin Heyat, 2019; Lai, Zhang, Zhang, Su, & Bin Heyat, 2019; Heyat et al., 2017; Mehdi et al., 2016; Pal et al., 2020) used the physiological signals to detect the sleep disorders based on power spectral density and machine learning techniques. Hasan et al. (2015) have approached analysis for classification of EEG signals combining with time and frequency analysis and leman network by the existing literature; it has been acquired that the classification of a network is more desirable to be materialized in most of the research work. It is rare of preexisting techniques related to a combination of neural network and DWT and FFT probability for EEG signal processing. In this research work, an original scheme utilizing DWT and FFT-based ANN approaches has been proposed to determine normal EEG and epileptic signals.

The EEG is the soundtrack of electrical activity within the mind (Bin Heyat et al., 2020c; Bin Heyat, Hasan, & Siddiqui, 2015; Bin Heyat, et al., 2019; Bin Heyat & Siddiqui, 2015; Guragai, Alshorman, Masadeh, & Bin Heyat, 2020). In most of the situations it is formed through electrodes that are placed directly on the brain from indicator probes which is implanted into the mind. Seizures are the result of a certain rise in frequency bands. The most recent work emphasizes five waves of the EEG signal generated in the brain as Delta wave, which has a frequency range 1–4 Hz, which occurs in the duration of sound slumber, through the early period and in very critical mind disease. Theta wave has a frequency range from 4 to 8 Hz. Alpha waves ranging from 8 to 13 Hz are in a sense and have a quiet rest. The persons who do not have any mental stress mean that is most common in ordinary people (Akhtar et al., 2020; Alshorman et al., 2021). Beta waves ranging from 13 to 30 Hz, people when they are in shock or alert situation. The person when he is in deep concentration. If a person stimulates in an alert manner, it will generate 30–60 Hz of waves.

The proposed model is applicable for the prediction of epileptic seizures and applying machine learning classifiers. It is used to detect performance evaluation. It is used for the detection of the type of disease and the stage of the disease. Based on the fundamental statistical features, it is used to classify accuracy. The present research is used to detect brain disorders and monitor the brain signals by detecting seizures. It is used to define the phase of a seizure by applying deep learning algorithms. The time-domain signals can be predicted and classified.

ANN is the tried replica of the human brain; an interconnected neural network is formed. This network system is self-updateable to surrounding parameters (Amato et al., 2013; Lopez-Garcia, Coronado-Mendoza, & Domínguez-Navarro, 2020). It is very much advanced and robust for complex linear mapping. However, it still has many weak points in the Back Propagation algorithm as it has a local minimum that is difficult to tackle. The core methodology of ANN is to factorize k number of classes in k diversity with two classifications. So, trying to mimic the human brain, ANN also tries to solve some complex problems in a much simplest manner. To elaborate on this, ensemble networking can be explained. In this, boosting and bagging are utilized to work together. In the solving technique, bagging tactic covers up the bagging mechanism. Here bagging is conducted for unsteady learning troubles. Considering this, bagging is studied in this study. ANN consists of a bundle of interconnected networks, every one of which holds subnets (P). This entire sample gives a set of priority classes belonging to particular group results. All the described groups are featured in the final predictions. In the present study, we detect normal EEG signals and epileptic EEG signals. Then evaluated the parameters of feature extraction and classification. The feature extraction has been performed by DWT and FFT and a combination of both. For classification, ANN has been used. The optimization technique used is PCA, which gives better accuracy. As per our knowledge, this is the first time that the combination and comparative analysis has been performed. The main contribution of this paper is.

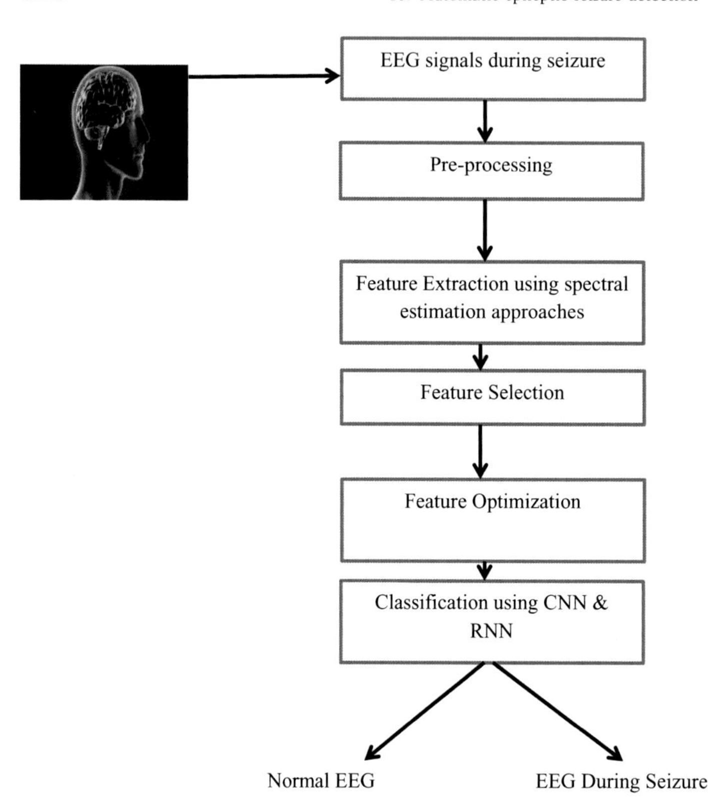

FIG. 10.1 Block diagram of the proposed study.

1. EEG signal analysis for the automatic detection of an epileptic seizure.
2. Automated classification and optimization of the epileptic seizure.
3. ANN for the automatic classification of the EEG signal.

This analogy is synchronized to each connection without each other's interference. The block diagram of the proposed model has been illustrated as follows in Fig. 10.1.

Materials and methods

The basic methodology used in the paper is the data that is first being extracted, and then classification performed using an ANN. The data has been taken from the open-access data set and used for the frequency 173.61 Hz. The data set and the methodology are given in the following.

Data set

The data set applied in the research has been taken from the open-access data at the Department of Epileptology, University of Bonn (www.meb.uni-bonn.de/epileptologie/science/physik/eegdata.html). The entire database contains 100 single-channel easy sections of a time duration of 23.6 s collected at the frequency rate of 173.61 Hz. The whole EEG signals were recorded by the 128-channel amplifier system. In this research work, to estimate the pursuance of the presented technique, the EEG signals are divided into two subsets one is normal data set, which is being taken by the five healthy human beings with their eyes open, and another one is collected from the patients having epileptic seizures, which is described as normal and epileptic data, respectively.

Proposed methodology

In the analysis here, the reprocessing system having low-pass filter has been implemented. Feature abstraction was applied for this study is DWT and Fast Fourier Transform based analysis. For the DWT, five structures were taken into account and three features were for FFT and used to input the ANN (Lai, Zhang, Zhang, Su, & Bin Heyat, 2019).

Discrete wavelet transform

DWT may be termed signal mirroring as it does not alter the content. It works with improvisation in the algorithm. Traditional Fourier transformation is referred to with this technique. Due to this, the basic lacking of WT is eliminated (Zhang et al., 2020). DWT can optimally perform with time-frequency-based signals with different resolutions. It can handle particular time and frequency details for high and low frequencies, respectively. This technique is a part of wavelet transformation. The second type is continuous wavelet transformation (CWT); as the name describes, CWT is not working for a particular domain, and DWT is more specific regarding signal processing. A particular case of DWT is CWT. CWT can be expressed by the following mathematical equation:

$$WT(a,b) = \int_{-\infty}^{\infty} x(t)\left(\frac{1}{\sqrt{a}}\right)\psi\frac{(t-b)}{a}dt \tag{10.1}$$

The above equation shows the Wavelet function ψ, scaling a, shifting b parameters, and signal x (t) is shown in Eq. (10.1). This fraction is unstable, so a and b need to be brought to an even number. This will result in the DWT shown here.

$$\psi(x) = \sum\nolimits_{k=-\infty}^{\infty}(-1)^k a_{N-1-k}\psi(2x-k) \tag{10.2}$$

Here "a" and "b" are substituted by 2^i2^i and 2^i2^ik.

Table 10.1 shows the algorithm of the proposed method using the EEG signal.

In biomedical engineering, DWT here can handle the precise domain of time and frequency aspects. This pulverizes the EEG signal in multiple frames. This work in stages in level by level for the alignment of coefficients. For DWT, these coefficients have great importance, which depends on wavelet function. The current research shows that EEG signals can be processed in five levels of order 4 (db4).

Classification method

To solve a problem, it is the most challenging task for intelligent entities. A machine is entitled to perform new tasks. Like human beings' machines can adapt to unique circumstances and capability of learning from the information stored. Machine learning approaches comprise of ANN, perceptron, metaheuristic, and swarm intelligence. Likewise, living being's neural networking, a machine also attain sense to take decision by analyzing the given data set. This

TABLE 10.1 Algorithm of the proposed method using the EEG signal.

Algorithm of decomposition of EEG signals into IMF
Step-1 Read EEG signal, x_i
Step-2 $e_0 = x_i$
Step-3 Identification of maximum and minimum of x_k, $K = K + 1$
Step-4 set upper envelopes, u_k and lower envelope, l_k
Step-5 Calculate the mean of the envelopes $m_k = [u_k + l_k]/2$
Step-6 Calculate $C_k = C_k - 1 - m_k$
Step-7 If C_k is an IMF: $I = i + 1$ $D_i = C_{k,\ xi} = x_i - 1 - x_i$ Go to step 2, otherwise End
Step-8 If more function cannot be extracted, residence $R_i = X_i$
Step-9 Define target data for ANN classification
Step-10 Choose a training function trainer, which is usually the fastest
Step-11 Create a pattern recognition network
Step-12 Setup division of delta for training, validation, testing
Step-13 Train and test the network
Step-14 Train the network using ANN
Step-15 View the network
Step-16 The validated results are displayed
Step-17 END

gets progressed under A.I. Today's era is using machine learning to design intellectual machinery. This kind of technique trillion commands in a fraction of seconds. Machine learning is used to upgrade the machinery using some basic learning techniques, as stated here.

A supervised learning approach is employed for various purposes. In supervised learning, the data set is generated by all introduced inputs. This process magically works on the strategy of regression. In regression, we can find the statistical relationship between different variables. Reinforcement learning is that in which data upgrade it according to the surrounding action. In this learning, there is no supervisor and controls data directly. An ML algorithm can learn from the framework to calibrate the system for upcoming situations and classified for delta or gradient descent rules. It is a supervised approach of learning varying on input sets. Just like the artificial neural networking to function like a human brain, ANN is brought to functioning.

The mathematical interpretation of a type of machine learning algorithm is as follows:

$$V_k(z, w) = \sum\nolimits_{j=0}^{n} z_j w_{jk} \tag{10.3}$$

The sigmoid function in respect of output function can be illustrated as

$$f_k(\overline{z}, \overline{w}) = \frac{1}{1 + e^{V(\tau, w)}} \tag{10.4}$$

Therefore neural error description is illustrated in the following:

$$R_k\left(\underline{z}, \underline{w}, d\right) = \left\{f_k\left(\underline{z}, \underline{w}\right) - d_k\right\}^2 \tag{10.5}$$

where d_k represents the kth expression expected and integrated error is given in the following:

$$R_k\left(\underline{z}, \underline{w}, d\right) = \left\{f_k\left(\underline{z}, \underline{w}\right) - d_j\right\}^2 \tag{10.6}$$

Since $\Delta w\beta - \frac{\partial R}{\partial w'}$ the complete error is decreased through the application of the gradient descent approach. By working with delta rule, derivation is made on error partially as

$$w_{jk} = -\eta \frac{\partial R}{\partial w_{jk}} \tag{10.7}$$

$$\frac{\partial R}{\partial f_k} = 2(f_k - d_k) \tag{10.8}$$

$$\frac{\partial R}{\partial w_{jk}} = \frac{\partial f_k}{\partial V_k}\frac{\partial V_k}{\partial w_{jk}} = f_k\,(1 - f_k) \tag{10.9}$$

From Eq. (10.6) and (10.7), we conclude that

$$\frac{\partial R}{\partial w_{jk}} = \frac{\partial R}{\partial f_k}\frac{\partial f_k}{\partial w_{jk}} = 2(f_k - d_k)f_k\,(1 - f_k)\,z_j \tag{10.10}$$

Here weight calibration is shown using Eq. (10.5)–(10.8):

$$\Delta w_{jk} = -2\eta(f_k - d_k)f_k\,(1 - f_k) \tag{10.11}$$

Results

Preprocessing

It is the approach that can make the extraction procedure easier. These approaches used in EEG signals are reliable on the task of the submissions. Some of the techniques are used generally to enhance the eminence of signal-to-noise ratio, namely, general regular referencing. Resampling the statistics, cleaning, depraved frequency recognition, autonomous element investigation, pooching unremitting statistics, and period refutation are the utmost general approaches in the preprocessing phase of EEG recordings. As we focus on the movable image, we need to deploy data intended for particular electrodes on the sufferer's scalp. The recordings are subjected to the prior cortex brain region. The goals of images can be summarized as unremitting images that can be detected as Mu, beta forms in the EEG recording. Table 10.2 gives a brief analysis of the frequency band of the EEG signal using fifth-level decomposition. In Fig. 10.2, feature extraction of normal and epileptic patients has been depicted with FFT and DWT, where time has been taken on the *x*-axis and amplitude has been taken on the *y*-axis with approximately 214 patients.

TABLE 10.2 shows the frequency band of the EEG signal using the fifth level decomposition.

Decomposition level	Subbands	Frequency range (Hz)
1	Sd1	43.40–86.80
2	Sd2	21.70–43.40
3	Sd3	10.80–21.70
4	Sd4	5.40–10.80
5	Sd5	2.70–5.40

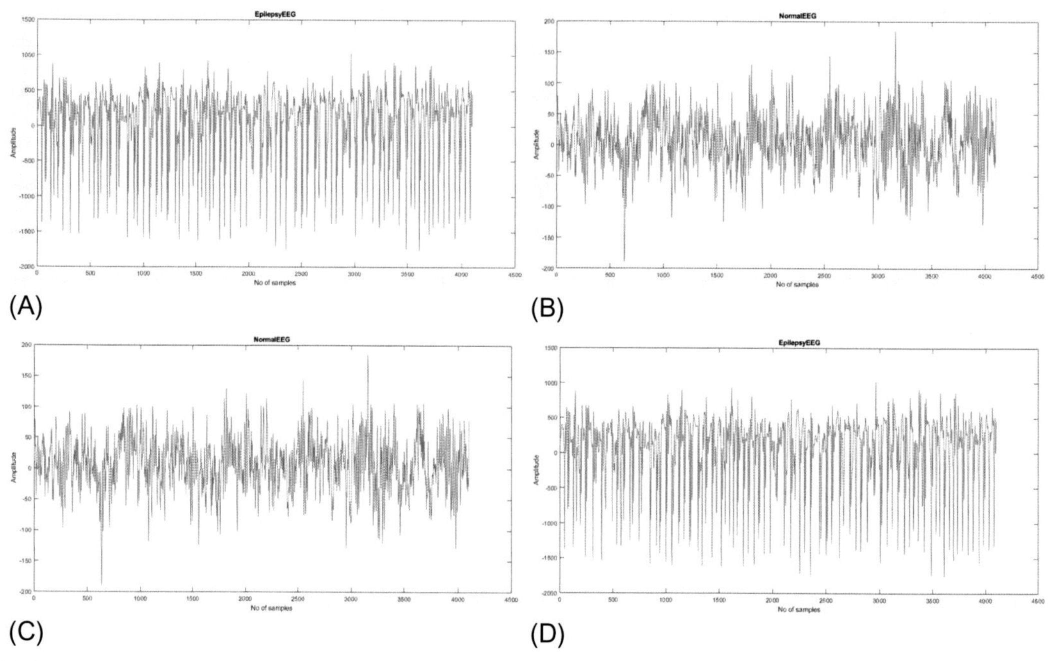

FIG. 10.2 EEG signal processing: (A) normal human with FFT processing, (B) epileptic seizure patients with FFT processing, (C) normal human with DWT processing, and (D) epileptic seizure patients with DWT processing.

Feature extraction and classification

It is also an essential stage in the procedure of EEG signal classification. Here the analysis has been performed to suggest a pattern recognition technique that differentiates EEG signals noted during the period of distinctive behavioral situations. The main focus has been done on the wavelet-based feature extraction in which the multiresolution disintegration into expanded and estimated constraints and relative wavelet energy is calculated. Then the extracted features are normalized to zero mean and unity variance, then optimized using PCA, that is, principal component analysis. The proposed approach (128 channels) by detecting two classifications, that is, ANN, gives the highest accuracy of 97.24%. The planned scheme resulted in crucially higher classification pursuance by applying machine learning classifiers as compared to statistical feature extraction. The consequences give the proposed feature extraction approach, which classifies the EEG signal tuned during behavioral goals with the highest range of precision. Tables 10.3 and 10.4 depict statistical parameters of features through FFT and a variety of features extracted from three data sets, respectively.

Discussion

The analysis used the pattern recognition basis technique to classify EEG signals tuned during the behavioral phase of consciousness. The relative fifth-level energy disintegration results give the highest performances for low frequency (0–3.90 Hz) and above

TABLE 10.3 shows the statistical parameters of features through FFT.

Parameter	Maximum value	Minimum value	Mean value	RMS	SD	Entropy
Normal	185	−190	15.844	113.7	168.15	0.86047
Epilepsy	800	−782	27.117	825.8	920.89	1.0132

TABLE 10.4 shows the variety of features extracted from three data sets.

Data	Feature	Sa5	Sd3	Sd4	Sd5	Raw EEG
Set A	Mean	2.45 ± 0.91	1.29 ± 0.49	1.67 ± 0.68	1.15 ± 30	0.55 ± 0.16
	Energy	$(1.12 \pm 0.75) \times 10^5$	$(1.25 \pm 0.9) \times 10^5$	$(1.11 \pm 0.93) \times 10^5$	$(2.48 \pm 0.13) \times 10^5$	$(2.11 \pm 1.17) \times 10^5$
	Standard deviation	1.19 ± 0.39	0.68 ± 0.22	0.91 ± 0.35	0.64 ± 0.17	0.61 ± 0.18
	Max value	6.59 ± 2.40	3.39 ± 1.31	4.95 ± 1.91	3.30 ± 0.95	2.46 ± 0.59
Set-B	Mean	3.65 ± 3.24	0.48 ± 0.31	1.19 ± 0.75	2.04 ± 1.74	0.51 ± 0.39
	Energy	$(4.08 \pm 0.17) \times 10^5$	$(3.17 \pm 0.68) \times 10^5$	$(7.92 \pm 0.75) \times 10^5$	$(1.38 \pm 4.32) \times 10^5$	$(3.36 \pm 9.09) \times 10^5$
	Standard deviation	1.93 ± 1.75	0.37 ± 0.37	0.78 ± 3.81	1.25 ± 1.26	0.5 ± 0.58
	Max value	9.65 ± 8.37	2.49 ± 2.81	4.32 ± 1.74	6.27 ± 5.67	3.01 ± 2.81
Set- C	Mean	8.16 ± 4.65	4.93 ± 3.33	8.42 ± 4.58	10.63 ± 5.89	2.44 ± 1.16
	Energy	$(1.64 \pm 1.83) \times 10^6$	$(2.48 \pm 3.28) \times 10^6$	$(2.98 \pm 2.94) \times 10^6$	$(2.36 \pm 2.41) \times 10^6$	$(4.77 \pm 4.13) \times 10^6$
	Standard deviation	5.17 ± 2.87	2.93 ± 2.01	4.21 ± 2.09	4.76 ± 2.47	3.06 ± 1.47
	Max value	28.87 ± 17.44	14.88 ± 9.17	21.74 ± 11.12	23.69 ± 12.18	10.67 ± 5.11

(3.90–7.81 Hz), which is an essential feature for EEG classification. These results give the DWT analysis to briefly specify EEG signals and compute the total energy levels for distinctive frequency bands. Additionally, the planned feature selection application minimized noncrucial features from a feature set before giving it to classifiers, which minimizes the computational cost. The results are compared to the previously publicly available data set, which offers the distinctive pursuance levels (precision) gained by statistical techniques to classify two distinctive behavioral events using EEG signals. Fig. 10.3 depicts the best-performance receiver operating characteristic (ROC) of FFT with ANN. Fig. 10.4 shows the confusion matrix of the training, testing, validation, and all, using FFT with ANN. Fig. 10.5 shows the training state of FFT with ANN, in which the epoch is 874. Fig. 10.6 gives the best performance of ROC of the FFT and DWT with ANN. Fig. 10.7 represents the confusion matrix of the training, testing, and validation, using FFT and DWT with ANN. Fig. 10.8 shows the training state of DWT

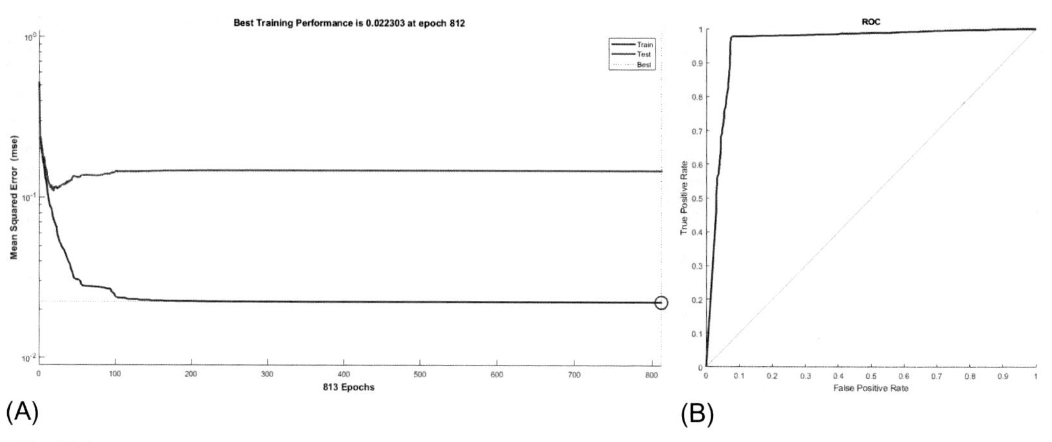

FIG. 10.3 Best performance receiver operating characteristic (ROC) of FFT with ANN.

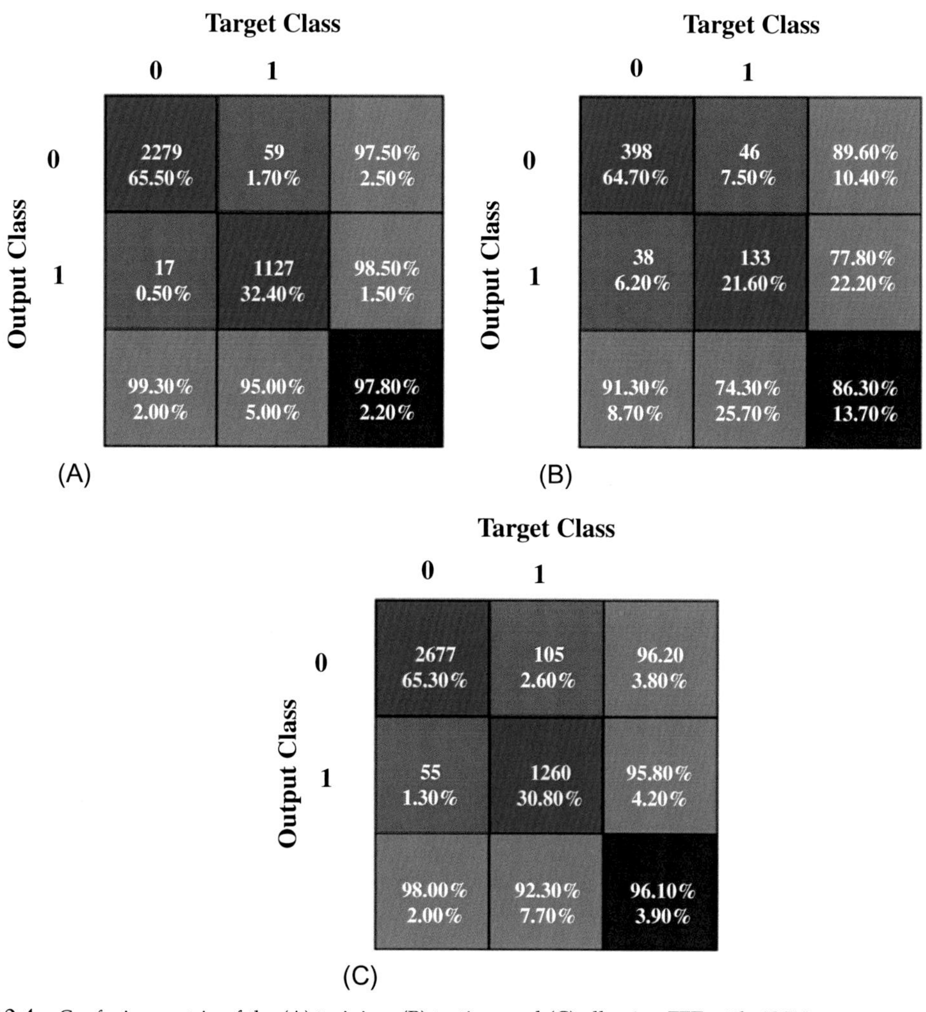

FIG. 10.4 Confusion matrix of the (A) training, (B) testing, and (C) all using FFT with ANN.

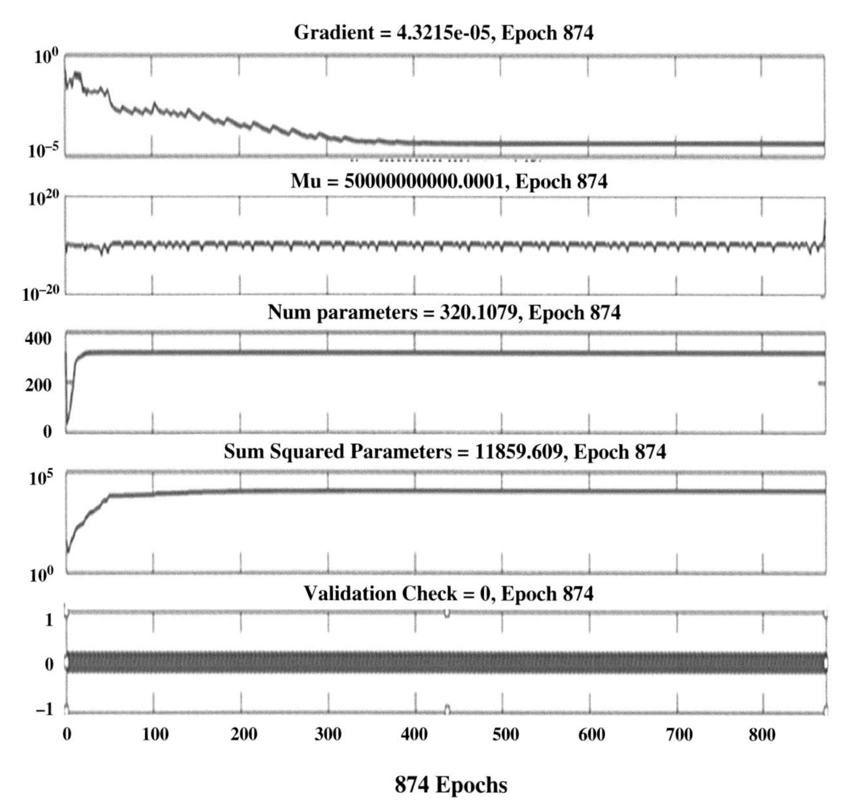

FIG. 10.5 Training state of FFT with ANN in which epoch is 874 (A) Gradient = 4.321e-05, Epochs 874; (B) $\mu = 5 \times 10^{10}$, epoch is 874; (C) numeric parameter is 320.1079, and epochs is 874; (D) validation check is 0, and epoch is 874; and (E) 874 epochs.

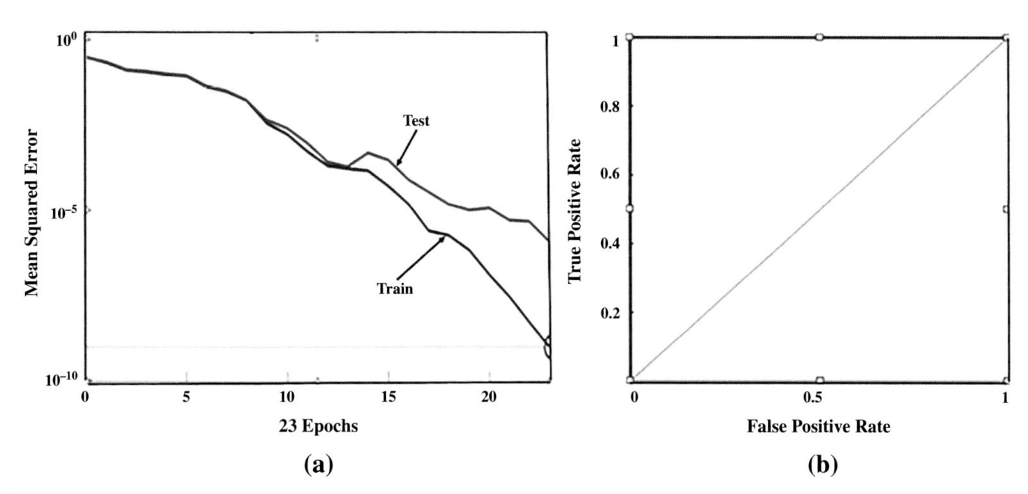

FIG. 10.6 Best performance. (A) ROC of the FFT and (B) DWT with ANN.

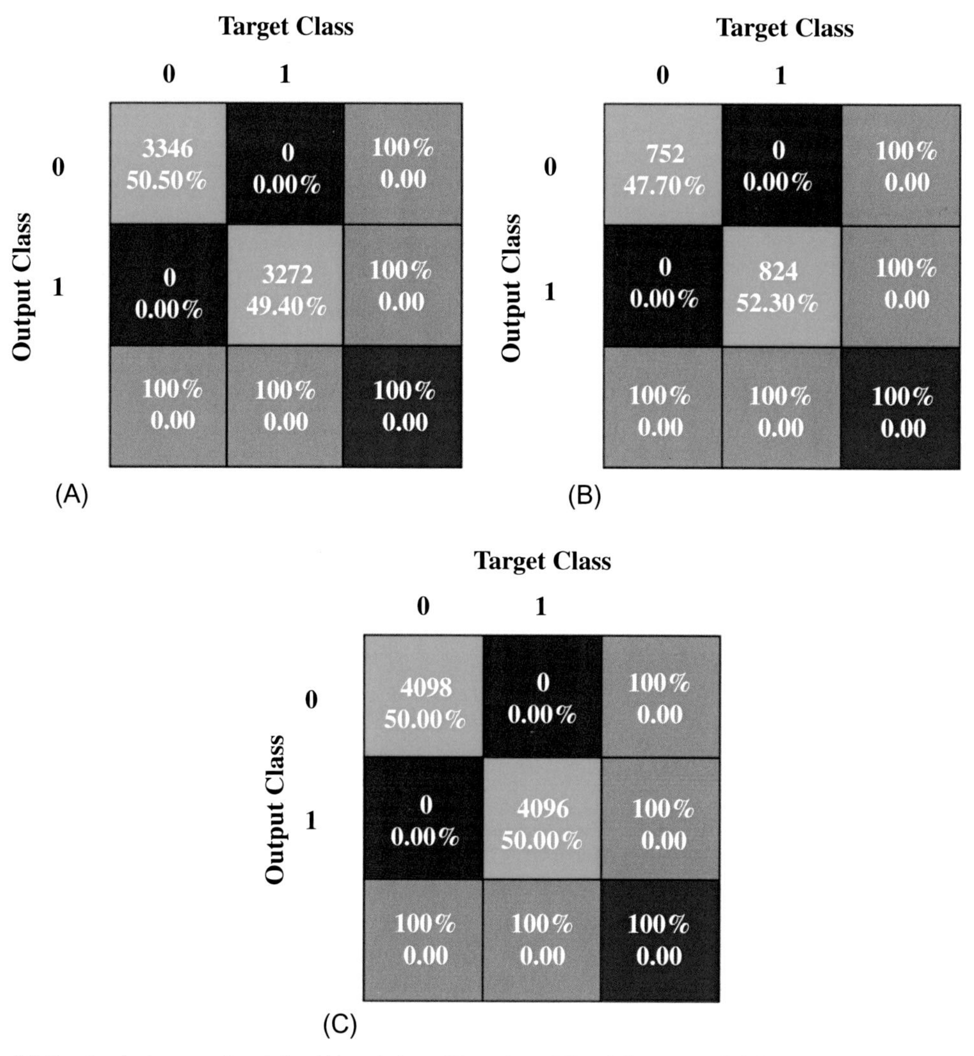

FIG. 10.7 Confusion matrix of the (A) training, (B) testing, (C) validation, and (D) all, using FFT and DWT with ANN.

and FFT with ANN, in which the epoch is 23. Table 10.5 gives the standard parameters taken for EEG data. Table 10.6 provides the classification with accuracy results for DWT. Fig. 10.9 depicts the comparative analysis between average accuracy vs epoch, average time vs epoch, and average accuracy vs average time.

Fig. 10.10 compares the previously selected and proposed system accuracy and previously selected and proposed ANN classifier. Table 10.7 gives the comparative analysis of the previously selected and proposed system accuracy. Table 10.8 shows the comparison between the previously selected and proposed ANN classifier.

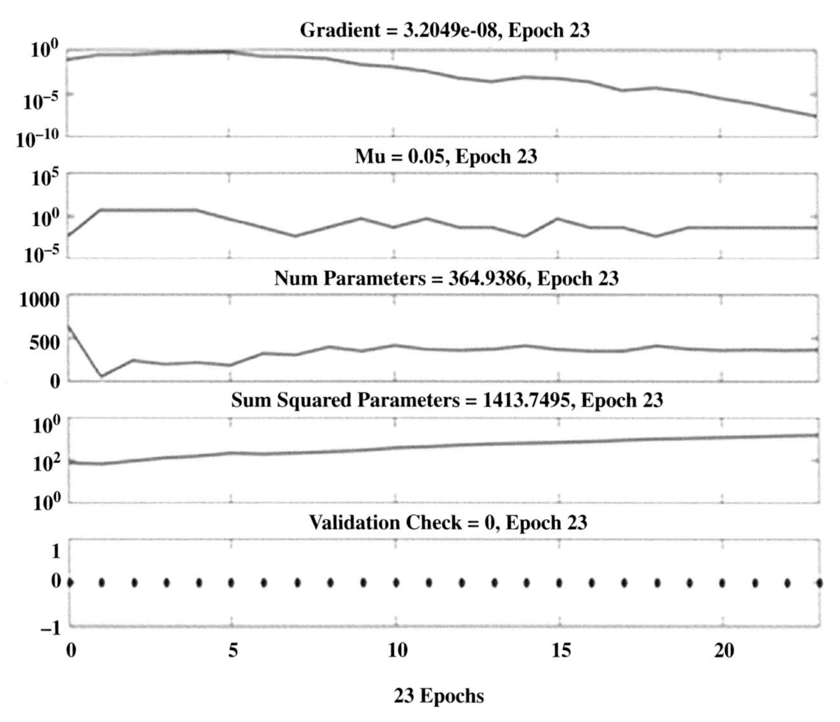

FIG. 10.8 Training state of DWT and FFT with ANN in which epoch is 23. (A) Gradient = 4.321e-05, epoch 874; (B) $\mu = 5 \times 10^{10}$, epoch is 874; (C) numeric parameter is 320.1079, and epoch is 874; (D) Validation check is 0, and epoch is 874; and (E) 874 epochs.

TABLE 10.5 shows the standard parameters taken for EEG data.

Decomposed signal (Haar)	Frequency range (Hz)	Decomposition signal level
D1	43.40–86.8	1
D2	21.70–43.40	2
D3	10.85–21.70	3
D4	5.43–10.85	4
A4	0–5.43	4

TABLE 10.6 Classification accuracy using ANN.

	Accuracy (%)		Time (min)	
Epoch	Overall	Average	Overall	Average
2500	99.8, 100, 99.9	97.2	0:42, 0:43, 0:40	0.41
5000	99.7, 100, 99.9	96.7	1:35, 1:31, 1:30	1.32
7000	99.9, 100, 100	96.9	2:25, 2:30, 2:40	2.31
8000	99.9, 100, 100	98.2	3:30, 3:60, 3:40	3.43
10,000	99.9, 100, 99.9	97.6	3:57, 3:58, 3:59	3.58
Mean		**97.24**	–	**2.21**

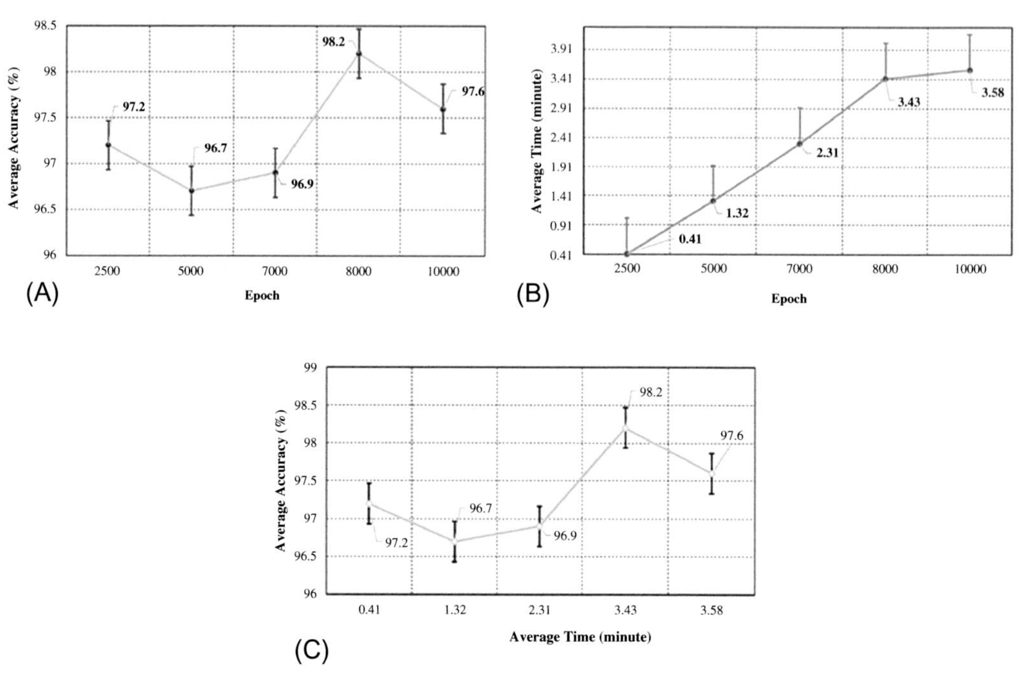

FIG. 10.9 Comparative analysis between (A) average accuracy vs epoch, (B) average time vs epoch, and (C) average accuracy vs average time.

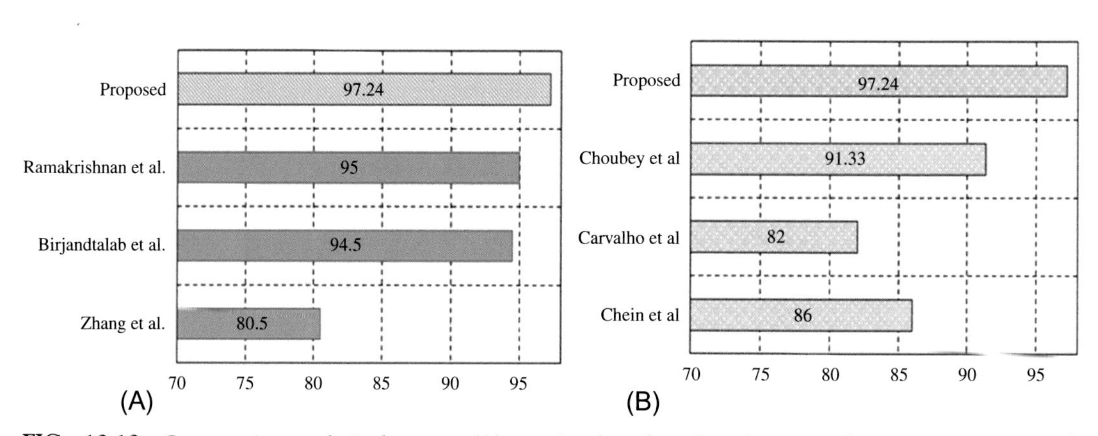

FIG. 10.10 Comparative analysis between (A) previously selected and proposed system accuracy and (B) previously selected and the proposed ANN classifier.

The proposed model is applicable for the prediction of epileptic seizures and applying machine learning classifiers. It is used to detect performance evaluation. It is used for the detection of the type of disease and the stage of the disease. Based on the fundamental statistical features, it is used to classify accuracy. The present research is used to detect the type of brain disorder and monitor the brain signals by detecting seizures. It is used to define the seizure

TABLE 10.7 Comparison analysis of the previously selected and proposed system accuracy.

Reference	Year	Signal	Classifier	Accuracy (%)
Zhang et al. (2020)	2020	EEG	CNN	80.50
Birjandtalab, Pouyan, & Nourani (2016)	2016	EEG	GMM clustering	94.50
Ramakrishnan & Muthanantha Murugavel (2019)	2018	EEG	LDAG-SVM	95.00
Proposed		**EEG**	**ANN**	**97.24**

TABLE 10.8 Comparison between previously selected and proposed ANN classifier.

Reference	Year	Detection	Accuracy (%)
Chien, Hsu, Lee, Sung, & Lin (2021)	2021	Parkinson	86
Carvalho et al. (2020)	2020	COVID-19	82
Choubey & Pandey (2020)	2020	Epilepsy	91.33
Proposed		**Epileptic Seizure**	**97.24**

phase by applying deep learning algorithms; the time domain signals can be predicted and classified.

The limitations of the present study are focused on brain signals. It is unable to detect heart signals, physical signals, insomnia signals, and several other disorders. It is unable to detect the X-rays, EOG, and EMG signals. It can only detect and classify seizures. It cannot be used to identify physiological signals. It cannot be used to classify ECG signals and MRI signals and other tumor detection techniques.

Conclusion

The seizure identification is of paramount medical importance; however, it is not simple to drag off. In this research work, we pursue to examine the potential of DWT using ANN to classify the signals and give the accuracy. The differentiation of the features is crucially enhanced by combining DWT and utilizing it for feature extraction. To design robust outcomes, various feature extracting and detection techniques are taken into consideration. The presented algorithm can discriminate among different families based on the outcomes with a medically important classification accuracy of 97.24%. In further research, we aim to authenticate the outcomes with massive data sets and enhance the precision by improving the system configuration.

Acknowledgments

We are thankful to Prof. Lai, Prof. Naseem, Prof. Singh, Prof. Umar, Prof. Tumrani, Dr. Gul, Dr., Zakaria, Ms. Hashmi, and Ms. Perween for the valuable discussion.

References

Akhtar, F., Bin Heyat, M. B., Li, J. P., Patel, P. K., Pal, R., & Guragai, B. (2020). *Role of machine learning in human stress: A review*. https://doi.org/10.1109/ICCWAMTIP51612.2020.9317396.

Alhagry, S., Aly, A., & El-Khoribi, R. (2017). Emotion recognition based on EEG using LSTM recurrent neural network. *International Journal of Advanced Computer Science and Applications*. https://doi.org/10.14569/ijacsa.2017.081046.

Alshorman, O., et al. (2021). Frontal lobe real-time EEG analysis using machine learning techniques for mental stress detection. *Journal of Integrative Neuroscience*, *21*, 1–11.

Amato, F., López, A., Peña-Méndez, E. M., Vaňhara, P., Hampl, A., & Havel, J. (2013). Artificial neural networks in medical diagnosis. *Journal of Applied Biomedicine*. https://doi.org/10.2478/v10136-012-0031-x.

Bin Heyat, M. B. (2016). *Insomnia: Medical sleep disorder & diagnosis* (1st ed.). Hamburg, Germany: Anchor Academic Publishing.

Bin Heyat, M. B., Akhtar, F., Ammar, M., Hayat, B., & Azad, S. (2016). Power spectral density are used in the investigation of insomnia neurological disorder. In *XL- pre congress symposium* (pp. 45–50).

Bin Heyat, M. B., Akhtar, F., & Azad, S. (2016). Comparative analysis of original wave and filtered wave of EEG signal used in the prognostic of bruxism medical sleep syndrome. *International Journal of Trend in Scientific Research and Development*, *1*(1), 7–9. https://doi.org/10.31142/ijtsrd53.

Bin Heyat, B., Akhtar, F., Singh, S. K., & Siddiqui, M. M. (2017). Hamming window are used in the prognostic of insomnia. In *International seminar present scenario future prospectives Res. Eng. Sci. (ISPSFPRES)* (pp. 65–71).

Bin Heyat, B., Hasan, Y. M., & Siddiqui, M. M. (2015). EEG signals and wireless transfer of EEG signals. *International Journal of Advanced Research in Computer and Communication Engineering*, *4*(12), 10–12. https://doi.org/10.17148/IJARCCE.2015.412143.

Bin Heyat, M. B., Lai, D., Akhtar, F., Ansari, M. A., Khan, A., & Alkahtani, F. (2020a). Progress in detection of insomnia sleep disorder: A comprehensive review. *Current Drug Targets*. https://doi.org/10.2174/1389450121666201027125828.

Bin Heyat, M. B., Lai, D., Khan, F. I., & Zhang, Y. (2019). Sleep bruxism detection using decision tree method by the combination of C4-P4 and C4-A1 channels of scalp EEG. *IEEE Access*, *7*, 102542–102553. https://doi.org/10.1109/access.2019.2928020.

Bin Heyat, M., & Siddiqui, M. M. (2015). Recording of EEG, ECG, EMG signal. *International Journal of Advanced Research in Computer Science and Software Engineering*, *5*(10), 813–815. [Online]. Available www.ivline.org.

Bin Heyat, M. B., et al. (2020b). Detection, treatment planning, and genetic predisposition of bruxism: A systematic mapping process and network visualization technique. *CNS & Neurological Disorders: Drug Targets*. https://doi.org/10.2174/1871527319666201110124954.

Bin Heyat, M. B., et al., Gupta, D., Bhattacharyya, S., & Khanna, A. (Eds.). (2020c). Bruxism detection using single-channel C4-A1 on human sleep S2 stage recording. In *Intelligent data analysis: From data gathering to data comprehension* (1st ed., pp. 347–367). John Wiley & Sons.

Bin Heyat, M. B., et al. (2020d). A novel hybrid machine learning classification for the detection of bruxism patients using physiological signals. *Applied Sciences*, *10*(21), 7410. https://doi.org/10.3390/app10217410.

Birjandtalab, J., Pouyan, M. B., & Nourani, M. (2016). *Unsupervised EEG analysis for automated epileptic seizure detection*. https://doi.org/10.1117/12.2243622.

Carvalho, A. R. S., et al. (2020). COVID-19 chest computed tomography to stratify severity and disease extension by artificial neural network computer aided diagnosis. *Frontiers in Medicine*, *7*(December), 1–11. https://doi.org/10.3389/fmed.2020.577609.

Chien, C., Hsu, S., Lee, T., Sung, P., & Lin, C. (2021). Using artificial neural network to discriminate Parkinson' s disease from other Parkinsonisms by focusing on putamen of dopamine transporter SPECT images. *Biomedicines*, *9*(1), 12. https://doi.org/10.3390/biomedicines9010012.

Choubey, H., & Pandey, A. (2020). A combination of statistical parameters for the detection of epilepsy and EEG classification using ANN and KNN classifier. *Signal, Image and Video Processing*. https://doi.org/10.1007/s11760-020-01767-4.

Direito, B., Ventura, F., Teixeira, C., & Dourado, A. (2011). *Optimized feature subsets for epileptic seizure prediction studies*. https://doi.org/10.1109/IEMBS.2011.6090472.

Donos, C., Maliia, M. D., Dümpelmann, M., & Schulze-Bonhage, A. (2018). Seizure onset predicts its type. *Epilepsia*. https://doi.org/10.1111/epi.13997.

Gadhoumi, K., Lina, J. M., & Gotman, J. (2012). Discriminating preictal and interictal states in patients with temporal lobe epilepsy using wavelet analysis of intracerebral EEG. *Clinical Neurophysiology*. https://doi.org/10.1016/j.clinph.2012.03.001.

Ghosh-Dastidar, S., Adeli, H., & Dadmehr, N. (2007). Mixed-band wavelet-chaos-neural network methodology for epilepsy and epileptic seizure detection. *IEEE Transactions on Biomedical Engineering*. https://doi.org/10.1109/TBME.2007.891945.

Guo, L., Rivero, D., Dorado, J., Rabuñal, J. R., & Pazos, A. (2010). Automatic epileptic seizure detection in EEGs based on line length feature and artificial neural networks. *Journal of Neuroscience Methods*. https://doi.org/10.1016/j.jneumeth.2010.05.020.

Guragai, B., Alshorman, O., Masadeh, M., & Bin Heyat, M. B. (2020). *A survey on deep learning classification algorithms for motor imagery*. https://doi.org/10.1109/ICM50269.2020.9331503.

Hasan, Y. M., Bin Heyat, B., Siddiqui, M. M., Azad, S., & Akhtar, F. (2015). An overview of sleep and stages of sleep. *International Journal of Advanced Research in Computer and Communication Engineering*, *4*(12), 505–507. https://doi.org/10.17148/IJARCCE.2015.412144.

Hasib, M. M., Nayak, T., & Huang, Y. (2018). *A hierarchical LSTM model with attention for modeling EEG non-stationarity for human decision prediction*. https://doi.org/10.1109/BHI.2018.8333380.

Heyat, B. B., Akhtar, F., Mehdi, A., Azad, S., Azad, S., & Azad, S. (2017). Normalized power are used in the diagnosis of insomnia medical sleep syndrome through EMG1-EMG2 channel. *Austin Journal of Sleep Disorders*, *4*(1), 2–4.

Lai, D., Bin Heyat, M. B., Khan, F. I., & Zhang, Y. (2019). Prognosis of sleep bruxism using power spectral density approach applied on EEG signal of both EMG1-EMG2 and ECG1-ECG2 channels. *IEEE Access*, *7*, 82553–82562. https://doi.org/10.1109/ACCESS.2019.2924181.

Lai, D., Zhang, X., Zhang, Y., & Bin Heyat, M. B. (Jul. 2019). Convolutional neural network based detection of atrial fibrillation combing R-R intervals and F-wave frequency spectrum. In *2019 41st annual international conference of the IEEE engineering in medicine and biology society (EMBC)* (pp. 4897–4900). https://doi.org/10.1109/EMBC.2019.8856342.

Lai, D., Zhang, Y., Zhang, X., Su, Y., & Bin Heyat, M. B. (2019). An automated strategy for early risk identification of sudden cardiac death by using machine learning approach on measurable arrhythmic risk markers. *IEEE Access*. https://doi.org/10.1109/access.2019.2925847.

Li, S., Zhou, W., Yuan, Q., & Liu, Y. (2013). Seizure prediction using spike rate of intracranial EEG. *IEEE Transactions on Neural Systems and Rehabilitation Engineering*. https://doi.org/10.1109/TNSRE.2013.2282153.

Lopez-Garcia, T. B., Coronado-Mendoza, A., & Domínguez-Navarro, J. A. (2020). Artificial neural networks in microgrids: A review. *Engineering Applications of Artificial Intelligence*. https://doi.org/10.1016/j.engappai.2020.103894.

Mehdi, S., Bin Heyat, B., Akhtar, F., Ammar, M., Heyat, B., & Gupta, T. (2016). Cure of epilepsy by different system of medicine. *International Journal of Technical Research & Science*, *1*(8), 244–247.

Moghim, N., & Corne, D. (2011). *Evaluating bio-inspired approaches for advance prediction of epileptic seizures*. https://doi.org/10.1109/NaBIC.2011.6089646.

Mormann, F., Andrzejak, R. G., Elger, C. E., & Lehnertz, K. (2007). Seizure prediction: The long and winding road. *Brain*. https://doi.org/10.1093/brain/awl241.

Pal, R., et al. (2020). *Effect of Maha Mrityunjaya HYMN recitation on human brain for the analysis of single EEG channel C4-A1 using machine learning classifiers on yoga practitioner*. https://doi.org/10.1109/ICCWAMTIP51612.2020.9317384.

Ramakrishnan, S., & Muthanantha Murugavel, A. S. (2019). Epileptic seizure detection using fuzzy-rules-based sub-band specific features and layered multi-class SVM. *Pattern Analysis and Applications*. https://doi.org/10.1007/s10044-018-0691-6.

Tripathi, S., Acharya, S., Sharma, R. D., Mittal, S., & Bhattacharya, S. (2017). Using deep and convolutional neural networks for accurate emotion classification on DEAP dataset. In *Proc. Thirty-First AAAI Conf. Artif. Intell.*

Tzallas, A. T., Tsipouras, M. G., & Fotiadis, D. I. (2007). Automatic seizure detection based on time-frequency analysis and artificial neural networks. *Computational Intelligence and Neuroscience*. https://doi.org/10.1155/2007/80510.

Zhang, X., Yao, L., Dong, M., Liu, Z., Zhang, Y., & Li, Y. (2020). Adversarial representation learning for robust patient-independent epileptic seizure detection. *IEEE Journal of Biomedical and Health Informatics*. https://doi.org/10.1109/JBHI.2020.2971610.

CHAPTER

11

Event identification by fusing EEG and EMG signals

Kashif Sherwani and Munna Khan

Department of Electrical Engineering, Jamia Millia Islamia, New Delhi, India

Introduction

One of the major causes of adult disability is stroke. Approximately 20 million people all over the world per year are affected by different types of stroke (MacMahon, 2002; Ward & Cohen, 2004). Among these around 30% are under the age of 65. There are other diseases also, which causes paralysis such as Multiple Sclerosis (MS) and Spinal Cord Injury (SCI). Statistically, around 2.5 million and 12.1–57.8 cases per million are affected by these diseases, respectively (Frohman, O'Donoghue, & Northrop, 2011; Martins, Freitas, Martins, Dartigues, & Barat, 1998; van Asbeck, Post, & Pangalila, 2000). There is an increase in motor disability due to increase in traumatic injuries, strokes, lifestyle, and so on. Assistive devices help to live a normal life, but their interaction with these rehabilitation devices is limited due to limited command control (Sherwani, Khan, & Kumar, 2015). To compensate for the lost or paralyzed limb, there is a major growth in the development of assistive devices such as versatile and lightweight prostheses and orthoses, but still reliable and intuitive control of these devices is a challenge (Atkins, Heard, & Donovan, 1996; Chiri et al., 2012; Iqbal, Tsagarakis, Fiorilla, & Caldwell, 2010; Pons, 2010; Vitiello et al., 2013). Surveys show that around 50% of amputees are having difficulty in operating their prosthetic hand because of low functionality and controllability. Electromyography (EMG) signals are used for the control of these prosthetic devices from the early days (Peerdeman et al., 2011; Zecca, Micera, Carrozza, & Dario, 2002). However, there are lots of practical reasons that affect this control such as effective and fast calibration system, good quality biosignal, and so on (Sherwani, Kumar, Chemori, Khan, & Mohammed, 2020). Adding more sensory systems improves the assistive system function (Weber, Friesen, & Miller, 2012). Also, hybridization of signals yields better and stable results as compared to the singular approach (Jiang, Zhou, Yin, Yu, & Hu, 2014; Sherwani & Kumar, 2016).

Computational Intelligence in Healthcare Applications
https://doi.org/10.1016/B978-0-323-99031-8.00020-X

Electroencephalography (EEG) signal can be used for predictive classification during different dynamic and static conditions (Kashif, Neelesh and Munna, 2018; Velu & de Sa, 2013). During treadmill walking, Fz, Cz, and C4 show high activity in both legs for the knee joint and Cz and C4 for the hip joint (Presacco, Forrester, & Contreras-Vidal, 2012). Anterior, right lateral, and right anterior occipital equally contribute during the activation of the distributed, sparse, and complex cortical network (Presacco, Goodman, Forrester, & Jose Luis, 2011).

In this work, the EEG and EMG signals are taken simultaneously. The main aim is to get the furnished and refined EEG signals. This is achieved by removing almost all the noises and artifacts. EEG gives the higher value of autocorrelation; this property of EEG along with EMG is exploited for validating the results.

Experimental procedure and data analysis

Experimental paradigm

Two dynamic tasks were selected for this experiment, that is, sit to stand and stand to sit. Several muscles are responsible for these movements, but only those muscles were selected which are responsible for the flexion and extension of the knee joints. Rectus femoris, semimembranosus, and gastrocnemius were the three selected muscles. EEG and EMG, both the biosignals, were taken simultaneously. EEG electrodes were placed using the "10–20" electrode placement system (Gómez-Herrero et al., 2006). The scalp positions taken for the recording of electrical activity were Fz, Cz, C4, Pz, P3, and P4. Electrode placement for the EEG is shown in Fig. 11.1. Initially, the subject sits on the chair in a relaxed position to acquire

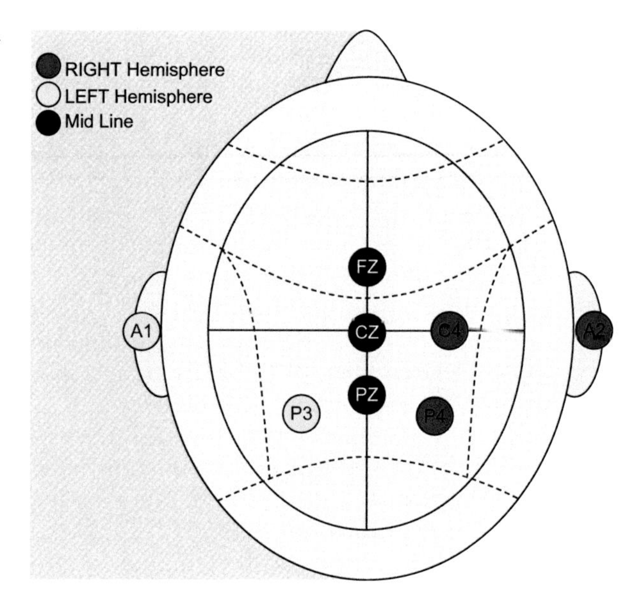

FIG. 11.1 EEG electrode positions. *No permission required.*

the baseline. He was asked to stand from the sit position at a normal self-selected speed, wait for 10 s, and then go to sit position. This task was repeated five times.

Data collection

Three healthy persons (male, age 24 ± 3 years; height 5′5″ ± 3″ (ft); weight 70 ± 4 kg) participated in these experiments with no known neurological or physical disorder. At the start of each experiment, EMG signal was set to base zero, when the subject stands straight. This was done to set the reference signal to zero. The signals at this position were taken as the reference for all measurements. For EEG signal acquisition, Biopac MP150 system which includes electroencephalogram amplifier (EEG 100C) was used. EEG100C is a single channel, differential input, high gain, and biopotential amplifier for monitoring the neuronal activity of the brain. A 16-channel EEG module in compliance with the international "10–20" electrode placement system is used for recording EEG signals with a sampling rate of 500 Hz. This module was designed to pass EEG signals with minimal distortion.

Data processing

EMG signal processing

The EMG signal was acquired with a sampling rate of 500 Hz. From this, raw signal DC offset was removed to remove the low-frequency noise. Then high-pass filtering was done with a cutoff frequency of 5–30 Hz. Butterworth filter was applied before rectifying the signal to get the absolute values at each point.

EEG signal processing

EEG was recorded with a sampling rate of 500 Hz. When the EEG data was recorded from any subject, then some noise components were originated with the signal. So, 35-Hz low-pass and 1-Hz high-pass filters with 50 Hz notch were used to remove the noise components, but still, there were some noise components. For reducing these noise components, the automatic blind source separation (BSS) method (autobss, a MATLAB command) was used (Presacco et al., 2011). This command was used for removing the artifacts from the original signal. It performs automatic electrooculography (EOG) artifact correction using the BSS and identifies the EOG components using fractal analysis. Figs. 11.2 and 11.3 show comparative results before and after the noise removal during stand to sit and sit to stand at Fz, Cz, and P3 electrode positions, respectively.

It shows that the sit position has the negative amplitude in all three given electrodes, whereas the stand position has positive amplitude. So, with that position, the muscle position for the sit to stand intention has been identified. Similarly, in the case of sit to stand position, the sit has the negative amplitude and the stand has the positive.

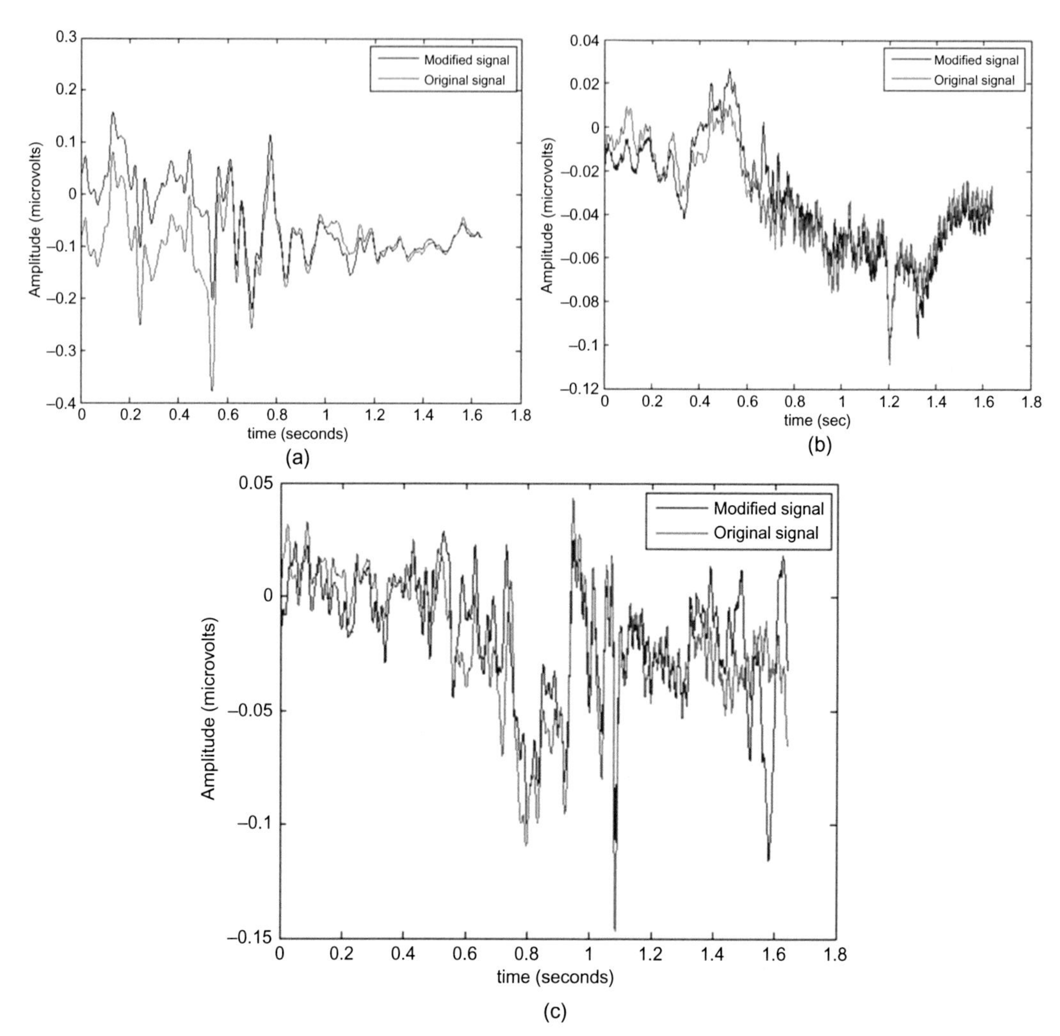

FIG. 11.2 Comparative results of the EEG signal before and after artifact removal during stand to sit of (A) P3, (B) Fz, and (C) Cz electrode positions, respectively. *No permission required.*

Canonical correlation analysis

Canonical correlation analysis (CCA) is a statistical method used to measure the linear relationship between two multidimensional variables A and B.

$$\text{Canonical correlation}, \rho = \frac{Cov[a, b]}{\sqrt{V[a]}\sqrt{V[b]}} \tag{11.1}$$

where *Cov* and *V* indicate the covariance and variance, respectively, and CCA is a multivariate extension of ordinary correlation analysis.

Neck muscles act as movement artifacts during gait and contaminate the EMG signal. So, CCA is used to detect and segregate EMG components (Maarten et al., 2010). It is based on an assumption that EEG signals have much higher autocorrelation than EMG signals. Basically,

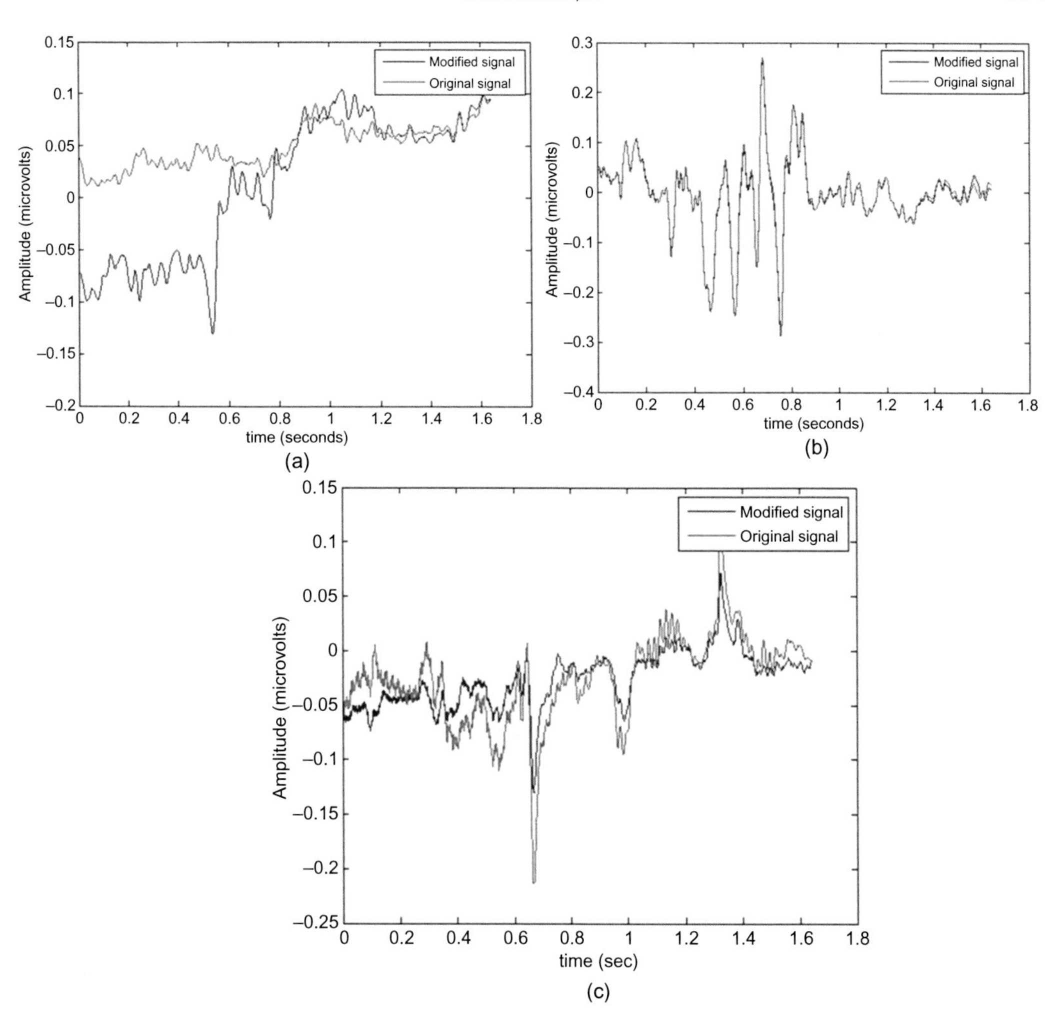

FIG. 11.3 Comparative results of the EEG signal before and after artifact removal during sit to stand of (A) P3, (B) Cz, and (C) Fz electrode positions, respectively. *No permission required.*

it is a decomposition method that removes the EMG components from the raw signals. It calculates the spectral power. EMG component is identified when its power is 4 times than the EEG. Hence it is considered as an artifact and removed.

Statistical analysis

The EEG and EMG signals were acquired concurrently at different instances, but for the same movement, that is, sit to stand and stand to sit. To quantify the muscle activation degree, the signal was processed with root mean square (RMS) (Wang, Li, Wang, & Meng, 2015). The RMS values of the three different muscles from the right and left lower limb positions and power spectral density (PSD) of the six different scalp positions were analyzed and are given

TABLE 11.1 RMS voltage and PSD of the right and left knee corresponding to their muscles and scalp electrode position.

	RMS voltage (mV)					
	Subject 1		Subject 2		Subject 3	
EMG	Sit to stand	Stand to sit	Sit to stand	Stand to sit	Sit to stand	Stand to sit
Right gastrocnemius	0.1132	0.092	0.098	0.104	0.104	0.118
Left gastrocnemius	0.0892	0.072	0.083	0.071	0.071	0.071
Right rectus femoris	0.1177	0.122	0.133	0.126	0.122	0.126
Left rectus femoris	0.1522	0.12	0.113	0.128	0.09	0.113
Right semimembranosus	0.057	0.048	0.051	0.048	0.048	0.051
Left semimembranosus	0.0832	0.077	0.076	0.074	0.071	0.076
Raw EEG	PSD (mV^2/Hz)					
Fz	3.12E-09	1.21E-8	1.48E-9	4.22E-10	5.09E-10	5.68E-10
Cz	5.40E-9	8.10E-8	2.01E-9	2.25E-08	1.02E-06	8.55E-09
C4	1.14E-8	3.24E-8	1.64E-8	1.67E-08	1.77E-08	8.54E-09
P3	1.08E-8	5.86E-8	1.11E-9	2.94E-08	9.88E-08	1.63E-08
Pz	4.02E-8	3.10E-7	2.48E-10	8.94E-09	1.98E-08	1.55E-09
P4	2.38E-8	8.31E-8	1.40E-9	8.68E-08	1.23E-07	6.00E-08

later in Table 11.1. The EEG and EMG signals were calculated for sit to stand and stand to sit positions by averaging the magnitude of the PSD values of the EMG and EEG data.

Correlation coefficient between sit to stand and stand to sit of raw EEG

The raw EEG data during sit to stand and stand to sit positions was analyzed. Here the PSD and correlation coefficient between sit to stand and stand to sit were calculated after applying the CCA.

The six scalp positions which are considered for the lower limb intention give the higher values of correlation coefficient, as shown in Table 11.2.

Delta wave lies within the range of 0.5–4 Hz. These waves often create confusion because of the movement artifacts caused by the neck and facial muscles. So, these waves were selected for validating the proposed methodology. Table 11.3 shows the higher value of the correlation coefficient in the delta band also.

Results

The PSD and correlation coefficient of the raw EEG signal were calculated at six different electrode positions during sit to stand and stand to sit positions. The comparative plot of PSD of EEG vs RMS value of EMG during sit to stand and stand to sit is shown in Fig. 11.4.

TABLE 11.2 Correlation coefficient between sit to stand and stand to sit at different scalp positions of raw EEG.

Electrodes	Subject	Sit to stand (PSD)	Stand to sit (PSD)	Correlation coefficient
Fz	1	7.36E-06	5.58E-06	0.7243
	2	7.48E-06	1.22E-05	
	3	1.01E-05	2.38E-06	
Cz	1	8.61E-06	1.20E-05	0.975841
	2	2.72E-05	2.63E-05	
	3	1.14E-05	1.06E-05	
C4	1	5.64E-06	1.17E-05	0.62831
	2	6.79E-06	2.33E-06	
	3	3.02E-05	1.00E-06	
P3	1	1.03E-05	2.91E-06	0.65151
	2	2.88E-06	2.97E-05	
	3	4.23E-06	3.35E-06	
Pz	1	2.64E-05	3.80E-06	0.98515
	2	6.11E-06	6.70E-06	
	3	7.16E-06	7.16E-06	
P4	1	1.77E-05	5.27E-06	0.99275
	2	1.85E-05	6.70E-06	
	3	9.75E-06	3.88E-05	

TABLE 11.3 Correlation coefficient of the average PSD values of sit to stand and stand to sit at different scalp positions of delta band.

Electrodes	Sessions	Sit to stand	Stand to sit	Correlation coefficient
Fz	1	6.57E-06	1.01E-05	0.99974
	2	4.05E-06	1.20E-05	
	3	1.96E-05	1.72E-06	
Cz	1	1.26E-05	8.97E-06	0.95716
	2	5.86E-06	1.33E-05	
	3	6.57E-06	6.57E-06	
C4	1	2.83E-06	5.57E-06	0.73872
	2	8.06E-06	3.02E-06	
	3	1.73E-06	1.95E-05	
P3	1	7.85E-06	2.39E-06	0.97947

Continued

TABLE 11.3 Correlation coefficient of the average PSD values of sit to stand and stand to sit at different scalp positions of delta band—Cont'd

Electrodes	Sessions	Sit to stand	Stand to sit	Correlation coefficient
	2	1.25E-06	1.74E-05	
	3	4.99E-06	6.16E-06	
Pz	1	1.53E-05	3.75E-06	0.54253
	2	3.47E-06	1.69E-05	
	3	2.89E-06	5.07E-06	
P4	1	1.11E-05	4.51E-06	0.908332
	2	5.93E-06	4.39E-06	
	3	1.29E-05	1.30E-05	

No permission required.

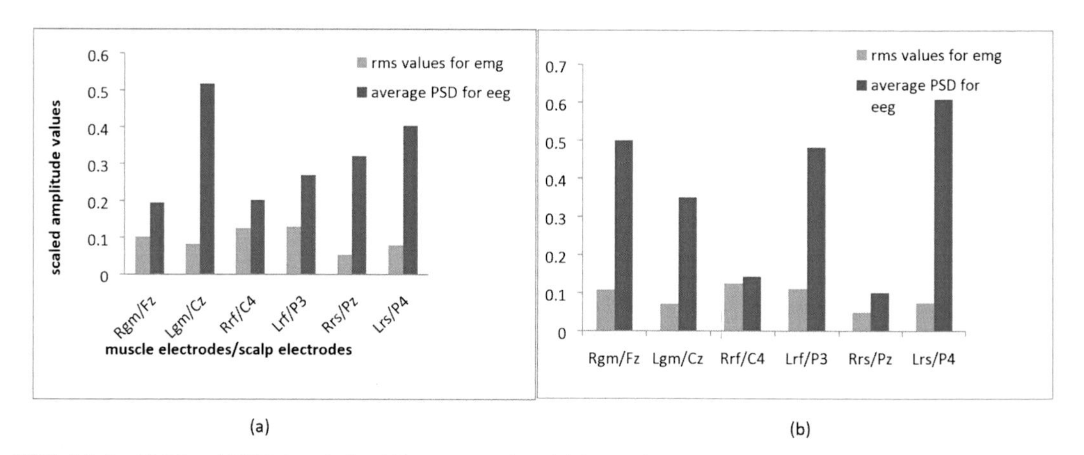

FIG. 11.4 EMG and EEG signals for (A) sit to stand and (B) stand to sit. *No permission required.*

The plot shows the correlation between the magnitude values of the muscle positions and the scalp positions. Lesser the difference, better the correlation between the muscle and the EEG electrode position. During this dynamic movement, the right leg muscle position gives consistent values and the C4 scalp position gives the best result for lower limb intention. RMS and PSD are in proximity.

Fig. 11.5 shows the results of EEG correlation coefficient after applying the CCA. It is observed that Fz, Cz, C4, P3, and P4 give higher values of the correlation coefficient.

The comparison between the correlation coefficient (r) before CCA and after CCA is tabulated in Table 11.4.

It is observed that after applying the CCA decomposition, the correlation coefficient has been increased not only in the raw EEG but also in the delta frequency region (Fig. 11.6).

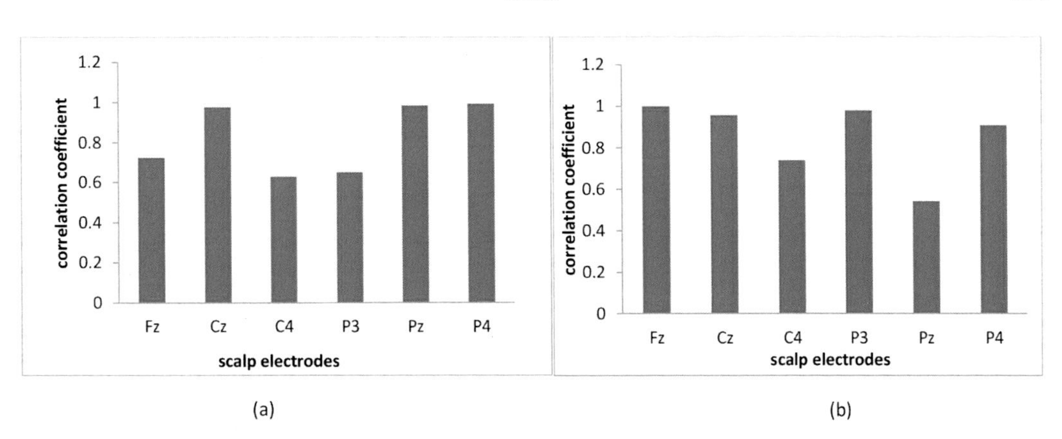

FIG. 11.5 Plot between different scalp positions and correlation values of (A) raw EEG and (B) delta band. *No permission required.*

TABLE 11.4 Comparison between r values after and before CCA of raw EEG.

	Raw EEG		EEG in delta band	
Electrodes	**Correlation coefficient before CCA**	**Correlation coefficient after CCA**	**Correlation coefficient before CCA**	**Correlation coefficient after CCA**
Fz	−0.03364	0.724296	−0.53245	0.999735
Cz	0.999664	0.975841	0.996095	0.957163
C4	0.383283	0.628309	0.493912	0.738724
P3	0.999952	0.651506	0.968929	0.979474
Pz	0.956679	0.985147	0.997266	0.542534
P4	0.985728	0.992748	0.825587	0.908332

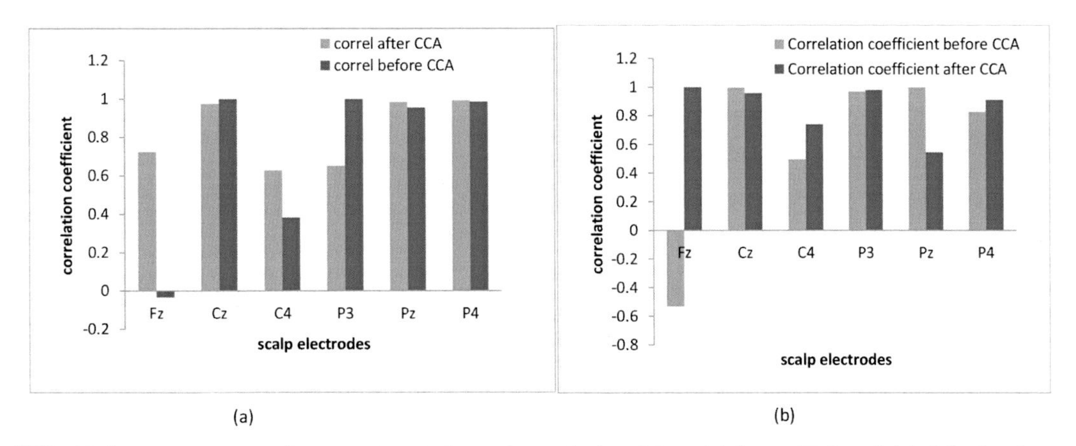

FIG. 11.6 Correlation coefficient values before CCA and after CCA of (A) raw EEG and (B) delta band. *No permission required.*

Conclusion

The EEG and EMG signals identify that which part of the scalp is used to detect the intention of sit to stand or stand to sit. The consistency of the values from the same muscle or scalp position can detect its intention for the controlled strategy. In this work, EEG signal refinement was done by removing all the unwanted noise and artifacts.

The EEG raw signal was passed through normal filters. Normal filters remove only 10%–20% of the noise. Therefore BSS methodology was used to remove the remaining noise. Results show that the BSS method improves the signal-to-noise ratio. When a person moves, there are EMG artifacts in the EEG signal; this is because of the movement of the neck and facial muscles. CCA decomposition was used to remove these EMG components from the EEG. Delta waves often create confusion because of the movement artifacts caused by the neck and facial muscles. So these waves were selected for validating the methodology with the EMG signals acquired from the muscles during sit to stand and stand to sit movements. It is observed that the C4 electrode position gives the best result for this activity.

EEG gives a higher value of correlation. So, more is the correlation coefficient better the EEG. It is concluded that correlation coefficient (*r*) values are higher after applying the CCA not only in the raw EEG but also in the furnished delta band region.

It is also concluded that Fz, Cz, C4, Pz, and P4 scalp electrode positions give the higher values of correlation coefficients. Compared to previous studies (Bulea, Prasad, Kilicarslan, & Contreras-Vida, 2013), the proposed methodology gives a better result.

References

Atkins, D. J., Heard, D. C. Y., & Donovan, W. H. (1996). Epidemiologic overview of individuals with upper-limb loss and their reported research priorities. *JPO Journal of Prosthetics and Orthotics*, *8*(1), 2–11. https://doi.org/10.1097/00008526-199600810-00003.

Bulea, T. C., Prasad, S., Kilicarslan, A., & Contreras-Vida, J. L. (2013). Classification of stand-to-sit and sit-to-stand movement from low frequency EEG with locality preserving dimensionality reduction. In *Annual international conference of the IEEE engineering in medicine and biology society. IEEE engineering in medicine and biology society. Annual international conference, 2013* (pp. 6341–6344). https://doi.org/10.1109/EMBC.2013.6611004.

Chiri, A., Vitiello, N., Giovacchini, F., Roccella, S., Vecchi, F., & Carrozza, M. C. (2012). Mechatronic design and characterization of the index finger module of a hand exoskeleton for post-stroke rehabilitation. *IEEE/ASME Transactions on Mechatronics*, *17*(5), 884–894. https://doi.org/10.1109/TMECH.2011.2144614.

Frohman, T. C., O'Donoghue, D. L., & Northrop, D. (2011). *A practical primer: Multiple sclerosis for the physician assistant*. Dallas, TX: Southwestern Medical Center.

Gómez-Herrero, G., De Clercq, W., Anwar, H., Kara, O., Egiazarian, K., Van Huffel, S., et al. (2006). Automatic removal of ocular artifacts in the EEG without an EOG reference channel. In *Proceedings of the 7th nordic signal processing symposium, NORSIG 2006* (pp. 130–133). https://doi.org/10.1109/NORSIG.2006.275210.

Iqbal, J., Tsagarakis, N. G., Fiorilla, A. E., & Caldwell, D. G. (2010). A portable rehabilitation device for the Hand. In *Annual international conference of the IEEE engineering in medicine and biology society. IEEE engineering in medicine and biology society. Annual international conference, 2010* (pp. 3694–3697). https://doi.org/10.1109/IEMBS.2010.5627448.

Jiang, J., Zhou, Z., Yin, E., Yu, Y., & Hu, D. (2014). Hybrid brain-computer interface (BCI) based on the EEG and EOG signals. *Bio-Medical Materials and Engineering*, *24*(6), 2919–2925. https://doi.org/10.3233/BME-141111.

Kashif, S., Neelesh, K., & Munna, K. (2018). Effect of voluntary and involuntary joint movement on EEG signals. *Journal of Scientific and Industrial Research (JSIR)*, *77*(12), 710–712. http://nopr.niscair.res.in/handle/123456789/45483.

Maarten, D. V., Riès, S., Vanderperren, K., Vanrumste, B., Alario, F.-X., Huffel, V. S., et al. (2010). Removal of muscle artifacts from EEG recordings of spoken language production. *Neuroinformatics*, *8*(2), 135–150. https://doi.org/10.1007/S12021-010-9071-0.

MacMahon, S. (2002). Introduction: The global burden of stroke. *Clinician's Manual on Blood Pressure and Stroke Prevention*, 1–6.

Martins, F., Freitas, F., Martins, L., Dartigues, J. F., & Barat, M. (1998). Spinal cord injuries–epidemiology in Portugal's central region. *Spinal Cord*, *36*(8), 574–578. https://doi.org/10.1038/SJ.SC.3100657.

Peerdeman, B., Boere, D., Witteveen, H., Rianne Huis in 't Veld, Hermens, H., Stramigioli, S., et al. (2011). Myoelectric forearm prostheses: State of the art from a user-centered perspective. *Journal of Rehabilitation Research and Development*, *48*(6), 719–737. https://doi.org/10.1682/JRRD.2010.08.0161.

Pons, J. L. (2010). Rehabilitation exoskeletal robotics. The promise of an emerging field. *IEEE Engineering in Medicine and Biology Magazine : The Quarterly Magazine of the Engineering in Medicine & Biology Society*, *29*(3), 57–63. https://doi.org/10.1109/MEMB.2010.936548.

Presacco, A., Forrester, L. W., & Contreras-Vidal, J. L. (2012). Decoding intra-limb and inter-limb kinematics during treadmill walking from scalp electroencephalographic (EEG) signals. *IEEE Transactions on Neural Systems and Rehabilitation Engineering*, *20*(2), 212. https://doi.org/10.1109/TNSRE.2012.2188304.

Presacco, A., Goodman, R., Forrester, L., & Jose Luis, C.-V. (2011). Neural decoding of treadmill walking from noninvasive electroencephalographic signals. *Journal of Neurophysiology*, *106*(4), 1875–1887. https://doi.org/10.1152/JN.00104.2011.

Sherwani, K. I. K., Khan, M., & Kumar, N. (2015). Assessment of postural stability using centre of pressure and knee progression during dynamic gait. In *2015 annual IEEE India conference (INDICON)* (pp. 1–5). IEEE. https://doi.org/10.1109/INDICON.2015.7443634.

Sherwani, K. I. K., & Kumar, N. (2016). Fusion of EEG and EMG signals for gait intent detection. *MMU Journal of Management & Technology*, *1*(1), 50–55.

Sherwani, K. I. K., Kumar, N., Chemori, A., Khan, M., & Mohammed, S. (2020). RISE-based adaptive control for EICoSI exoskeleton to assist knee joint mobility. *Robotics and Autonomous Systems*, *124*. https://doi.org/10.1016/j.robot.2019.103354.

van Asbeck, F. W., Post, M. W. M., & Pangalila, R. F. (2000). An epidemiological description of spinal cord injuries in The Netherlands in 1994. *Spinal Cord*, *38*(7), 420–424. https://doi.org/10.1038/SJ.SC.3101003.

Velu, P., & de Sa, V. R. (2013). Single-trial classification of gait and point movement preparation from human EEG. *Frontiers in Neuroscience*, *0*(7), 84. https://doi.org/10.3389/FNINS.2013.00084.

Vitiello, N., Lenzi, T., Roccella, S., De Rossi, S. M. M., Cattin, E., Giovacchini, F., et al. (2013). NEUROExos: A powered elbow exoskeleton for physical rehabilitation. *IEEE Transactions on Robotics*, *29*(1), 220–235. https://doi.org/10.1109/TRO.2012.2211492.

Wang, L., Li, H., Wang, Z., & Meng, F. (2015). Study on upper limb rehabilitation system based on surface EMG. *Bio-Medical Materials and Engineering*, *26*(Suppl 1), S795–S801. https://doi.org/10.3233/BME-151371.

Ward, N. S., & Cohen, L. G. (2004). Mechanisms underlying recovery of motor function after stroke. *Archives of Neurology*, *61*(12), 1844–1848. https://doi.org/10.1001/ARCHNEUR.61.12.1844.

Weber, D. J., Friesen, R., & Miller, L. E. (2012). Interfacing the somatosensory system to restore touch and proprioception: Essential considerations. *Journal of Motor Behavior*, *44*(6), 403–418. https://doi.org/10.1080/00222895.2012.735283.

Zecca, M., Micera, S., Carrozza, M. C., & Dario, P. (2002). Control of multifunctional prosthetic hands by processing the electromyographic signal. *Critical Reviews in Biomedical Engineering*, *30*(4–6), 459–485. https://doi.org/10.1615/CRITREVBIOMEDENG.V30.I456.80.

CHAPTER

12

Hand gesture recognition for the prediction of Alzheimer's disease

R. Sivakani[a] *and Gufran Ansari*[b]

[a]Computer Science and Engineering, B.S. Abdur Rahman Crescent Institute of Science and Technology, Chennai, India [b]Department of Computer Applications, B.S. Abdur Rahman Crescent Institute of Science & Technology, Chennai, India

Alzheimer's disease and gestures

Alzheimer's disease (AD) is a neurological disorder that affects people older than 65 years. It causes memory impairment, among other neurological symptoms. More than 5.8 million people aged 65 years or older in the United States are affected by AD, 80% of whom are older than 75 years. Worldwide, more than 50 million people are affected by AD (https://www.mayoclinic.org/diseases-conditions/alzheimers-disease/symptoms-causes/syc-20350447). In the initial stages of the disease, the patient cannot recall recent events and incidents. As the disease progresses, the patient begins to have trouble remembering how to complete daily tasks. AD medications provide only temporary relief; there is no cure for AD. Patients with AD cannot live independently, thus require a caregiver to monitor and assist them. Researchers are putting much effort into the study of AD. One area of research that is attracting attention is gesture recognition, in which the computing process takes place with the help of human–computer interaction. The computer captures the patient's gestures and interprets them as commands using deep learning algorithms. Gestures can be recognized from the face or hand. Recently, researchers have been focusing on emotion recognition as well.

Computational Intelligence in Healthcare Applications
https://doi.org/10.1016/B978-0-323-99031-8.00019-3

Types of gestures

Online gesture

An online gesture is the gesture generated by the computer in real time.

Offline gesture

An offline gesture is the gesture generated by the computer after interaction with objects.

Input devices for gesture recognition

Data gloves

Data gloves send input signals to the computer based on bodily motion.

Cameras

Cameras generate input images or videos based on bodily motion and send the data to the computer.

Gesture-based controllers

These controllers track the movement of the body and give input to the computer.

Sensors

Sensors detect bodily movement and give input to the computer.

Literature survey

According to some estimates, the number of patients with AD is expected to increase to 13.8 million by 2050 (Alzheimer's Association Report, 2020). To help care for these patients, mobile applications and wearable devices used in conjunction with the Internet of Things (IoT) have been introduced. For example, there are smartwatch apps that provide information about the patient's heart rate and their location so that caregivers can monitor the patient. These apps can also send reminders to the patient about daily activities that need to be completed (Aljehani et al., 2018). An example of a wearable device for AD patients is a smart cap developed using deep learning methods that helps patients recognize their relatives and caregivers. This cap also sends information to caregivers about the patient's location (Boppana et al., 2019). A tablet-based application has been developed for comparing the speech responses of AD patients to daily life questions to differentiate mild cognitive impairment (MCI) from cognitively normal (CN) (Yamada et al., 2021). A smart home automated intelligent system has been proposed for predicting the multimodal symptoms of AD patients to manage critical situations using actions. Contextual-based analysis has been done using ontology technology. A random forest algorithm has been used for the prediction of critical situations (Harish & Gayathri, 2019).

A hand gesture recognition system using the leap motion device has been developed. In this system, the sequential data is collected using a long short-term memory (LSTM) algorithm that exploits the unidirectional LSTM and bidirectional LSTM separately. The prediction is done using the hybrid model of bidirectional and unidirectional LSTM and specified as HBU-LSTM. Results show the recognition rate of the HBU-LSTM was 90%. Because this system is time-consuming, the researchers plan to use a Graphics Processing Unit (GPU) in future work (Ameur et al., 2020). A review has been done to analyze works on M-health applications for AD (Elfaki & Alotaibi, 2018). A smart home-based IoT device application has been developed for monitoring the health condition of AD patients. This device recommends appropriate treatment cased on the behavior and movement of the patient. Blood pressure and body temperature information is collected via smartwatch and sensors. The security for the data was provided using the Message Queue Telemetry Transport and WebSocket protocols. Also, the location of the doctors, patients, and the ambulance are recorded and maintained by an admin at the backend (Oskouei et al., 2020). A data glove with a 3D flex sensor has been designed to record arm and knuckle motion. In this system, a convolutional neural network (CNN) detects the deep features, which are then classified using an LSTM model. The accuracy of hand gesture recognition using this wearable device was 99.93% for American Sign Language and 96.1% for Chinese Sign Language (Yuan et al., 2020). The recognition of daily life activities has been accomplished using multimodal wearable sensors and Bluetooth beacons. In one study, the classification accuracy for 19 activities was 80%. In the future, the study authors plan to design a wearable prototype for the remote assessment of older people with dementia (De et al., 2015).

A model using biometric monitoring devices (BMDs) and artificial intelligence (AI) technology for describing patient perceptions was developed in one study. In this study, researchers recruited a group of 1183 adult patients from the Community of Patients for Research (ComPaRe) in France. Of these patients, 20% found the technology to be useful, 3% had a negative experience, and 35% refused to participate (Tran et al., 2019). A review of wearable IoT devices for health care was conducted and the challenges of the research explained in Surantha et al. (2021). An A-to-Z guide of wearable technology can be found in Godfrey et al. (2018). A review of IoT and wearable technologies for the assessment of elderly people was conducted by Stavropoulos et al. (2020). A model using CNN for real-time gesture recognition has been applied for AD. In this model, inputs were given in the format of images (Kibbanahalli Shivalingappa et al., 2020). AD has also been predicted using machine learning algorithms (Sivakani & Ansari, 2020). The missing values are generated using the machine learning methods in the Alzheimer's dataset (Sivakani & Ahmad Ansari, 2020). A CNN model with two stages was proposed for hand gesture recognition. Segmentation was done in the first stage and gesture identification was done in the second stage (Dadashzadeh et al., 2019). Two other methods of hand gesture recognition have been implemented using RNN and CNN algorithms. In the first method, only the RNN algorithm was used, and in the second method, both the RNN and CNN algorithms were implemented in combination for hand gesture recognition (Shin & Sung, 2016).

A novel framework has been designed for real-time hand gesture recognition. Images of the gestures are given as inputs, and the background is subtracted from the hand region. Segmentation was done for the recognition of the fingers. A rule classification was implemented for predicting gestures (Chen et al., 2014). CNN architecture has been designed for detecting and classifying gestures. In this study, the architecture was evaluated using the Ego Gesture and NVIDIA Dynamic Hand Gesture datasets. Results show the model achieved classification accuracy of 94.04% and 83.82% for each dataset, respectively (Kopuklu, Gunduz, Kose, & Rigoll, 2019). A new method for recognizing hand gestures was introduced by Freeman and Roth (1995). For gesture classification, researchers selected the feature vector using the orientation histogram method. Special hardware was used for digitizing the images before classification (Freeman & Roth, 1995). A new lightweight YOLOv3 and CNN model was designed for hand gesture recognition by Mujahid et al. (2021). The accuracy of the models was 82% and 85%, respectively. A review of the different techniques used for hand recognition was done by Oudah et al. (2020). The study authors also highlighted the types of gestures, segmentation process, classification models, and limitations of the models.

From this literature survey, it is apparent that many kinds of IoT-based smart wearable devices have been introduced for the benefit of older people. LSTM or CNN machine learning models are used to process image inputs to predict AD. In the subjective as well as machine learning methods, past patient data has to be processed to predict AD, whereas in deep learning methods, real-time data is used for prediction.

In "Proposed model" section, we describe our proposed model of hand gesture recognition for the prediction of AD. We present our results and discussion in "Result and discussion" section, and we conclude the chapter with "Conclusion and future work" section.

Proposed model

Fig. 12.1 shows the architecture of the proposed model. Inputs are given in the format of real-time videos captured via a web camera. The videos are then subjected to pre-processing and segmentation. Segmented images are checked for recognition; if an image is recognized, it is given for processing in the deep learning model. The deep learning model will learn the hand gesture and the prediction will be done. The deep learning model uses RNN, CNN, and LSTM algorithms and classification is done using the training data already stored in the dataset. Then, based on the gesture, the annotation is specified and the AD prediction will be done.

Result and discussion

We conducted two experiments: one using static images and one using dynamic images

4.1. Steps in the experiment for the static images
- Step 1: Input: Static images
- Step 2: Pre-processing
- Step 3: Segmentation
- Step 4: Hand gesture
- Step 5: Implementation using deep learning models
- Step 6: Output: Annotation and the prediction of Alzheimer's disease

4.2. Steps in the experiment for the dynamic images
- Step 1: Input: Dynamic images
- Step 2: Recording the input video for 2 s
- Step 3: Storing the images by the frame sequence of 30

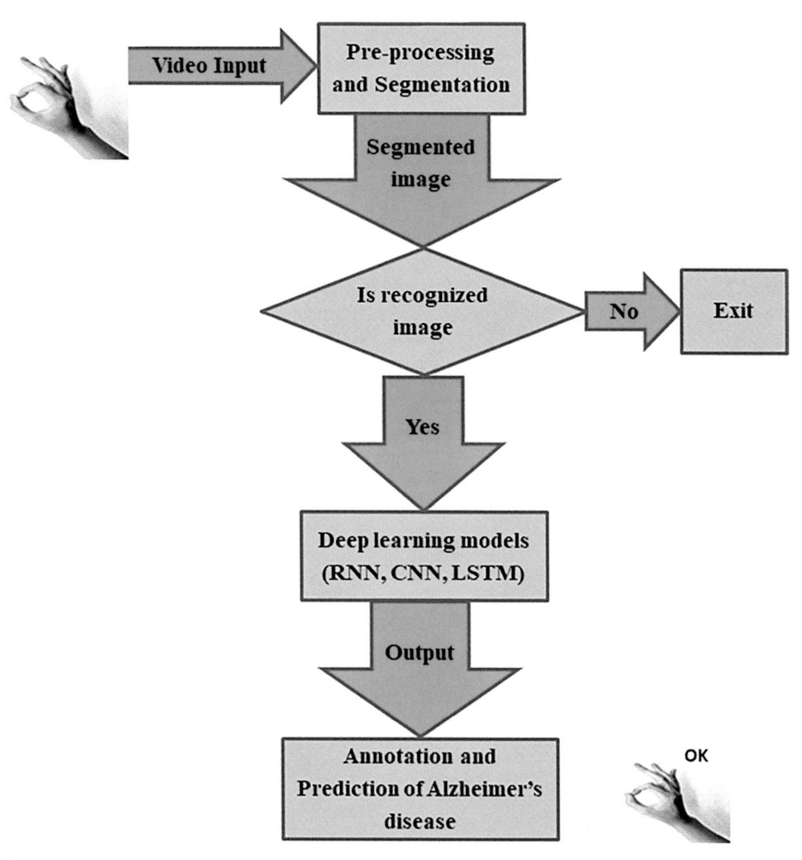

FIG. 12.1 Architecture of the proposed model.

Step 2: Pre-processing
Step 3: Segmentation
Step 4: Hand gesture
Step 5: Implementation using deep learning models
Step 6: Output: Annotation and the prediction of Alzheimer's disease

Fig. 12.2 explains the workflow of the proposed model. The input can be in the form of an image or video, which will be segmented and checked for the recognized gesture. If it is a recognized gesture, then it will be fed to the deep learning model for prediction.

First, the static images were given as inputs, processed for gesture recognition, and then annotated. Second, the dynamic images with video inputs were processed. The video input was preprocessed and segmented. The video was recorded for 2 s and then stopped. Then, the video was converted to a sequence of 30 frames and stored for processing. In the preprocessing step, each frame was resized to 120×120. Some frames were cropped and thus edge detection was employed. Training and testing data were stored for validating the images. Each video was stored as 30 frames and 10 gesture classes were considered for hand

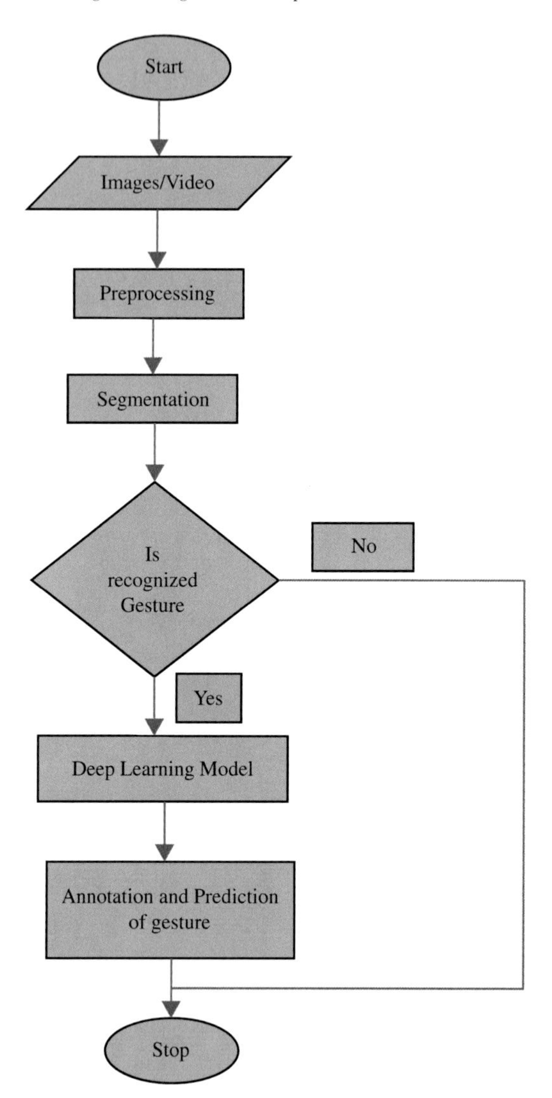

FIG. 12.2 Workflow of the proposed model.

gesture recognition: Ok, Good job, Hai (Hi), Dislike, Timeout, Collab (Collaboration work - handshaking), You, Me, Call me, and Alzheimer.

We evaluated hand gestures using a dataset of 10 gestures and 10 videos for each gesture. Each video was sequenced to 30 frames and stored as 30 images. Thus, a total of 3000 images were stored for processing.

The segmented images were subjected to prediction using deep learning models. A hybrid deep learning model was constructed using CNN, RNN, and LSTM algorithms.

The CNN model consists of the input layer, convolutional layers, pooling layers, fully connected layers, and the output layers. The RNN model consists of the input layer, hidden layer, and output layer. The LSTM model consists of the input layer, hidden layer, and output layer. LSTM can support hand gesture recognition for the unsegmented images as well. The current state and the previous state must be maintained to predict the correct output.

The current state can be generated using the formula:

$$h_t = f(h_{t-1}, X_t) \tag{12.1}$$

where h_t is the current state value, h_{t-1} is the previous state value, and X_t is the input state value.

The activation function formula is given as:

$$h_t = \tanh(W_{hh}h_{t-1} + W_{xh}X_t) \tag{12.2}$$

where W_{hh} is the weight at the recurrent neuron and W_{xh} is the weight at the input neuron.

The formula for calculating the input is given as:

$$Y_t = W_{hy}h_t \tag{12.3}$$

where Y_t is the output and W_{hy} is the weight of the output layer.

Fig. 12.3 shows the architecture of the deep learning model. In this model, the segmented image is given to the hybrid CNN–RNN–LSTM model for prediction purposes.

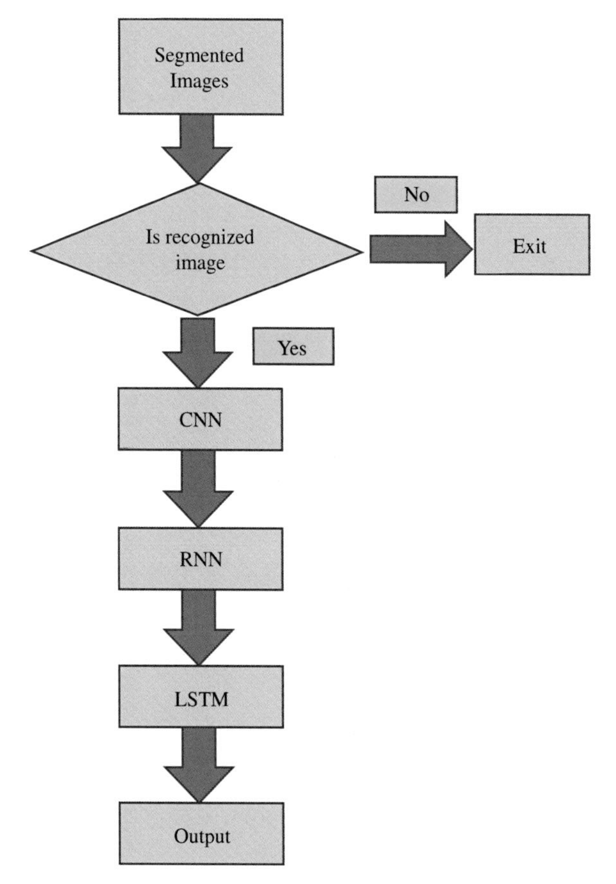

FIG. 12.3 Architecture of deep learning model.

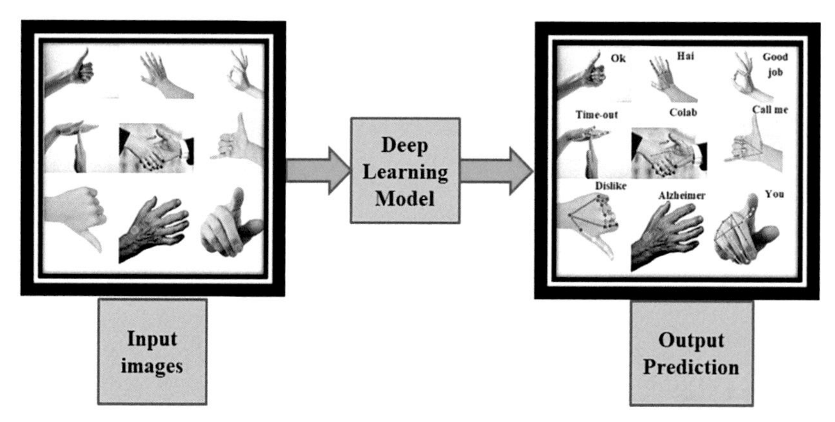

FIG. 12.4 Prediction of result.

The annotation specified for the images is based on the prediction. The segmented image is processed using a CNN, then given to the RNN layers, and finally processed in the LSTM layers to output the prediction. The processing is done in a hidden layer and the output is then predicted.

Fig. 12.4 shows the predicted outputs; the input images are processed using the deep learning model and the output is predicted with the annotation.

Fig. 12.5 is a graphical representation of the prediction accuracy achieved by the hybrid deep learning model. For static images, the model predicted AD with an accuracy of 90%. For dynamic images, the model predicted AD with an accuracy of 87%. Some of the images from the Alzheimer gesture class were unusable, thus we plan to rectify this in future work.

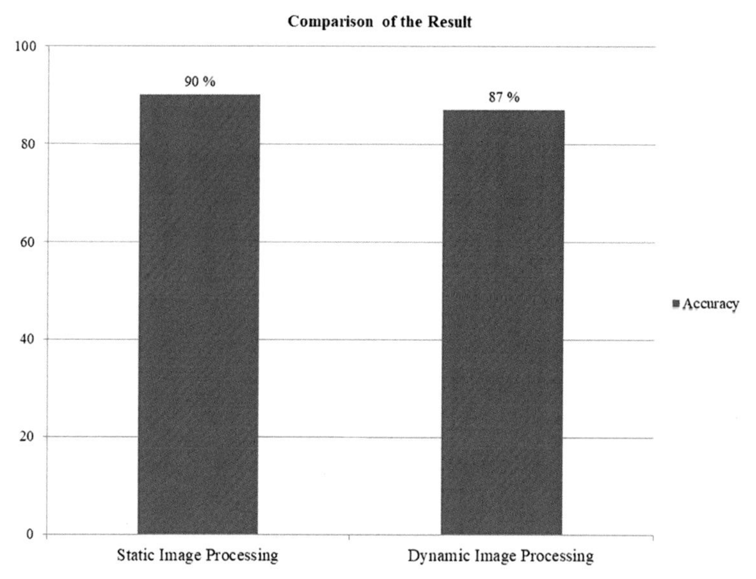

FIG. 12.5 Graphical representation of the result.

Conclusion and future work

In this chapter, we used hand gesture recognition and deep learning to predict AD. We constructed a hybrid model of CNN, RNN, and LSTM algorithms. Real-time hand gesture videos were recorded and annotated to identify AD. Results show the proposed system predicted AD with an accuracy of 90% using static images and an accuracy of 87% using dynamic images. In future work, we plan to use sensors for collecting data to achieve even greater accuracy.

References

Aljehani, S. S., Alhazmi, R. A., Aloufi, S. S., Aljehani, B. D., & Abdulrahman, R. (2018). iCare: Applying IoT technology for monitoring Alzheimer's patients. In *1st International conference on computer applications & information security (ICCAIS)* (pp. 1–6).

Alzheimer's Association Report. (2020). *2020 Alzheimer's disease facts and figures.* Alzheimer's & Dementia, Wiley Publisher.

Ameur, S., Khalifa, A. B., & Bouhlel, M. S. (2020). A novel hybrid bidirectional unidirectional LSTM network for dynamic hand gesture recognition with Leap Motion. *Entertainment Computing, 35,* 1–20. Elsevier Publisher.

Boppana, L., Kumari, P., Chidrewar, R., & Gadde, P. K. (2019). Smart cap for Alzheimer's patients using deep learning. In *2019 IEEE Region 10 Conference (TENCON)* (pp. 2466–2471). IEEE.

Chen, Z.-h., Kim, J.-T., Liang, J., Zhang, J., & Yuan, Y.-B. (2014). Real-time hand gesture recognition using finger segmentation. *The Scientific World Journal, 2014,* 1–9. https://doi.org/10.1155/2014/267872.

Dadashzadeh, A., Targhi, A. T., Tahmasbi, M., & Mirmehdi, M. (2019). HGR-Net: A fusion network for hand gesture segmentation and recognition. *IET - The Institution of Engineering and Technology, 13,* 700–707. https://doi.org/10.1049/iet-cvi.2018.5796.

De, D., Bharti, P., Das, S. K., & Chellappan, S. (2015). Multimodal wearable sensing for fine-grained activity recognition in healthcare. *IEEE Internet Computing, 19,* 26–35.

Elfaki, A. O., & Alotaibi, M. (2018). The role of M-health applications in the fight against Alzheimer's: Current and future directions. *Mhealth, 4,* 1–13.

Freeman, W. T., & Roth, M. (1995). Orientation histograms for hand gesture recognition. In *IEEE Intl. Wkshp. on Automatic Face and Gesture Recognition.*

Godfrey, A., Hetherington, V., Shum, H., Bonato, P., Lovell, N. H., & Stuart, S. (2018). From A to Z: Wearable technology explained. *Maturitas,* 40–47.

Harish, S., & Gayathri, K. S. (2019). Smart home-based prediction of symptoms of Alzheimer's disease using machine learning and contextual approach. In *International conference on computational intelligence in data science (ICCIDS), 2019* (pp. 1–6).

Kibbanahalli Shivalingappa, M. S., Ben Abdessalem, H., & Frasson, C. (2020). Real-time gesture recognition using deep learning towards Alzheimer's disease applications. In C. Frasson, P. Bamidis, & P. Vlamos (Eds.), *Lecture notes in computer science. Brain function assessment in learning, BFAL* Springer.

Kopuklu, O., Gunduz, A., Kose, N., & Rigoll, G. (2019). Real-time hand gesture detection and classification using convolutional neural networks. *2019 14th IEEE International Conference on Automatic Face & Gesture Recognition (FG 2019)* (pp. 1–8). IEEE.

Mujahid, A., Awan, M. J., Yasin, A., Mohammed, M. A., Damaševičius, R., Maskeliunas, R., et al. (2021). Real-time hand gesture recognition based on deep learning YOLOv3 model. *Applied Sciences, 11,* 1–15.

Oskouei, R. J., Lou, Z. M., Bakhtiari, Z., & Jalbani, K. B. (2020). IoT-based healthcare support system for Alzheimer's patients. *Wireless Communications and Mobile Computing.*

Oudah, M., Al-Naji, A., & Chahl, J. (2020). Hand gesture recognition based on computer vision: A review of techniques. *Journal of Imaging, 6,* 1–29.

Shin, S., & Sung, W. (2016). Dynamic hand gesture recognition for wearable devices with low complexity recurrent neural networks. *2016 IEEE International Symposium on Circuits and Systems (ISCAS)* (pp. 2274–2277). IEEE.

Sivakani, R., & Ahmad Ansari, G. (2020). Imputation using machine learning techniques. In *2020 4th International conference on computer, communication and signal processing (ICCCSP)* (pp. 1–6).
Sivakani, R., & Ansari, G. A. (2020). Machine learning framework for implementing Alzheimer's disease. In *2020 International conference on communication and signal processing (ICCSP)* (pp. 0588–0592).
Stavropoulos, T. G., Papastergiou, A., Mpaltadoros, L., Nikolopoulos, S., & Kompatsiaris, I. (2020). IoT wearable sensors and devices in elderly care: A literature review. *Sensors, 20*, 1–22.
Surantha, N., Atmaja, P., & David, M. W. (2021). A review of wearable internet-of-things device for healthcare. *Procedia Computer Science*, 936–943.
Tran, V. T., Riveros, C., & Ravaud, P. (2019). Patients' views of wearable devices and AI in healthcare: findings from the ComPaRe e-cohort. *npj Digital Medicine, 2*, 1–8.
Yamada, Y., Shinkawa, K., Kobayashi, M., Nishimura, M., Nemoto, M., Tsukada, E., et al. (2021). *Tablet-based automatic assessment for early detection of Alzheimer's disease using speech responses to daily life questions*. Frontiers in Digital Health.
Yuan, G., Liu, X., Yan, Q., Qiao, S., Wang, Z., & Yuan, L. (2020). Hand gesture recognition using deep feature fusion network based on wearable sensors. *IEEE Sensors Journal, 21*, 539–547.

CHAPTER

13

A frequency analysis-based apnea detection algorithm using photoplethysmography

G. Gaurav[a] *and R.S. Anand*[b]

[a]Faculty of Engineering and Technology, Datta Meghe Institute of Medical Sciences University, Wardha, Maharashtra, India [b]Department of Electrical Engineering, Indian Institute of Technology–Roorkee, Roorkee, Uttarakhand, India

Introduction

Breathing is the autonomous and continuous process controlled by the autonomic nervous system and is vital as apart of respiratory process. Generally, breathing is involuntary according to body metabolism feedback and does not require conscious effort. However, sometimes it ceases temporarily due to conscious or involuntary effort. Breathing when stopped involuntarily continuously for more than 20s then is termed as apnea. Other than apnea, there can be hypopnea where breathing becomes shallow and very slow and does not fulfill the body oxygen demand (JayBlock, Boysen, Wynne, & Hunt, 1979). Along with breathing stop, apnea is accompanied with slowing of heart rate termed as bradycardia. Apnea when occurs in sleep is called sleep apnea. In case of adults, mostly it is sleep apnea and obstructs sleep, but in case of neonates, it may lead to life threatening complications. The result in case of neonates can cause stroke or sudden infant death syndrome (Hunt, 2007; Moon et al., 2011). Underdevelopment of the nervous system may be the reason in prematured neonates for apnea. About 80% of neonates born before 30 weeks of gestation suffer some level of apnea (Apnea of prematurity, n.d.).

Earlier methods developed for apnea detection are diaphragmatic surface electromyography (Ochoa, Osorio, Torres, & McLeod, 2009), high-resolution accelerometer, and EM fit pressure sensor (Reinvuo, Hannula, Sorvoja, Alasaarela, & Myllylä, 2006). The earlier methods for apnea detection were reported to involve monitoring of heart rate, breathing rhythm, arterial blood

https://doi.org/10.1016/B978-0-323-99031-8.00017-X

oxygen and saturation level (Dransfield & Fox, 1980). The application of numerous sensors creates complexity in the monitoring of cardiovascular conditions along with discomfort to the neonate. For long-term respiratory, activity of neonates has become popular due to its simplicity (Fouzas, Priftis, & Anthracopoulos, 2011; Johansson, Öberg, & Sedin, 1999). Photoplethysmography (PPG) has been used for detecting apnea and hypopnea (Lazazzera et al., 2020).

In this work, we establish the presence of respiratory-induced rhythm (RIR) in the PPG signal using Fourier transform analysis. A simple averaging and thresholding-based method is developed to detect apnea using single wavelength or single-channel PPG.

Method of obtaining a PPG

We propose a method to monitor apnea using a single PPG. PPG is a noninvasive optoelectronic method that measures the blood volume change close to the skin. PPG is obtained by optically illuminating a portion of body. Generally, the sensor probes are attached to the peripheral region of the body-like fingers of the limbs, earlobe, and toe. The output is collected from reflected or transmitted light back through a photodetector. The intensity of transmitted or the reflected light depends majorly on the bone, skin layers (dermis and epidermis), and volume of the blood flowing through arteries and veins near the portion of body. The reflected or transmitted light is converted to electrical voltage or current to receive a time-varying signal, termed as PPG (Webster, 1997). Fig. 13.1A shows a transmittance-type PPG probe.

(a) PPG

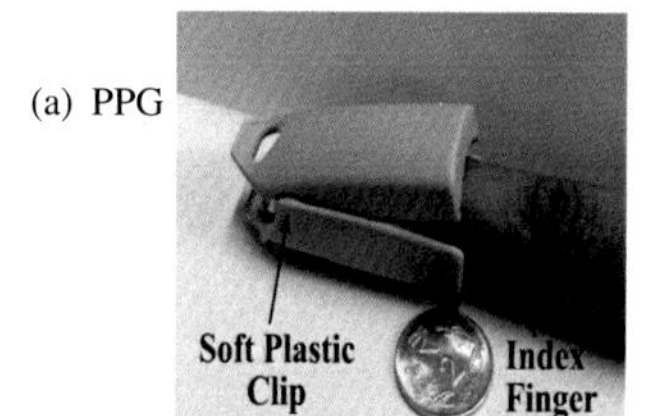

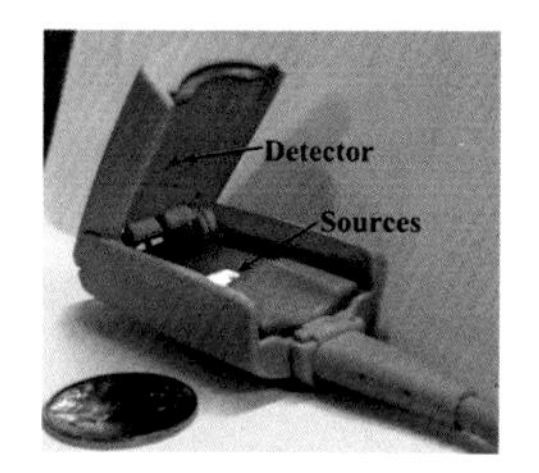

(b)

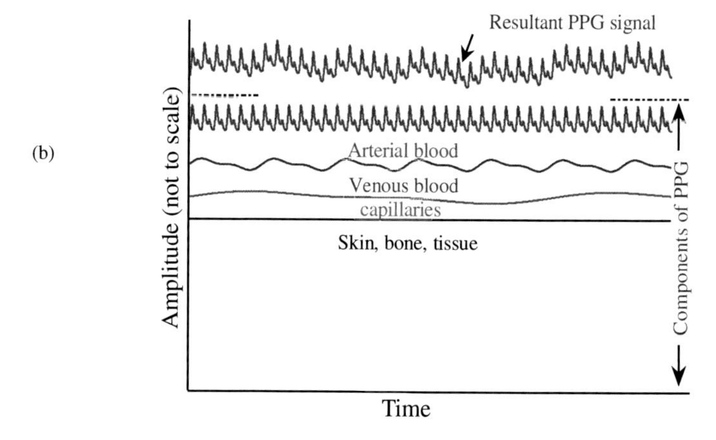

FIG. 13.1 PPG probe. (A) PPG probe (i) Nellcor transmittance probe showing detector and sources. (B) Typical PPG signal and its components. *No permission required.*

Components and uses of a PPG

Around nine-tenth of PPG is the DC component due to time-uniform regions like epidermis, bloodless skin, and bones. Due to the pumping of the blood through the heart which flows through arteries, time-varying response indicating heart pulse intensity is present. There is a slow time-varying component typically frequency lying between 0.05 and 0.5 Hz due to venous blood flow, which is aided by the breathing activity, so this portion of the signal corresponds to the respiratory signal. A very small portion around 1% of the signal is due to arterial blood perfusion. These significant frequency components can be observed in the power spectrum plot of the PPG signal. Fig. 13.1B shows DC and pulsatile components of the PPG signal due to various physiological sections (Barker & Tremper, 1987).

In general, PPG is used as a pulse oximeter to extract the arterial blood oxygen saturation level using two wavelengths (red: 660 nm and infrared: 940 nm) by the empirical comparative method (Reddy, George, Mohan, & Kumar, 2011). Also, heart pulse rate is possible to extract from arterial blood pulsation. It is hypothesized to have a relationship between venous blood pulsation and gaseous exchange in lungs (Brecher & Hubay, 1955). This hypothesis is used to extract breathing activity from single wavelength PPG.

PPG and respiratory-induced rhythm

Fig. 13.2 shows a pictorial schematic of a subject with sensors attached, describing the experimental setup to acquire the PPG signal and respiratory rhythm from the accelerometer

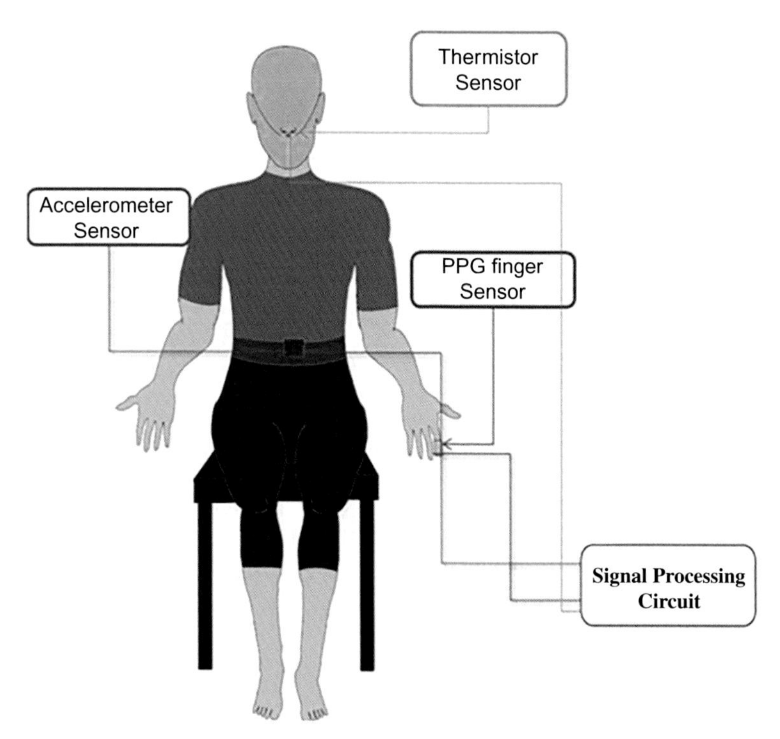

FIG. 13.2 Sensors placement. Sensors placement and alignment during data extraction from a volunteer. *No permission required.*

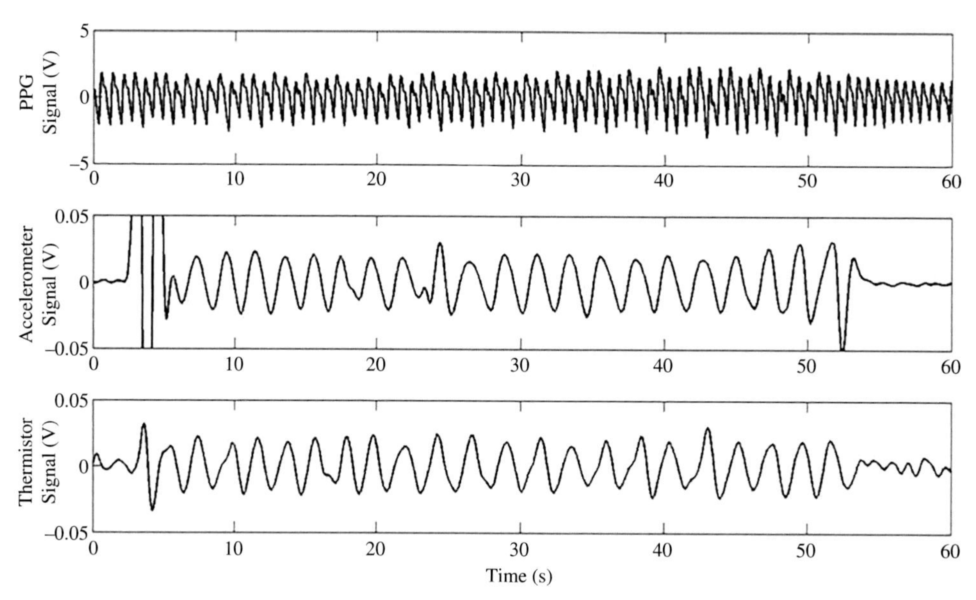

FIG. 13.3 Plot of the sensor signal. Plot of the 60-s segments. (A) PPG signal, (B) accelerometer signal, and (C) thermistor signal extracted of a volunteer. *No permission required.*

signal on abdomen and the thermistor signal on nostril. The PPG signal, accelerometer signal, and thermistor signals are shown in Fig. 13.3 for a typical volunteer. It is easily seen that the PPG does contain the breathing signal. In order to make the comparison more objective, the frequency spectrum of all the three signals are extracted and compared in Fig. 13.4.

It is observed in Fig. 13.4 that the PPG signal does possess a significant level of the frequency components matching the frequencies in the breathing signal.

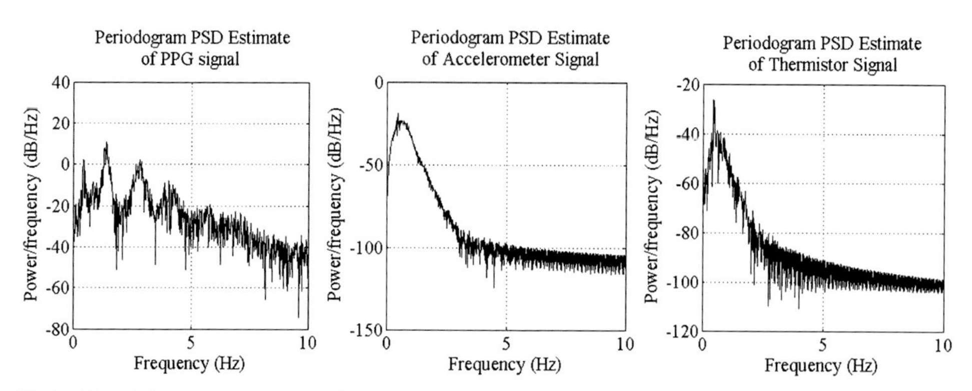

FIG. 13.4 Plot of the power spectrum of sensor signals. Plot of the power spectrum of various signals extracted of a volunteer: (A) PPG signal, (B) accelerometer signal, and (C) thermistor signal. *No permission required.*

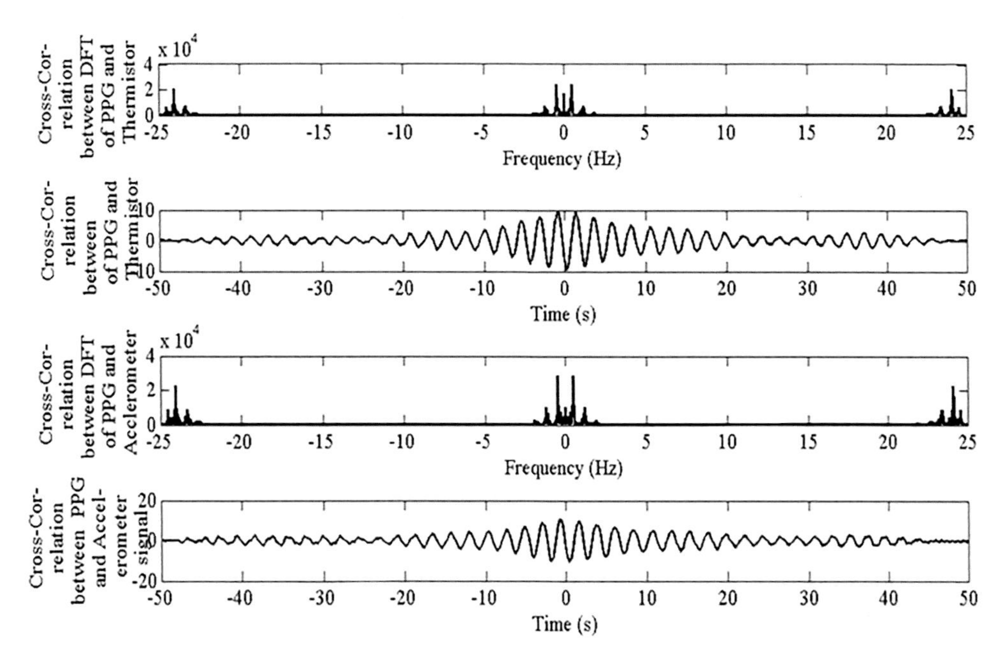

FIG. 13.5 Cross-correlation between the PPG signal and thermistor signal and the PPG signal and accelerometer signal. Plot of cross-correlation (B) between the PPG signal and thermistor signal generated due to breathing and (D) between the PPG signal and accelerometer signal generated due breathing and then cross-correlation of the discrete Fourier transform of (A) between the PPG signal and thermistor signal generated due to breathing and (C) between the PPG signal and accelerometer signal generated due to breathing. *No permission required.*

The cross-correlation between the PPG signal and accelerometer signal and the PPG signal and thermistor signal and the correlation of their respective Fourier transform is shown in Fig. 13.5. It is quite evident that the PPG signal and respiratory rhythm captured through an accelerometer or a thermistor are highly correlated.

Steps involved in apnea detection through PPG

PPG signal collection and Fourier transform extraction of the segment collected rhythm

From the PPG signal *ppg(i)*, a sufficiently large sample of length N_w is collected *[ppg(1):ppg* $(N_w)]$ denoted by *p(i,1)*, where *i* is the discrete time interval of the sample varying from $i=1$ to $i=N_w$ and *p(i,j)* is the *j*th sample of ppg, where each *p(i,j)* is of N_w length.

N_w depends upon the sample frequency f_s and the expectable range of breathing frequency *B*, where *B* is the least possible breathing frequency to be detected. To get the proper information of the breathing pattern inscribed in *p(i,j)*, a sample of at least *1/B* time length is taken. So visually a length of N_w equals to f_s/B is taken for analysis in each cycle. A discrete Fourier transform (DFT) of the collected sample is taken as *f (i,j)*, which is of the same length as the *p(i)* sample, that is, N_w.

Apnea detection method

The *f(i)* is processed to get the strength of the respiratory signal. Now from the *f(i)*, the frequency band that lies in the breathing frequency *[Bmin, Bmax]* is integrated and divided by the integration of the *f(i)* into the band of the second overtone of the heart pulse frequency, that is, *[0,2H]*; this ratio is termed as *Q(j)*.

$$R(j) = \sum_{i=Bmin*N_w/f_s}^{Bmax*N_w/f_s} f(i,j),\ H(j) = \sum_{i=0}^{2H*N_w/f_s} f(i,j)$$

$$Q(j) = \frac{R(j)}{H(j)}$$

Once *Q(j)* is computed, first *Nd* samples from *ppg(i)* are deleted and *Nd* new samples are appended, which creates new strand as *[ppg(1+Nd): ppg(Nw+Nd)]* denoted by *p(i,2)*; similarly, *p(i,j)* is *[ppg(1+j*Nd):ppg(Nw+j*Nd)]*.

An average of the *Q(j)* is computed as *(j)* after sufficient samples being collected, given by

$$\overline{Q}(j) = \frac{1}{N} \sum_{k=j-N_a-1}^{j-1} Q(k)$$

The computed is compared with *Q(j)*. If the *Q(j)* computed at any instant is less than 0.5 times, that is, if $Q(j) < 0.5*\overline{Q}(j)$, then apnea alarm is flagged on and a visual/audio signal is set, else flagged off. If the apnea alarm is on, then the last $\overline{Q}(j)$ value is kept intact for the next sample detection, that is, the $\overline{Q}(j)$ is kept frozen for the next time sample apnea detection, but further processing is continued. To check the efficacy of the mentioned method of apnea detection, the experiment was conducted on volunteers; details of the experiment are mentioned later. Fig. 13.6 illustrates the steps involved in apnea detection as a flow chart.

Results and discussion

Checking the sensitivity of the apnea detection algorithm based on the strand length of PPG used

Fig. 13.4 illustrates the Fourier spectrum curve of a strand sample of the PPG signal acquired. The portion of this spectrum from 0.05 to 0.6 Hz frequency, which lies in the respiratory frequency region, is integrated and divided by another portion which is integration of the spectrum from 0 to 2 Hz; at full band till heart rate frequency, this ratio is termed as *Q*. The *Q* will be larger if there is rapid or intense breathing and will be smaller if there is shallow breathing, cessation in breathing or apnea. Now to determine what length of the PPG signal is appropriate for apnea detection different strand length was analyzed with the apnea algorithm.

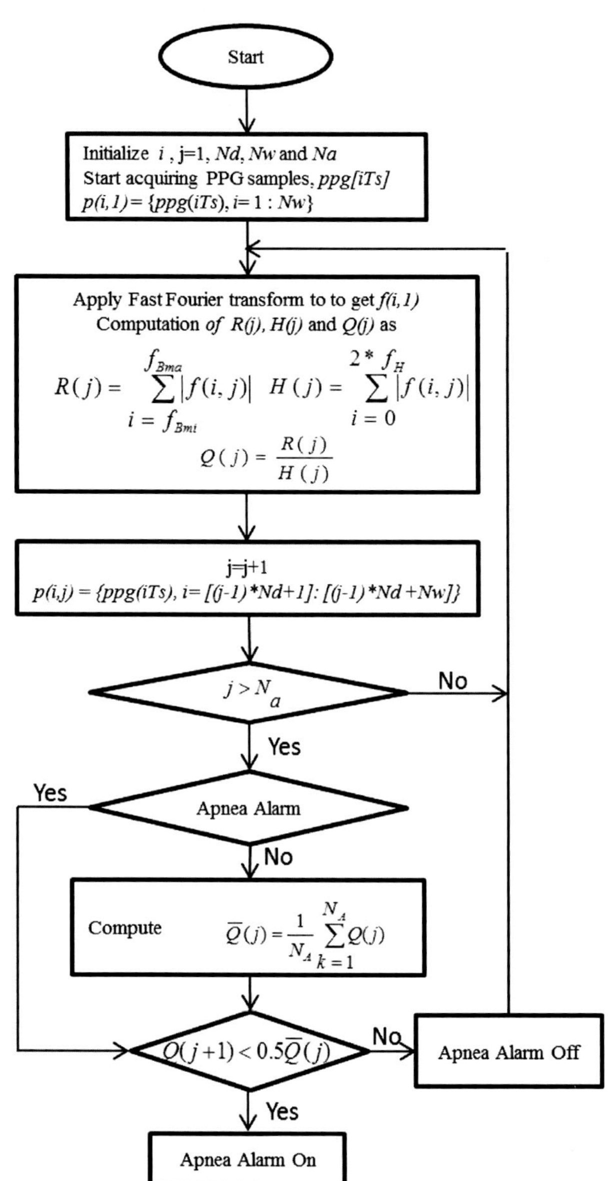

FIG. 13.6 Apnea detection algorithm. Flow chart of the apnea detection algorithm. *No permission required.*

Sensitivity of the apnea detection algorithm based on the strand length of PPG used

Fig. 13.7 illustrates the Fourier spectrum curve of a strand sample of the PPG signal acquired. The portion of this spectrum from 0.05 to 0.6 Hz frequency, which lies in the respiratory frequency region, is integrated and divided by another portion which is integration of the

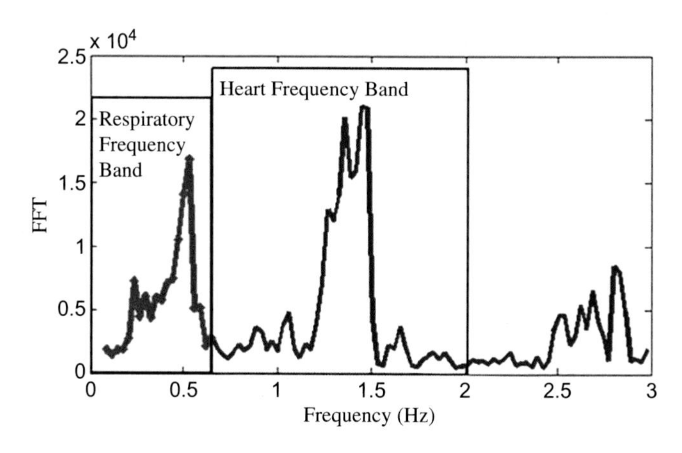

FIG. 13.7 Fourier transform spectrum. Fourier transform spectrum of 10s strand of the PPG signal. *No permission required.*

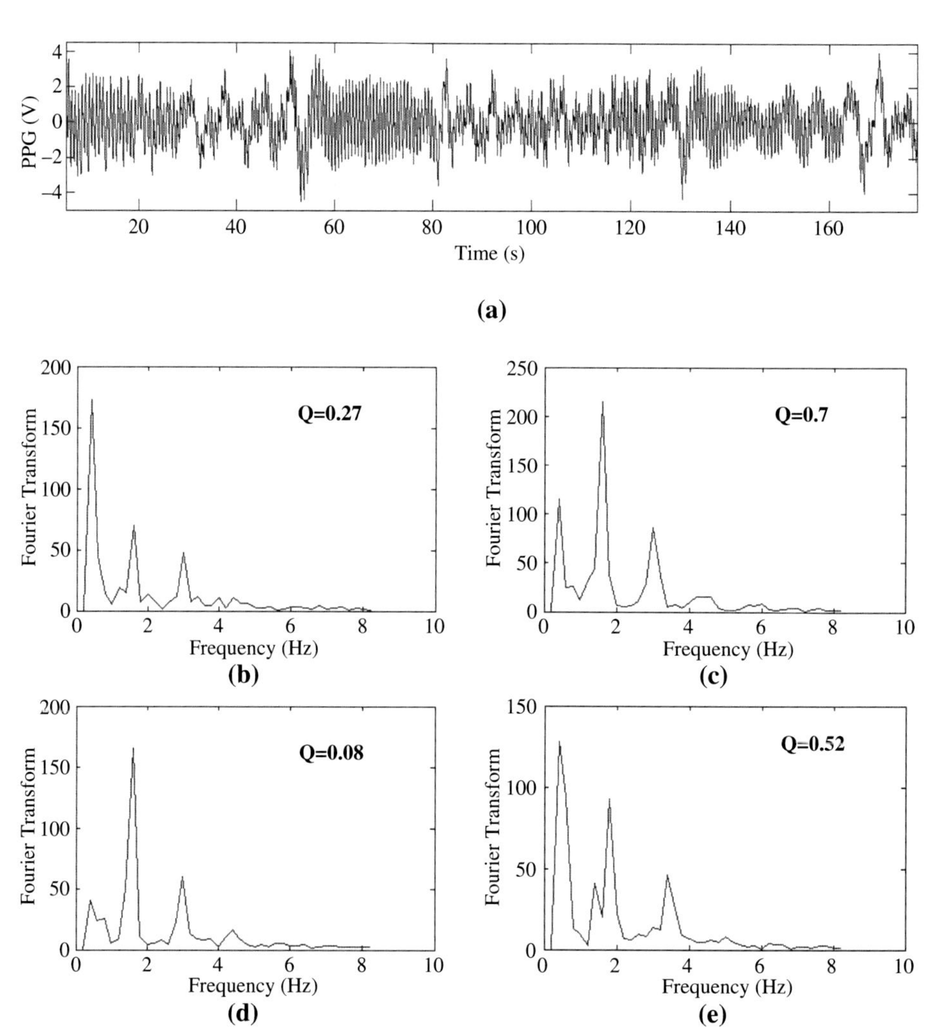

FIG. 13.8 Fourier transform-based time strand. Fourier transform-based frequency spectrum of different time strands, where (A) is the full segment of the PPG signal, (B) is the frequency spectrum for the PPG segment between 30 and 35s, (C) is the frequency spectrum for the PPG segment between 50 and 55s, (D) is the frequency spectrum for the PPG segment between 75 and 80s, and (E) is the frequency spectrum for the PPG segment between 90 and 95s. *No permission required.*

spectrum from 0 to 2 Hz; at full band till heart rate frequency, this ratio is termed as Q. The Q will be larger if there is rapid or intense breathing and will be smaller if there is shallow breathing, cessation in breathing or apnea. Fig. 13.8 shows Fourier transform for 10 s segment of PPG at different time intervals and variation in Q during normal breathing and apnea intervals. Fig. 13.8 clearly explains how the value of $Q(i)$ changes during breathing and apnea; decreases when there is apnea and returns to a larger value when the breathing restarts. Each segment of PPG taken for determining Q is of 5 s time duration. Added to it, the heart pulse rate and the blood oxygen saturation level (% SpO_2) are also recorded.

To determine the suitable strand length of the PPG signal for the apnea detection algorithm, different sample lengths were analyzed with the apnea algorithm. Fig. 13.9 shows the result of apnea algorithm on different strand length visually 5, 6, 8, and 10 s as apnea 1, 2, 3, and 4. The apnea algorithm output is compared to the accelerometer response for different strand lengths. It is observed that 5 s strand length is faster and accurate than other strand length. Whereas for strand length less than 5 s, the resolution of DFT becomes very less for respiratory frequency band analysis.

Experiment to determine the latency in apnea detection

To check the validity of the algorithm, the test was conducted on volunteers after taking informed permission from them. The algorithm was tested with 16 volunteers by simulating apnea. First, they were guided to take continuous, smooth, and deep breaths for 1 min, then were asked to stop breathing 20 s, and then again were asked to breathe in the previous pattern. This cycle was repeated 2 or 3 times. Since two PPG signals red and IR are available so along with heart rate, arterial blood oxygen saturation level is also monitored and recorded (Johansson et al., 1999).

Fig. 13.10 shows the data recorded from a volunteer. There is a time delay in apnea detection and accelerometer response showing the respiratory signal. The figure clearly shows that the apnea detected is not evident from the recording of the blood oxygen saturation level and or heart rate variability. Fig. 13.10 also shows that the apnea detection signal derived from the apnea algorithm is sensitive to apnea without any ambiguity. The time taken by the apnea algorithm to detect apnea for different volunteers was recorded and tabulated in Table 13.1.

Incapability of the PPG signal to detect obstructive apnea

Fig. 13.11 shows situation where there is cessation in air flow shown in the thermistor signal from time 130 to 190 s and from 310 to 380 s, whereas an accelerometer shows rhythm due to effort to breathe in the form of abdominal movement. $Q(i)$ plot in such a situation is not able to detect apnea (obstructive) as seen in Fig. 13.11, whereas %SpO_2 continuously decreases and even reaches the risk value less than 85%. Thus it validates that RIR due to the PPG signal is not sufficient to detect apnea.

Conclusion

A simple method to detect apnea based on the frequency component strength of the PPG signal is developed. PPG is a very commonly used optoelectronic method in the form of a

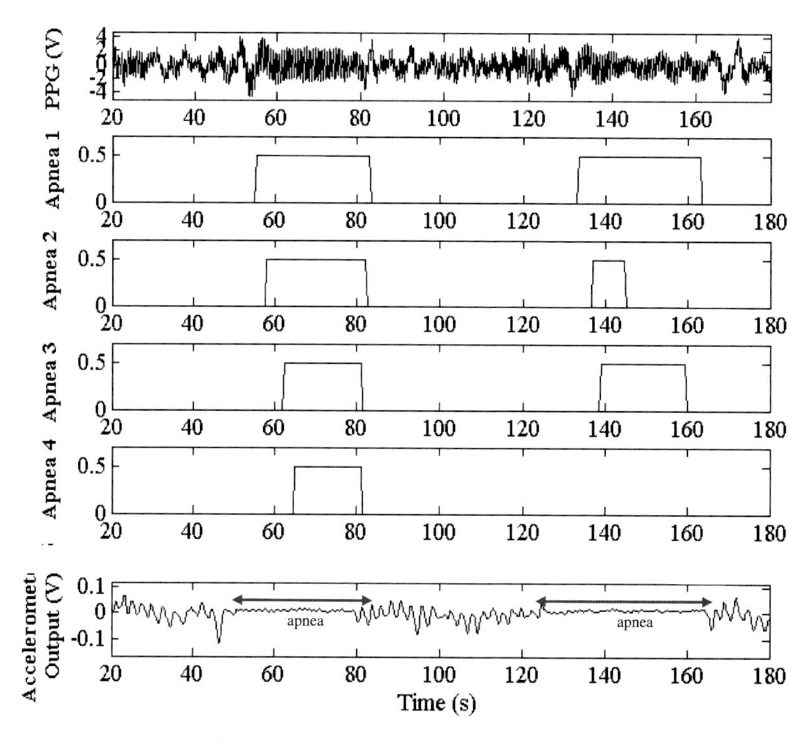

FIG. 13.9 Apnea detection for different time strands. Apnea detection apnea 1, 2, 3, and 4 using different time strands of 5, 6, 8, 10, and 20s. An accelerometer signal used as a reference with apnea at $t = 50–80$ s and $t = 125–165$ s. *No permission required.*

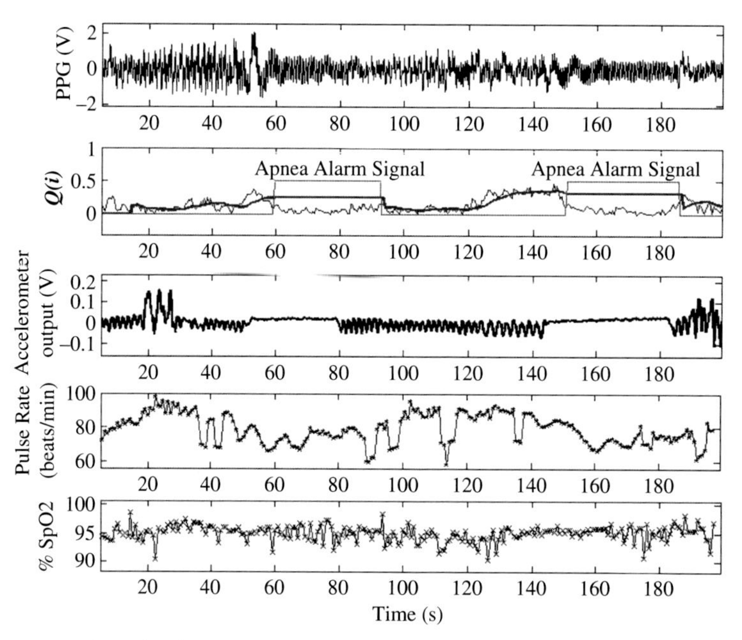

FIG. 13.10 Experimental results showing the PPG signal. Experimental results showing the PPG signal, apnea detection signal $Q(j)$, accelerometer output, pulse rate, and $\%SpO_2$. *No permission required.*

TABLE 13.1 Time taken to detect apnea and recovery

Volunteer no.	Time taken to detect apnea (s)	Time taken to recovery (s)	Volunteer no.	Time taken to detect apnea (s)	Time taken to recovery (s)
1	6.9	3.7	9	10.1	1.1
2	8.3	1.4	10	5.5	2
3	9.1	4	11	8.5	3
4	6.1	6.3	12	13	1.6
5	6.3	3.2	13	3.2	2.2
6	7.7	1.6	14	9.7	0.8
7	7.6	3.8	15	11.3	1.5
8	13.7	1.4	16	13	1.2

No permission required.

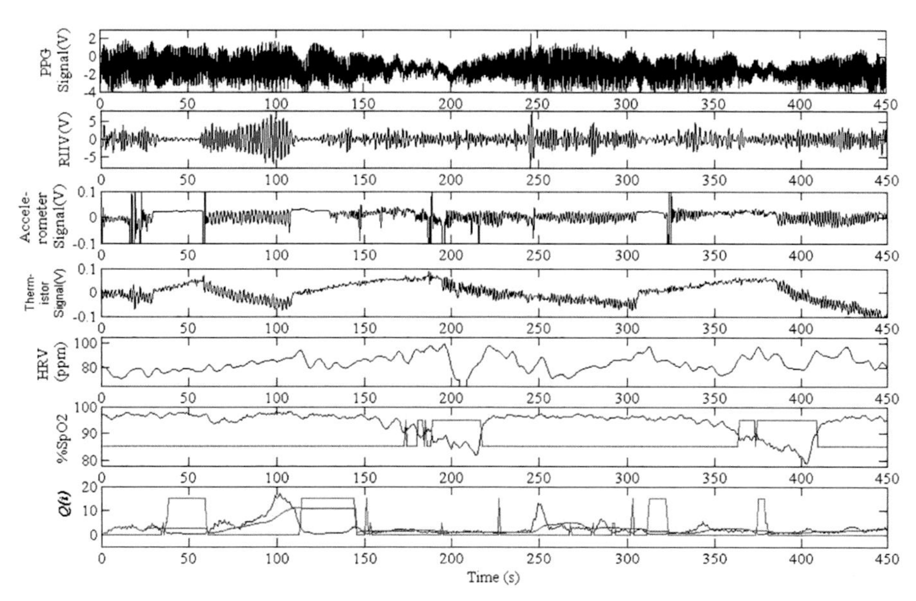

FIG. 13.11 Experimental results of subject B. Experimental results of subject B: (A) IR PPG, (B) RIIV through PPG signal, (C) accelerometer output, (D) thermistor signal reference to nasal airflow, (E) heart rate variability, (F) $\%SpO_2$ and asphyxia signal showing $\%SpO_2$ less than 90%, and (G) apnea signal. *No permission required.*

pulse oximeter for long-duration continuous cardiovascular health monitoring; this method can add the feature of apnea detection to the commercial pulse oximeter for a clinical use at a very low cost. Due to its simplicity and no ambiguity, it can be used to monitor respiratory rhythm and apnea along with other parameters like heart rate, blood oxygen saturation level. The experimental result from 16 volunteer establishes the efficacy of the method. Due to its noninvasive use, it is secured and nonhazardous to the patient being monitored, hence can also be use d to monitor the apnea in neonates. On the contrary, the proposed method is incapable in detecting the obstructive apnea condition.

References

Apnea of prematurity. (n.d.). http://www.aboutkidshealth.ca.

Barker, S. J., & Tremper, K. K. (1987). Pulse oximetry: Applications and limitations. *International Anesthesiology Clinics, 25*(3), 155–175.

Brecher, G. A., & Hubay, C. A. (1955). Pulmonary blood flow and venous return during spontaneous respiration. *Circulation Research, 3*(2), 210–214. https://doi.org/10.1161/01.RES.3.2.210.

Dransfield, D. A., & Fox, W. W. (1980). A noninvasive method for recording central and obstructive apnea with bradycardia in infants. *Critical Care Medicine, 8*(11), 663–666. https://doi.org/10.1097/00003246-198011000-00015.

Fouzas, S., Priftis, K. N., & Anthracopoulos, M. B. (2011). Pulse oximetry in pediatric practice. *Pediatrics, 128*(4), 740–752. https://doi.org/10.1542/peds.2011-0271.

Hunt, C. E. (2007). Small for gestational age infants and sudden infant death syndrome: A confluence of complex conditions. *Archives of Disease in Childhood: Fetal and Neonatal Edition, 92*(6), F428–F429. https://doi.org/10.1136/adc.2006.112243.

JayBlock, A., Boysen, P. G., Wynne, J. W., & Hunt, L. A. (1979). Sleep apnea, hypopnea and oxygen desaturation in normal subjects: A strong male predominance. *New England Journal of Medicine, 300*(10), 513–517. https://doi.org/10.1056/NEJM197903083001001.

Johansson, A., Öberg, P. A., & Sedin, G. (1999). Monitoring of heart and respiratory rates in newborn infants using a new photoplethysmographic technique. *Journal of Clinical Monitoring and Computing, 15*(7–8), 461–467. https://doi.org/10.1023/a:1009912831366.

Lazazzera, R., Deviaene, M., Varon, C., Buyse, B., Testelmans, D., Laguna, P., … Carrault, G. (2020). Detection and classification of sleep apnea and hypopnea using PPG and SpO_2 signals. *IEEE Transactions on Biomedical Engineering, 68*(5), 1496–1506. https://doi.org/10.1109/TBME.2020.3028041.

Moon, R. Y., Darnall, R. A., Goodstein, M. H., Hauck, F. R., Willinger, M., Shapiro-Mendoza, C. K., et al. (2011). SIDS and other sleep-related infant deaths: Expansion of recommendations for a safe infant sleeping environment. *Pediatrics, 128*(5), 1030–1039. https://doi.org/10.1542/peds.2011-2284.

Ochoa, J. M., Osorio, J. S., Torres, R., & McLeod, C. N. (2009). Development of an apnea detector for neonates using diaphragmatic surface electromyography. In *Proceedings of the 31st annual international conference of the IEEE engineering in medicine and biology society: engineering the future of biomedicine, EMBC 2009* (pp. 7095–7098). IEEE Computer Society. https://doi.org/10.1109/IEMBS.2009.5332906.

Reddy, A. K., George, N., Mohan, V., & Kumar. (2011). A novel method for the measurement of oxygen saturation in arterial blood. In *Proceedings of the IEEE I2MTC2011* (pp. 1627–1630).

Reinvuo, T., Hannula, M., Sorvoja, H., Alasaarela, E., & Myllylä, R. (2006). Measurement of respiratory rate with high-resolution accelerometer and EMFit pressure sensor. In *Proceedings of the 2006 IEEE sensors applications symposium* (pp. 192–195).

Webster, J. G. (1997). *Design of pulse oximeters*. Boca Raton: CRC Press.

CHAPTER

14

Noninvasive health monitoring using bioelectrical impedance analysis

Mahmood Aldobali[a], *Kirti Pal*[b], *and Harvinder Chhabra*[c]

[a]Department of Electrical Engineering, Gautam Buddha University, Greater Noida, Uttar Pradesh, India [b]Department of Electrical Engineering, School of Engineering, Gautam Buddha University, Greater Noida, UP, India [c]Indian Spinal Injuries Centre, New Delhi, India

Introduction

Bioelectrical impedance analysis (BIA) is a simple, fast, portable, noninvasive, safe, and inexpensive method and easy to use. BIA signifies a novel approach to evaluating body composition (BC) (Lukaski, Johnson, Bolonchuk, & Lykken, 1985). Consequently, BIA acts suitable for the laboratory, clinical, therapist, and field evaluation of a human body of tissue and fluid. BIA has parameters compartment models such as "Body Cell Mass (BCM), Body Mass Index (BMI), Fat Mass (FM), Fat-Free Mass (FFM), Total Body Water (TBW), Extracellular Water (ECW), Intracellular Water (ICW), Basal Metabolic Rate (BMR), Phase Angle (PhA), Resistance (R), Reactance (Xc), Body Capacitance (C)" (Brožek, Grande, Anderson, & Keys, 1963; Kushner & Schoeller, 1986; Siri & Lukaski, 1993; Wang et al., 1992). BC is an essential part of health and nutrition evaluation. Therefore, BIA is applied in health-related examinations, clinical laboratory medicine, and nutritional status in many patients or populations (Aldobali & Pal, 2021). However, BIA depends on the electrical conductivity used in the body. A low-level constant alternating current is presented as the resistance that depends on the frequency of current flow in the body and measures parameters such as (R, Xc, PhA, Z, and C). So, two or four surface electrodes are used to conduct a BIA instrument (Lukaski, 1987). Accordingly, Electrical bioimpedance varies with the signal frequency (SF), and multifrequency (MF) produces more information regarding tissue properties, which supports improving the characterization of tissue (Mikes et al., 1999). Nevertheless, BIA permits early detection of an inaccurate balance in BC, enhancing intervention and prevention initiatives. Regardless, BIA provides a measure of fluid and mass that may be a critical evaluation tool for the present health condition. Consequently, the most common BIA application

Computational Intelligence in Healthcare Applications
https://doi.org/10.1016/B978-0-323-99031-8.00008-9

regions include targeted control of treating weight methods, like patient hydration evaluations in intensive care units, including dialysis (González-Correa, 2018). This chapter aims to provide a summary of the fundamentals of a variety of compartment body composition models developed based on noninvasive clinical and nutritional effective treatments utilizing bioelectrical impedance analysis.

Principles of bioelectrical impedance analysis (BIA)

From the principle of a physical property characterized by mineral conductors in electrical circuits, it is well known that the resistance of the conducting materials is straightforwardly relative to the length of the conductor (L), including inversely proportionate into the "cross-sectional area" (A), as given in Fig. 14.1. Even though the body is not a single cylinder and its access is limited. Thus, an experiment may be conducted to determine the impedance Z coefficient (length2/R) and the amount of water that serves as an electrolyte that conducts electrical current throughout the body (Kyle et al., 2004b). To better grasp BIA performance and body composition evaluations, it is necessary to understand better the underlying physical described by impedance. The conductor volume is determined by calculating the right side of this equalization.

$$Z = \rho \frac{L*L}{LA} \tag{14.1}$$

where Z = impedance of a biological and geometrical system in ohm Ω.

ρ=Volume resistivity in ohm (Ω-cm) depending on the material, A = Conductor cross-sectional area (m^2). L = Conductor length in (cm). As Volume Where, V = Volume of cylinder $= A * L$.

$$\text{Then}, V = \rho L^2/Z \tag{14.2}$$

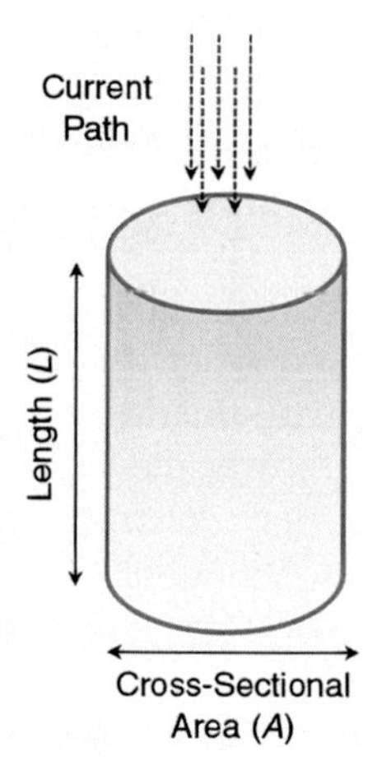

FIG. 14.1 The cylinder shape of the relation between impedance and geometry.

If ρ is constant (Kushner, Schoeller, Fjeld, & Danford, 1992).

Hence, from Eq. (14.2) concerning a single-cylinder, the conductor volume is straight proportionate to the square of the conductor length and inversely proportionate to the conductor impedance. The theoretical relation between impedance and electrical volume was suggested by Nyober, who proved that electrically determined biological volume is inversely related to the triangular system's impedance (Nyboer, 1960).

$$\text{Then}, z^2 = R^2 + X^2 \tag{14.3}$$

If X is supposed to be negligible, then

$$\sqrt{Z^2} = \sqrt{R^2} = \text{and } Z = R.\text{Hence}, V = L^2/Z \tag{14.4}$$

In practice, height is simply determining more than the conductive length from the wrist to the neck. The experimental relationship between lean body mass (LBM) (usually 73% water) and (H^2/R). Most of the suggested methods for measuring TBW were tested. It consisted of a regression equation for the (H^2/R) resistance index, which needs an obligation to match the current geometry through a proper coefficient. This parameter depends on many aspects, including the shape of the components under examination. Most BIA instruments use a single frequency (SF) at 50 kHz and are produced on the report of the assumption of continuous hydration of the body fat mass for FFM estimation. Consequently, some errors happen once there are conductive changes, differences in height compared to conductor length, and dissimilarities in body figures and body fragments. (The body parts act as if they were chained together, giving shorter and thicker layers more minuscule than the total resistance (R)). An additional challenge observed in BIA is that the body consists of two different types of resistance: capacitive resistance and resistive resistance (Kyle et al., 2004a). Capacitive resistance appeared from the cell membrane, while there was slight resistance due to the fluid inside the cells and between cells. Another complication is that the body provides resistance (R) electric current: resistive resistance and capacitive reactance. Capacitance (C) is issued from cell membranes and resistance R from fluid outside and inside the cells. Impedance is the word utilized to represent the compound of capacitance and resistance. Many electrical circuits were used to characterize the performance of biological membranes in vivo (Gudivaka, Schoeller, Kushner, & Bolt, 1999). Including one R arrangement and capacitance, respectively, and the other in parallel Fig. 14.2, while others are difficult. The circuit generally utilized to describe biological membranes in vivo is the circle in which the resistance R is regulated for the extracellular fluid ECF in parallel with another arm of the circle, which involves the capacitance C and resistance R to the intracellular Fluid ICF, respectively. The resistance and capacitance across the range of frequencies can be measured. The BIA operates at a single frequency SF at (50 kHz). At low frequencies, the current does not affect the performance of the cell layer, which works as insulation. Hence, the ECF's current paths are accountable for determining resistance R for the body R0.

The capacitor is an ideal (or almost perfect) capacitor at a constant frequency (or too high frequency). Therefore, a combination of ICW and ECF indicates whole-body resistance ($R\infty$). Due to the limitations of work and the phenomenon of multiple distributions that preclude the principal usage of direct current low or very high-frequencies and AC, it is assumed that resistance values at specific frequencies apply the Cole-Cole plot.

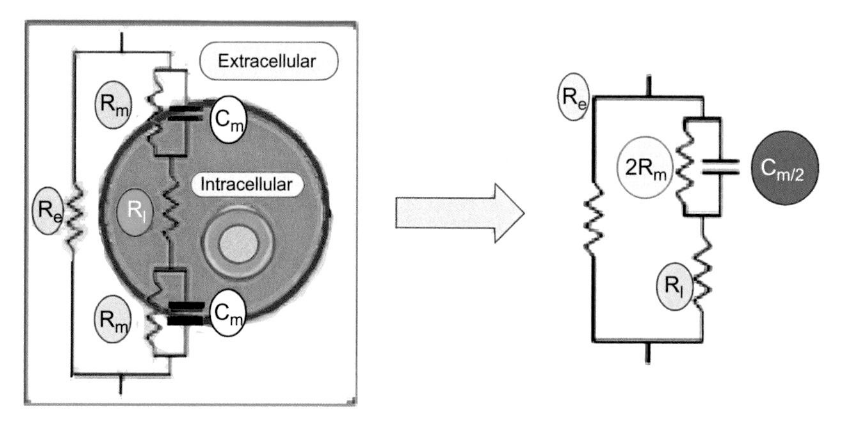

FIG. 14.2 Frick's circuit: Equivalent Circuit of a Cell.

Biological tissues of bioimpedance analysis

There are two parts to biological tissue impedance: resistance and responsiveness (capacitive). While only cell layers offer capacitive reactors, TBW and ECW give resistance to electric current. In dividing the cell, the membrane performs a passive function in transferring and adjusting ICW and ECW media. It also has a vital role in exchanging chemical varieties. The negative cell membrane is referred to as the BLM. According to the illustration (Fig. 14.1), this layer (7 nm thick) serves to let the movement of fat and water molecules, and likewise, it entirely prevents the passage of ions (Tuorkey, 2012; Lukaski, 1999). The essential electrical conductivity is relatively low and can be deemed insulation. The structure consists of an extracellular medium, BLM, and the intracellular medium is the dielectric conductor structure, and it works as a capacitor ($\sim 1\ \mu F/cm^2$). Fig. 14.3 shows the two-layer lipid membrane and its equivalent electrical circuit. Hence, cell membranes provide capacitive reactance. As a cell membrane does not enclose fat tissue cells, reactors are not influenced by fat in the body. Parallel to BLM, there are implanted proteins, carrier organelles, ion channels, and ion

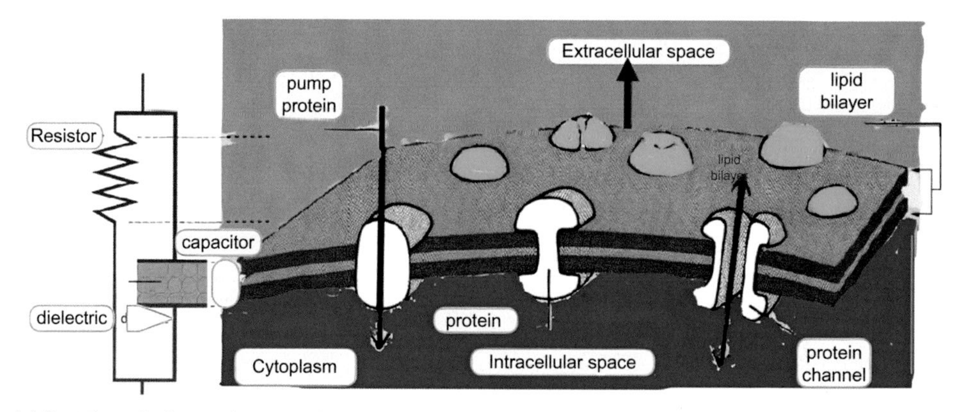

FIG. 14.3 The cell electrical circuit plasma membrane and its counterpart. *From http://www.rjlsystems.com.*

pumps. These structures are the essential components of the active role of the membrane. Ion channels are an acceptable structure that permits some ions to pass from the external into the cell or vice versa. These structures are eclectic to the atoms and can be opened or closed by a few electrical or chemical signals (Chertow et al., 1995; Kremer, Latin, Berg, & Stanek, 1998).

How BIA works

BIA works with an alternating current (AC) range of 800 μA to 10 mA at 50 kHz that flows through the human body's water. The voltage is measured, and the characteristics of body composition are computed. Muscles store the majority of the body's water. For this reason, if an individual has more muscle mass, they will have less body water, which decreases resistance (Lichtenbelt, 2010). Hence, Ohm's law, which says that the R of a component is proportional to the voltage reduction of the current as it runs through the resistive element, is applied to measure the whole-body or resistance:

$$R = \frac{V}{I} \tag{14.5}$$

where V = applied voltage drops (V), current (A).

Mainly all investigators have adopted a right-sided tetrapolar surface electrode technique since the benchmark research of Hoffer (Hoffer, Meador, & Simpson, 1969). After placing the subject on a nonconductive bed, separate the arms from the trunk and the legs. As shown in Fig. 14.4, two distal electrodes or an input signal (I) are located on the dorsal surfaces of a single hand and foot near the dearest to the metacarpal bones, sequentially. Two polar voltages (V) are connected to the wrist's granular projection and between the ankle's foot and center side. The impedance analyzer produces flows between 100 and 800 μA, over a scale of frequencies from 1.0 to 1000.0 kHz, and is established among the surface terminals.

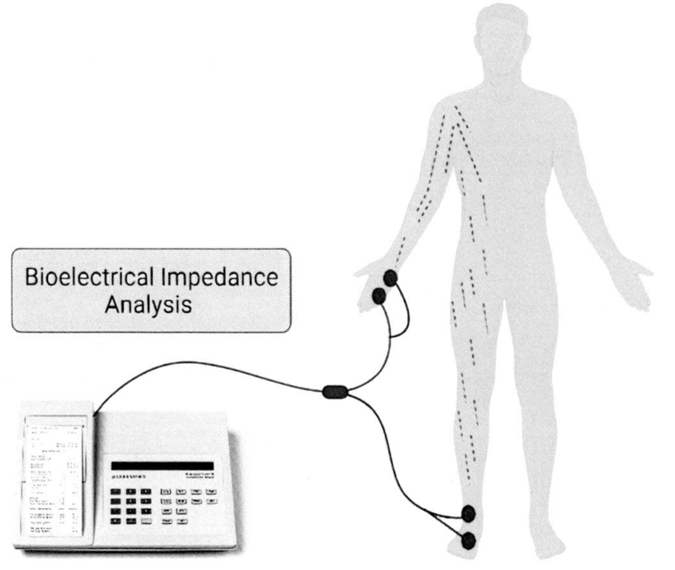

FIG. 14.4 The BIA configuration with four electrodes.

The voltage drop is estimated within the internal poles. The application of this 4-electrode design enables the high surface impedance to be bypassed, and then the impedance (R) and Xc are measured (Foster & Lukaski, 1996).

Bioimpedance analysis: Bioimpedance analyzers use a regression formula to calculate body FFM, TBW, ICW, BCM. The values from the bioimpedance analyzer are sent to the microprocessor in the analyzer, where the computations are performed. The computer-based analyzer uses formulas that encode FFM, BCM, TBW, and ICW, shown in Fig. 14.5.Methods of BIA

BIA has two types of frequencies: single and multiple frequencies depend on the electrical current utilized to determine the impedance (resistance) of body tissues shown in Fig. 14.6. Hence, BIA operating frequencies are different. The mass of tissues (particularly muscle and fat) is responsive to lower electric current frequencies, whereas the tissues of the neurological system require a higher frequency.

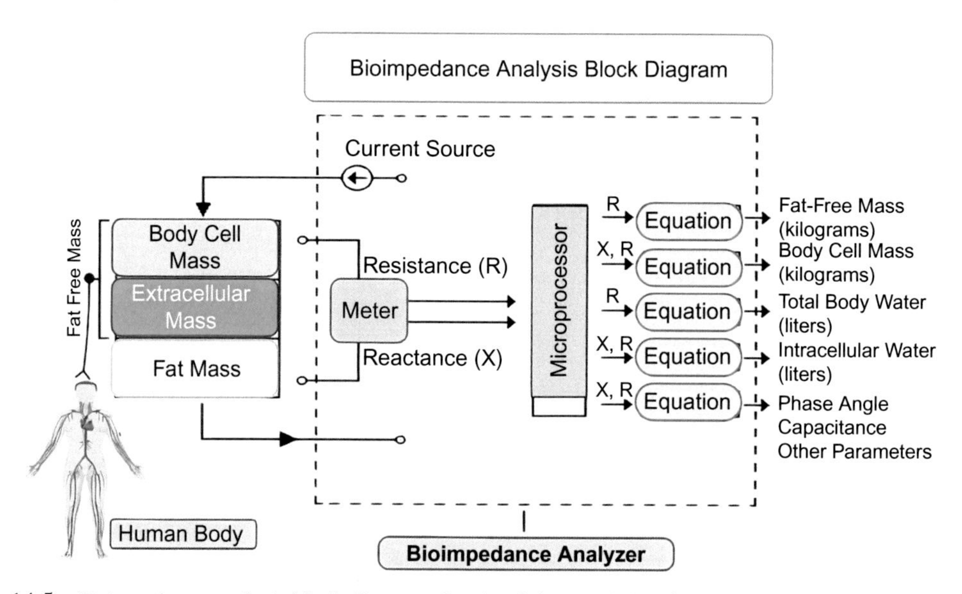

FIG. 14.5 Bioimpedance analysis block diagram. *Reprinted, by permission, from Biodynamics Corporation, Seattle, WA, U.S.A.*

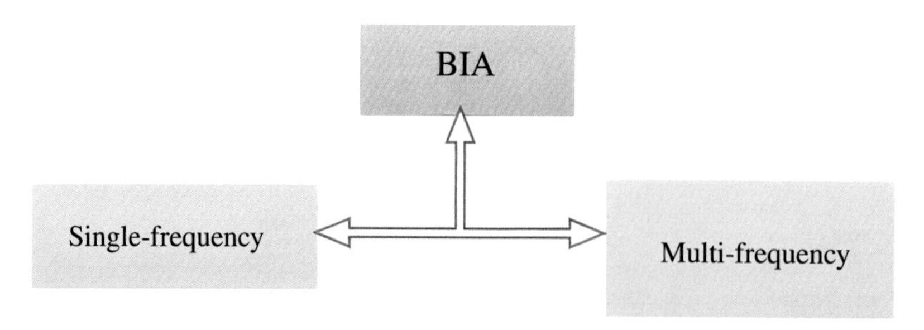

FIG. 14.6 Types of BIA.

Single-frequency BIA (SF-BIA) The SF-BIA is known for electrical current transfer within hand and foot surface terminals at 50 kHz. SF-BIA is considered the commonly utilized and oldest advanced technique of evaluation of body compartments. It defines an electrical current's conductive route, which has the opposite relationship between rated impedance and TBW (Kyle, Bosaeus, et al., 2004b; Martinsen & Schwan, 2015). BIA permits observing body fluids (extracellular/intracellular proportion) and subjects' nutritional status (Barbosa-Silva & Barros, 2005; Chen et al., 2016). SF-BIA tools evaluated TBW and FFM by applying derivative equations individually for healthy, hydrated individuals. SF-BIA presents limitations in different intracellular fluid ICF predictions. However, many investigations report a satisfying relationship in ICF evaluation (Olde Rikkert, Deurenberg, Jansen, Van't Hof, & Hoefnagels, 1997).

Multifrequency BIA (MF-BIA) Analysis of BIA can be conducted using concurrently with the electric current at various frequencies. MF-BIA processes multiple regression calculations evaluations of the TBW, ICW, ECW, FFM compartments by identifying them and low 5 kHz to above 500 kHz high-frequency electrical currents, respectively (Thomasset, 1962). It was developed in the 1990s when scholars employed low and high frequencies from 50 to 500 kHz to determine similar parameters (Ellis et al., 1999). On the contrary, at larger frequencies (>50 kHz), current flows through the cell layers and relates to intracellular and extracellular fluid cells (Deurenberg, 1996). According to "Patel et al.," MF-BIA remained more stable and less tendentious than SF-BIA to predict ECW. In contrast, SF-BIA was, contrasted to MF-BIA, extra reliable and less inclined to TBW in people with serious illnesses. However, in subsequent years, Jafrin et al. declared that BIA should technically employ a frequency range between 5 kHz and 1 MHz. They have indicated that MF-BIA, related to "bioelectric spectroscopy BIS," produces more reliable predictions of TBW and the exact estimate of ECW in clinical inpatients (Bera, 2014) (Hannan, Cowen, Freeman, & Wrate, 1993).

Factors of predicted body composition

Bioimpedance analysis is a broadly practiced technique to assess BC for epidemiological and clinical healthcare systems, evaluate body compartments, estimate the consistent variety of a nutritional state in a population, and control health opportunities for patients. The human body is formed from the intricate combination of various body membranes consisting of water, protein, fats, and minerals called (BC). The human body involves 20.0% protein, 64.0% water, 10.0% fat, 5.0% minerals, and almost 1.0% carbohydrate shown in Fig. 14.7 (Janaway, Percival, & Wilson, 2009).

Many techniques use different assumptions to calculate (BC) depending on the number of elements. At the moment, immediate in vivo assessment of BC is not achievable. Using indirect techniques and models were created. The "World Health Organization" (WHO) defines "nutritional status" in this context as the state of the body as a result of equilibrium of intake, absorption, and nutrients interacting with the individual's physiologic conditions. The study found that BC is directly linked to a series of health-related problems and the difference in mortality, including injury immunity, longevity, high functionality, and implementation of sports shown in Fig. 14.8. As we know from the tissue properties, the body has more fat. High

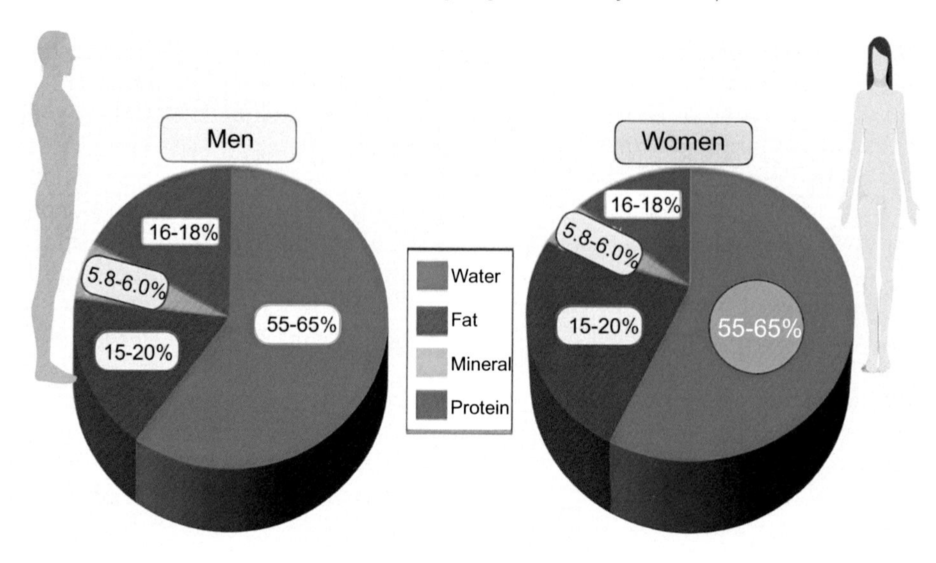

FIG. 14.7 The distribution of body composition (anatomy).

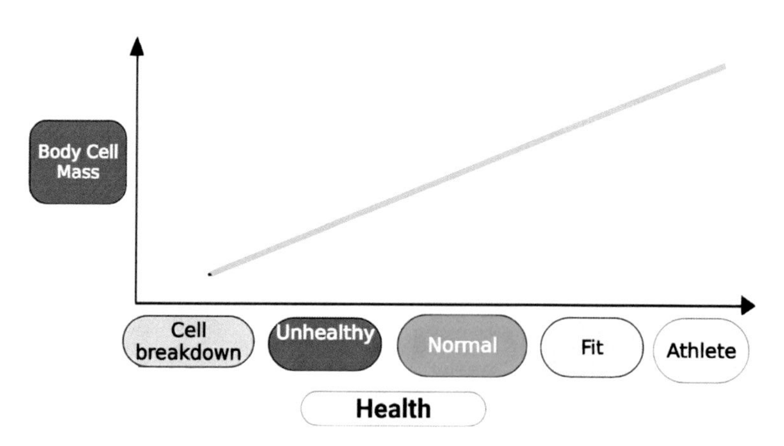

FIG. 14.8 Body composition and health. *Reprinted, by permission, from Biodynamics Corporation, Seattle, WA, U.S.A.*

resistance (impedance) flows. Nonconductor means noncurrent flow. Nevertheless, the human body has more muscle, which means lower impedance.

Consequently, the current flows smoothly. For more understanding of (BC) and fluid and mass partitions, BC is a medical evaluation of the distribution of tissues and fluids in the body. The body is composed of a series of membrane and fluid compartments shown in Fig. 14.9. It is vital to know better the body elements that BIA models intend to analyze (Bera, 2014; Kyle, Genton, Gremion, Slosman, & Pichard, 2004).

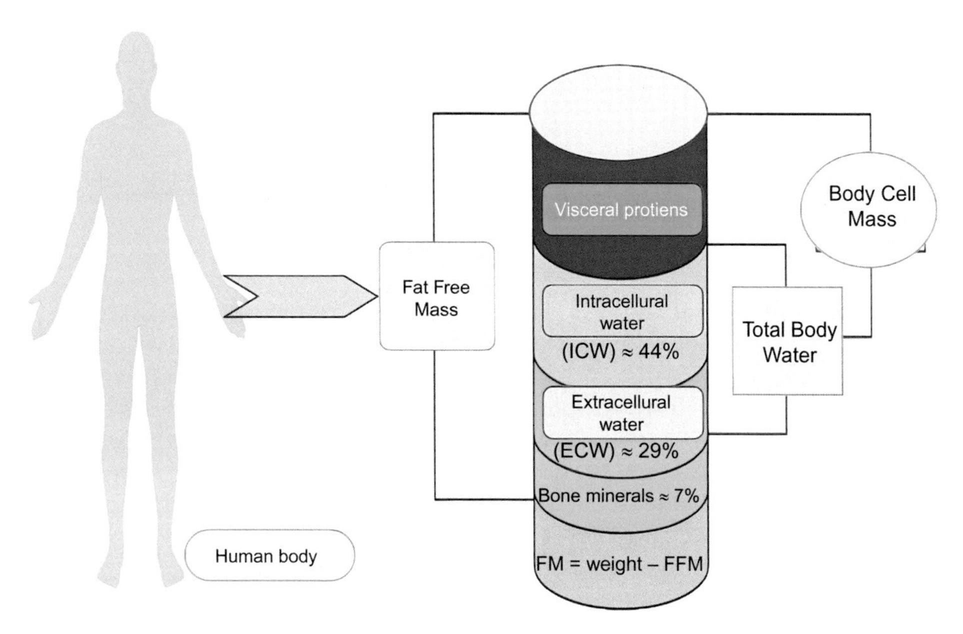

FIG. 14.9 The body composition graph illustrates fat mass, FFM, TBW, ICW, ECW, BCM, Visceral proteins, and Bone minerals. *Reprinted, by permission, from Biodynamics Corporation, Seattle, WA, U.S.A.*

Here we are evaluating the body composition compartments as following:

- **Fat mass (FM)** is the primary influence between individuals from the BC and can be calculated using an unintended BIA prediction (Bertemes-Filho & Simini, 2018). So, fat mass is the whole amount of fat deposited in the body, and it comprises the subsequent categories of fats:
- **Subcutaneous fat:** is placed merely below the skin. So, it acts as an energy resource and is covered by the external cold.
- **Visceral fat** is found more profound in the body, so visceral fat acts as a source of energy and pads the organs.
- **"Fat-free mass (FFM)":** It is also termed LBM. It comprises fat-free (nonfat) physical components. It is made up of nearly 73.0% water, 20.0% proteins, 6.0% minerals and 1% ash. Many equations have been suggested to determine the fat-free mass. Body composition diagrams such as FFM, TBW (ICW), (ECW), and BCM are displayed in Fig. 14.9 (Bera, 2014).

The literature predicts many BIA equations; most of these various equations have multiple regression equations in many subjects. In early 1987, BIA equations included height^2/resistance. After that, some parameters were involved, such as gender, age, weight, reactance, and body composition BC (Kyle, Bosaeus, et al., 2004b). Subsequent equivalences produce low standard errors SE to predict FFM, which is suitable for the public (Foster & Lukaski, 1996) - (Lohman et al., 2000).

$$\text{For male: FFM (kg)} = 0.485 \times \left[\frac{(ht(cm))^2}{R(\Omega)}\right] + 0.338 \times wt\,(kg) + 3.52 \tag{14.6}$$

$$\text{For females: FFM (kg)} = 0.475 \times \left[\frac{(ht(cm))^2}{R(\Omega)}\right] + 0.295 \times wt\,(kg) + 5.49 \tag{14.7}$$

where FFM is distributed into BCM and ECM.

- **Body cell mass (BCM)** contains all body cells, involving muscle cells, organic cells, plasma cells, and immune cells. Hence, BCM comprises the live cells in fat tissue, ensuring the body's fat is not processed. BCM also absorbs water from the interior of the cells. This water is called the ICW. However, potassium is the main electrolyte in ICW.
- **Extracellular mass (ECM) comprises** all metabolically stable "nonliving" body components; for example, blood plasma and bone minerals are shown in Fig. 14.10. ECM involves water that is outside living cells. It is termed ECW. The principal electrolyte of ECW is sodium [37] (Ott et al., 1995)

Body fat percentage (BFP)

Body fat (BF) is the number of fatty muscles in the body, which represents the total fatty tissue that does not contain any muscle mass MM, electrolyte, or body fluids (Kyle, Bosaeus, et al., 2004b) (Bera, 2014). It is the (TBM — FFM).

$$\text{"Body Fat (kg)} = \text{TBM(kg)} - \text{FFM (kg)"} \tag{14.8}$$

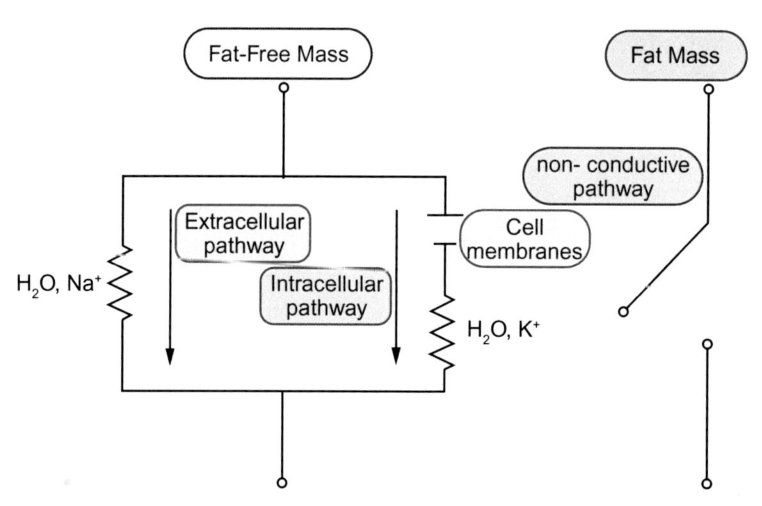

FIG. 14.10 Circuit model. *Reprinted, by permission, from Biodynamics Corporation, Seattle, WA, U.S.A.*

On the other hand, "the BFP of a person or other organism is (the total mass of fat divided by the total body mass, multiplied by 100)". Body fats compose primary physique lipids and stored body fats (Riyadi, Muthouwali, & Prakoso, 2017).

$$BF = \frac{\text{FM}(\text{kg})}{\text{Weight}(\text{kg})} \times \%100 \tag{14.9}$$

where BF = Body Fat, Fat Mass FM (kg), Weight (kg) (Dehghan & Merchant, 2008).

Body fat (BF) involved density 0.9 g/cm^3. Therefore, the fat performance as insulation for AC Fat cells cannot occupy the natural characteristics of BCM and consequently have little capacitive resistance (reactance) (Popa, Sirbu, Curseu, & Ionutas, 2006). There is a BFP for a particular type of athlete compared to the average for men and women shown in Table 14.1.

It is undoubtedly that confirmations body fat amount is a very relevant parameter, overall population groups. An increase in BFP is the initial and standard significant indicator of obesity; moreover, its threat like atherosclerosis and supplementary serious sicknesses are associated with obesity Table 14.2 (Haslam & James, 2005; Zingaretti, Nuñez, Gallagher, & Heymsfield, 2000).

TABLE 14.1 Compares to the average population, body fat scores, and essential body fats in some athletic and healthy males and females.

"Group	Female (fat %)	Male (fat %)
Essential fat	10–13	2–5
Athletes	14–20	6–13
Fitness	21–24	14–17
Average	25–31	18–24
Obese	32 and higher	25 and higher"

From WHO Consultation on Obesity (1999: Geneva, Switzerland) & World Health Organization. (2000). Obesity: Preventing and managing the global epidemic: Report of a WHO consultation. World Health Organization. https://apps.who.int/iris/handle/10665/42330.

TABLE 14.2 BFP for women and men, and the elemental distribution of groups by age and body fat%.

Women				
"Age	**Under fat**	**Healthy**	**Overweight**	**Obese**
20–40 years	Under 21%	21–33%	33–39%	Over 39%
41–60 years	Under 23%	23–35%	35–40%	Over 40%
61–79 years	Under 24%	24–36%	36–42%	Over 42%."
Men				
"Age	**Under fat**	**Healthy**	**Overweight**	**Obese**
20–40 years	Under 8%	8–19%	19–25%	Over 25%
41–60 years	Under 11%	11–22%	22–27%	Over 27%
41–60 years	41–60 years	41–60 years	41–60 years	41–60 years
61–79 years	Under 13%	13–25%	25–30%	Over 30%."

From Haslam, D., & James, W. (2005). Obesity. Lancet. *67483–1. https://doi.org/10.1016/S0140-6736(05)67483-1.*

Body mass index (BMI)

As we know from the studies, the free and publicly available body composition assessment is BMI, since the index is easy compared to overweight. Moreover, BMI is widely used as the only parameter showing the rate of obesity or thinness in people. In the 1840s, a Belgian polymath, Adolphe Quetelet, invented the BMI parameter. BMI is an association between weight (W) and height (H) that is correlated through body fat BF and fitness risk (Kg/m^2). Hence, BMI has a basic equation to calculate BMI in the human body that is as follows:

$$BMI = \frac{Weight(Kg)}{Height^2(m)} \tag{14.10}$$

Bodyweight in (Kg) is known as body weight, split into meters by the square of body height (H^2).

The WHO and other parties have approved BMI to practice as the standard for reporting obesity statistics and describing BC's status. However, the BMI parameter cannot mention anything regarding the BC and fat combination of the tested body. Recent research studies have discovered that BMI is a nonparticular prediction of BC in healthful adults (Lichtenbelt, 2010; McCarthy, Samani-Radia, Jebb, & Prentice, 2014; Stolarczyk, Heyward, Hicks, & Baumgartner, 1994) (Lichtenbelt, 2010; Stolarczyk et al., 1994) (McCarthy et al., 2014). The WHO provided a table with the captured value that correlated with the global standard of adults for various classified categories: overweight, underweight, and obesity shown in Table 14.3.

BMI is a practical analytical analysis for assessing numerous examined subjects. This system allows a surveyor to apply relationship standards within groups associated with the common mass and assess adiposity shown in Fig. 14.11 (Jeukendrup, 2017). Research has classified the health risk correlated as a full spectrum of BMI values:

Muscle mass (MM)

Muscle mass (MM), or "skeletal muscle mass" (SMM), is the muscle that attaches to your bones and enables you to move. It is considered an essential index of total physical health. Some muscles can be improved and strengthened during exercise (chest muscles, quadriceps, biceps). For an adult male, the expected value is 42% of body mass, and for normal women, it

TABLE 14.3 Distribution of BC release on BMI rate.

"BMI (kg/m²)	Nutritional status
Below 18.5	Underweight
18.5–24.9	Normal weight
25.0–29.9	Pre-obesity
30.0–34.9	Obesity class I
35.0–39.9	Obesity class II
Above 40	Obesity class III."

From WHO. (2008). WHO global report on falls prevention in older age. Geneva: World Health Organization. https://www.who.int/publications/i/item/9789241563536.

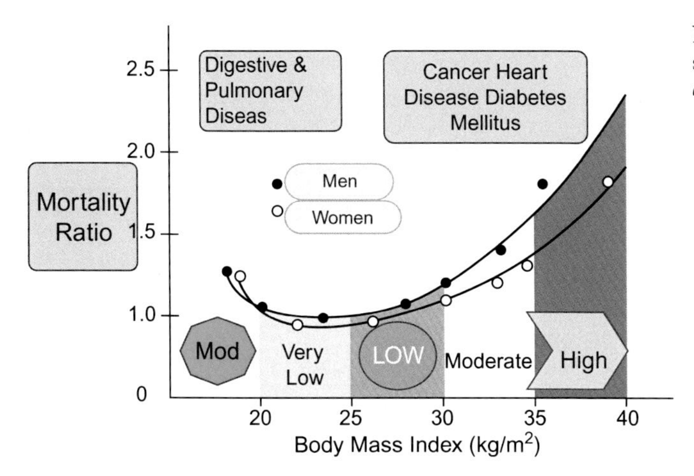

FIG. 14.11 Classified the health risk associated. *Reprinted, by permission, from Biodynamics Corporation, Seattle, WA, U.S.A.*

is 36% of body mass (Roberts et al., 2009). There are differences in equations for determining muscle mass. It mainly depends on the cases measured. Lee et al. used specific values to measure anthropometric and bioimpedance (Lee et al., 2000).

The anthropometric established predictive equalization can be utilized:

$$\text{SMM (Kg)} = \text{H}_\text{m}\,(0.244 \times \text{BM}_\text{Kg}) + (7.8 \times \text{H}) + (6.6 \times \text{G}) - (0.98 \times \text{Age}) + (\text{Ethnicity} - 3.3) \tag{14.11}$$

where, SMM in (Kg); H height (cm); Gender: male = 1, female = 0; (BM) body mass (kg).

Ethnicity: White = 0, $R^2 = 0.86$; African-American = 1.2; SEE = 2.6; Asian = 1,4.

The muscles comprise a large volume of ECW as the sodium and chloride ions are separated. That is an ideal tool for present transmission. Therefore, MM has less impedance than FM, which is extremely valuable in BC analysis of bioimpedance. So, because of certain principles of electrical properties of muscle mass (mm), BIA may be utilized to perform the impedance Z of a body and anthropometric physique estimations Janssen, Heymsfield, Baumgartner, and Ross, (2000).

A prediction equivalence was obtained by applying SF resistance as follows:

$$SMM\,(Kg) = \left[\left(\frac{H_t}{R} \times 0.401\right) + (Gender \times 3.825) + (Age \times (-0.0710))\right] + 5.1020 \tag{14.12}$$

Anywhere, R is resistance Ω in kHz; Ht is the height (cm), Gender, male = one female = 0, age is in years. The R2 and SE of the regression prediction's termination remained 2.7 kg and (9%) 0.86, individually. Muscle mass is an important marker used in obesity treatment. The purpose of losing body weight is to reduce the patient's body fat, reducing the total value. Differences in body MM should be mentioned if there is no operating error and an increment in body MM.

Basal metabolic rate (BMR)

BMR is the number of calories metabolized through rest for ("24 h"), which is the key to efficient weight. The average patient represents the BMR of more than 90% of the total daily spending and burned more than 90% of the calorie intake burned during the patient's comfort, as shown in Fig. 14.12. Hence, it is determined that BMR by FFM and FFM is the only one that is metabolized. The formula is natural. More patients' FFM increased calorie expenditure (Deye et al., 2016; Ozeki et al., 2018).

The equation that the analyst uses to predict BMR, depending on Grande and Keys, is the following:

$$BMR\left(\frac{\text{calories}}{\text{day}}\right) = 3.2 \times FFM(kg) \tag{14.13}$$

While increasing the calorie expenditure through exercise spending and professional practice, the exercise's central feature is maintaining the FFM, which means more calories burned on the FFM throughout the day. At the same time, the patient is relaxed.

Total body water (TBW)

TBW consists of two variables: ICW (approximately 44%) and ECW (about 29%) of the whole-body water FFM volume. Moreover, it is similar to 73.2% in healthy hydration patients, as in Equalization (Jaffrin & Morel, 2008).

$$\text{TBW} = 0.73\text{FFM} \tag{14.14}$$

From these values, a simple formula for calculating TBW can be mentioned as follows:

$$TBW = ECW + ICW \tag{14.15}$$

Fig. 14.13 shows the compartments of the water models.

ICW is water held inside the cell. Healthy cells preserve their veracity and maintain their fluid inside. At the same time, ECW is water positioned outside the cell. Jaffrin and Morel

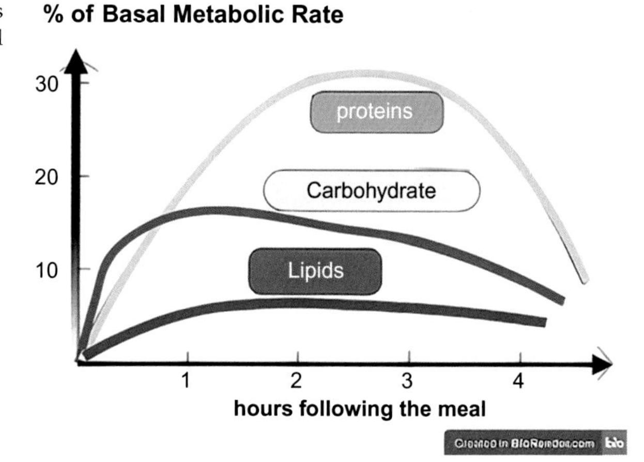

FIG. 14.12 Postprandial thermogenesis occurs in the basal metabolic rate at various levels based on the composition of the food consumed.

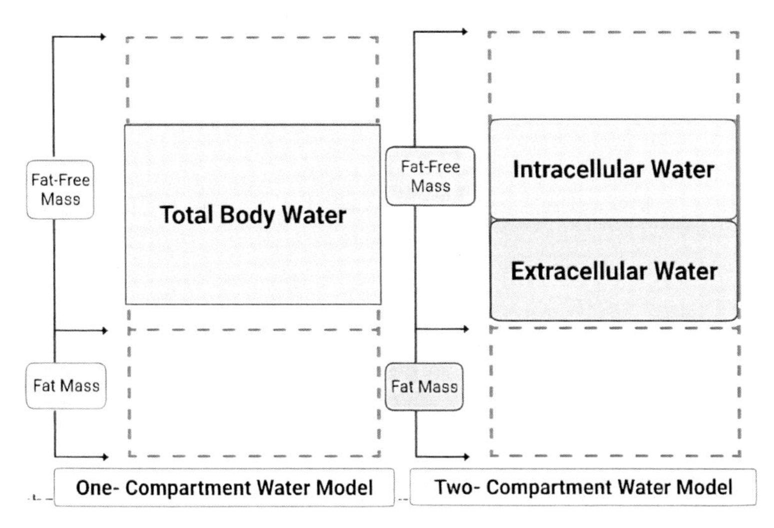

FIG. 14.13 The Compartments of the water models. *Reprinted, by permission, from Biodynamics Corporation, Seattle, WA, U.S.A.*

observed that from 1985 to 1994, they were considered the most TBW calculation equations depending on conditions predicted using H^2/R. That was clarified by Kyle et al. (Khalil, Mohktar, & Ibrahim, 2014; Kyle, Genton, et al., 2004). Thomasset was the first to use the equation to study the variance of total body water and multiple and single frequencies (Thomasset, 1962). To measure total body water TBW first, we must apply high-frequency current penetrating cells to conduct the ICW and ECW of the human body (Lorenzo, Andreoli, & Deurenberg, 1997; Patel, Matthie, Withers, Peterson, & Zarowitz, 1994). The efficacy of (MF) impedance to distinguish between TBW, ICW, and ECW may be essential to describe fluid dislocation and explore potential differences in hydration levels in particular exceptional clinical cases, such as kidney disease (Hwang et al., 2010).

L. Schoeller and R. F. Kushner have approved some formulations for total body water (Kushner et al., 1992).

$$\text{TBW (kg)} = 1.84 + 0.45\left(\frac{H^2}{R}\right) + 0.14(\text{W}) \tag{14.16}$$

where the H^2 = height square, R = resistance, and W = weight.

Hence, "According to Lukaski et al. And Jawon Medical Inc. In South Korea, the equation is given by":

$$\text{TBW(kg)} = 1.84 + 0.377\left(\frac{H^2}{R}\right) + 0.11(\text{W}) - 0.08\text{Age} + 2.9\text{Gender} + 4.56 \tag{14.17}$$

$$\text{TBW (kg)} = \text{A} \times \left(\frac{H^2}{R}\right) + \text{B} \times (\text{W}) - \text{C} \times \text{Age} + D\,\text{Gender} + E \tag{14.18}$$

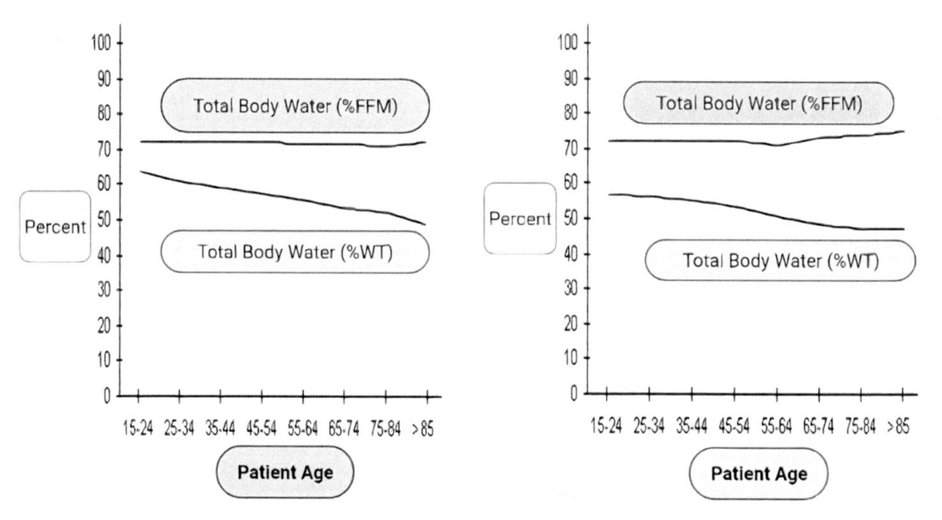

FIG. 14.14 Natural water distribution for males and females. *Reprinted, by permission, from Biodynamics Corporation, Seattle, WA, U.S.A.*

where A, B, C, D, and E are constant values.

Other BIA equations can be initiated in numerous works published over a lengthy period (Cox-Reijven & Soeters, 2000; Hannan et al., 1993; Hlubik, Stritecka, & Hlubik, 2013; Sun et al., 2003; Wang et al., 1992). Hence, Fig. 14.14 shows TBW natural water distribution for a male and female.

In BIA, the standard error of evaluating the TBW <2 L of water or $<4\%$ error for the pure 50 L of TBW contributed to some study equations (Conference, 1994).

$$W = a\,Ht/R + b\,Wt + c \tag{14.19}$$

Extracellular water (ECW)

ECW is part of TBW and regularly eliminates body fluids. This liquid comprises blood plasma, fluid interfaces, supracellular juices, and a high $Na+$ quantity and is low in $K+$ ions (Sargent & Gotch, 1989). The ECW resistance is approaching that of saline, and the cumulative strength is nearly 40 cm. Usually, the ECW amount is 60% of TBW. This body liquid further manages the flow of water and electrolytes in the body, shown in Fig. 14.15 (Kyle, Genton, Karsegard, Slosman, & Pichard, 2001). As we know from the literature, researchers proved that for calculating ECW, it must provide multiple frequencies because, at low frequencies, the current flows through the ICW only. Therefore, MF-BIA was introduced to enhance the accuracy of calculating TBW, ECF, and ICF. Hence, the equation has also changed. Due to technical purposes, estimates of the impedance regularly exerted by surface electrodes at a frequency (5-1 K) kHz are limited. Individually, the essential ECW Re account and one TBW ($R\infty$) by frequency extrapolation are endless and unlimited. Behind this hypothesis, the ECW is determined using the

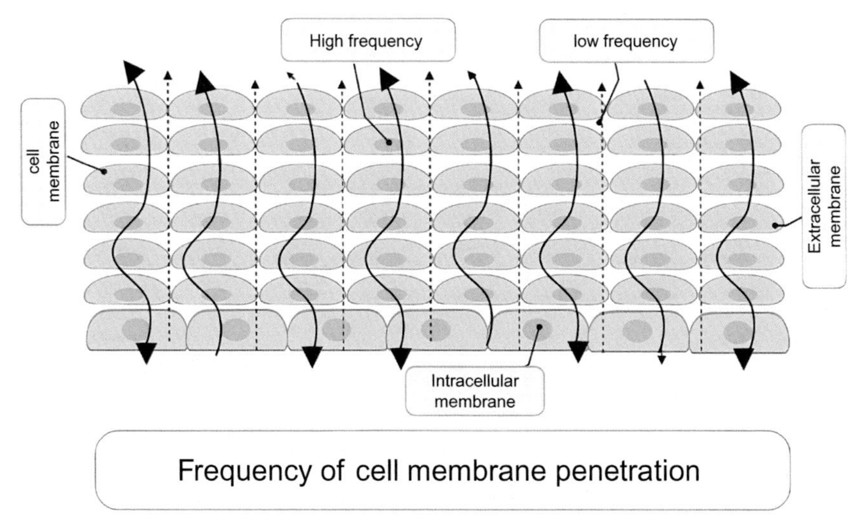

FIG. 14.15 Frequency of cell membrane penetration and current.

impedance at the frequency of zero and approach the individual body as the amount of 5-cylinder (limbs and torso).

Jaffrin et al. have defined the final equation is the following Jaffrin and Morel (2008).

$$Ve = (H2 + W12/\mathrm{Re})23 \tag{14.20}$$

$$V_e = K_e \left(\frac{H^2 + WW^{1/2}}{R_e} \right)^{2/3} \tag{14.21}$$

Van Loan et al. recommend "ke = 0.306" for males and "0.316" for females, with "Ve in a liter, H in cm, Re, and weight in (kg)."

Pichler et al. observed the BIS system utilizing Implement equipment (SFB7) in FFM, ECF, and TBW, addressing the deuterium space system (DSS), the sodium bromide space approach, and the DXA system sequentially. The research was used on 32 healthy cases and 84 subjects, including numerous varieties of illnesses (Pichler, Amouzadeh-Ghadikolai, Leis, & Skrabal, 2013).

$$V_{ECF\_MALE} = 0.11 + 0.11Wt + 0.24\frac{H^2}{R_{ECF}} \tag{14.22}$$

$$V_{ECF\_FEMALE} = 1.24 + 0.09Wt + 0.28\frac{H^2}{R_{ECF}} \tag{14.23}$$

For ECF, it obtained 0.87 = male and 0.89 = female, as in Eqs. (14.11) and (14.12) (Khalil et al., 2014; Pichler et al., 2013).

All these equations can be utilized to estimate the size of ECW. The examiner must decide which is better for his particular conditions and data acquisition.

Intracellular water (ICW)

ICW commonly refers to entire body fluids within cells. Ionic composition depends on ICW on the cells in which they are deemed in the ICW. Hence, the ionic properties of ICW are unknown, and the resistivity cannot be defined, as is the case with ECW. However, it could be stated that ICW mainly includes K^+ ions by a concentration of approximately160 (mequiv./l) (Hlubik et al., 2013). In other words, ICW contains almost 70% of the cytosol, such as a mix of water and some different dissolved components. (Inbody, compony). The parameter of impedance may be used to calculate the ICW volume equation for TBW estimation. To an assessment volume of ICW through impedance values, ICW must be calculated by using resistance Ri. This method is considered to be Re and Ri parallel to a symbol (0). Following this supposition, Ri could be determined to use total resistance "RTot" and EC resistance Re as the following:

$$R_i = \frac{(R_{e\times} R_{tol})}{(R_e - R_{tol})} \tag{14.24}$$

where the total resistance isR_{tol} and extracellular resistance is R_e.

Fig. 14.16(A) shows the membrane of intercellular water, and (B) shows the model of ICW calculation considering R_e.

Ri describes the "ICW part and the layer capacitance Cm Ri could not be utilized as ICW resistance unequal Re for ECW." De Lorenzo et al. have studied the investigation of formal analysis and a derivative equalization as follows (De Lorenzo et al., 1998):

$$V_i = 12.20 + 0.370650 \times \frac{H_T}{R_{ICW}} - 0.1320 \times age + 0.105\,weight \tag{14.25}$$

where R_{ICW} = intracellular resistance; 0 for women and 1 for men.

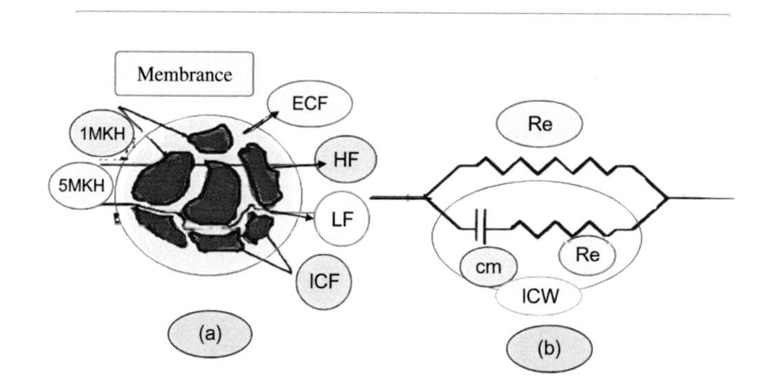

FIG. 14.16 Frick's circuit: Equivalent Circuit of a Cell.

Resistance (R)

Resistance (R) is the current flow through a substance, measured in volts and amperes. Consider a material with a high resistance that needs a considerable voltage (V) to create a particular current. The material has low resistance, R needs low potential (v) to perform an equal current value. Perhaps the most straightforward approach relative to this is to realize that a material with a low R conducts well. In contrast, a material with a higher R performs inadequately. In the human body, the low impedance is correlated with high values of FFM. Significant R is associated with fewer values than lower FFM shown in Fig. 14.17. Hence, FFM is a stable transmitter of electrical current, as it comprises more considerable amounts of water and electrolytes (Corporation, 2021) (Formenti, Bolgiaghi, & Chiumello, 2018).

Target's resistance is dedicated by shape, which is defined as length (L), "surface area (A)," and material form, which is represented by resistivity (ρ), as revealed in the Eq. (14.18) (Ghassemlooy, 1997). In biological systems, resistance occurs due to entire water through the body, and resistance happens due to cell membrane capacitance (Khalil et al., 2014; Kyle, Bosaeus, et al., 2004a).

$$\text{Resistance}\,(R_{\Omega}) = \rho_{m.\Omega}\,\frac{L_{(m)}}{A_{(m^2)}} \tag{14.26}$$

The relation between FFM and resistance R (high conductivity and low conductivity).

Reactance (Xc)

Reactance (Xc) is the impact on the electric current produced by a substance's safe energy ability. On the other hand, reactance is a period of delay between voltage and current used. Therefore, it describes the resistance the capacitor exerts to the alternating current. Due to the lipoprotein membranes, all the cell layers behave similarly to small capacitors, so the reaction

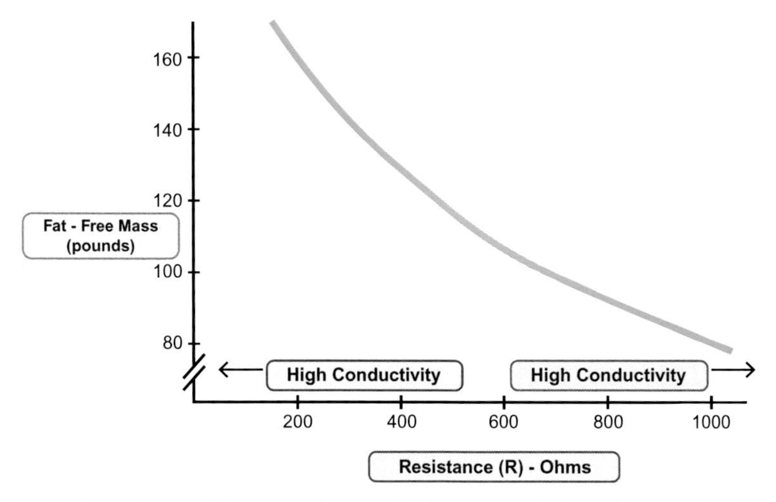

FIG. 14.17 The relation between FFM and resistance R (high conductivity and low conductivity). *Reprinted, by permission, from Biodynamics Corporation, Seattle, WA, U.S.A.*

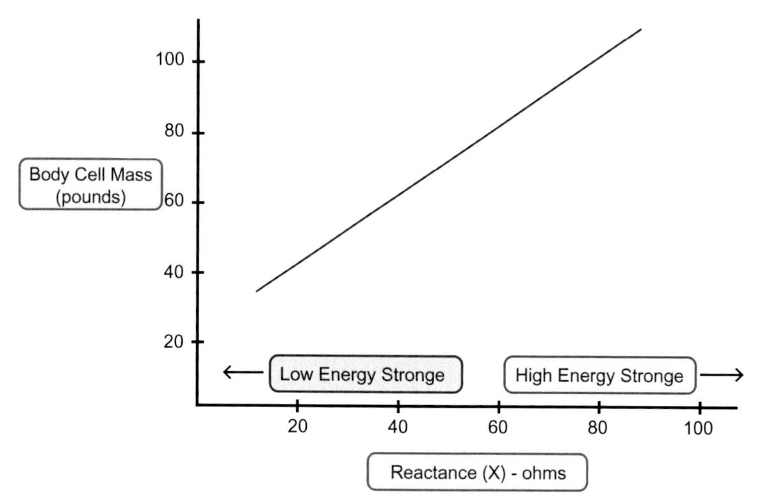

FIG. 14.18 The association between BCM and reactance(X) (high energy storage and low energy storage). *Reprinted, by permission, from Biodynamics Corporation, Seattle, WA, U.S.A.*

estimates the body's cell mass. Consequently, "In the human body," large reactance is correlated with high volumes of BCM (intracellular mass ICM). Lower reactance is connected with fewer amounts of BCM, shown in Fig. 14.18 (Corporation, 2021; Formenti et al., 2018).

In biological systems, reactance occurs due to cell membrane capacitance.

$$Reactance(X_C) = \frac{1}{2\pi f_{(Hz)} C_F} \tag{14.27}$$

From Eq. (14.20), reactance is interpreted as the resistance to changing the voltage crossing the body and is inversely correlated to the signal frequency (f) and capacitance (De Lorenzo et al., 1998; Khalil et al., 2014)

Capacitance (C)

"Body capacitance" (C) is the total capacity of potential storage in the BCM. An indicator of the presence of healthy cell layers is the presence of large capacitance. Lower capacitance means that there are fewer amounts of healthy cell layers shown in Fig. 14.19. Their amount and quality determine the capacity of the cell layers in the cell block segment.

Overall, (C and Xc) follow each other. A significant C or high Xc symbolizes copious amounts of intact cellular layers. It is imperative to remark that the capacity is linked to the quantity of BCM and its quality. A high BCM with a healthy low cell density may have equal capacitance as a small BCM with an intact high cell density. Therefore, Reasonable notes a decrease in capacity due to weight loss in a healthy body (Ghassemlooy, 1997). The equation for capacitance (C) for the duration of determined resistance and reactance is:

$$\text{Capactance}(\text{C}) = 10^7/n^* \times /(X^2 * R^2) \tag{14.28}$$

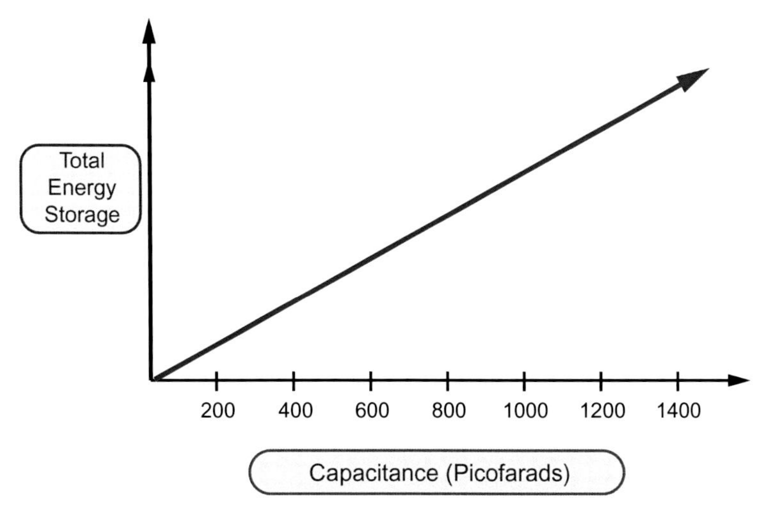

FIG. 14.19 The association between total energy storage and capacitance (pF). *Reprinted, by permission, from Biodynamics Corporation, Seattle, WA, U.S.A.*

The capacitance analyzer affords a valuable substitute for the reactance Xc for physicians to utilize body capacitance C in their observations.

Impedance (Z)

"Total impedance (Z) is the vector quantity of the impacts of resistance (R) and reactance (x) on a flow in the human body." Theoretically, the impedance (Z) is the rate of voltage (V) to current (I) and is occasionally utilized in biological impedance study analyses. Therefore, it depends on the frequency of the practical flow, specified the amount of impedance ($|Z|$) and the phase angle (φ), as noted in Eqs. (14.22)–(14.24) (Ghassemlooy, 1997). Bioimpedance is a complicated amount consisting of resistance (R) created by TBW and reactance (Xc) affected by cell membrane capacitance shown in Fig. 14.20 (Khalil et al., 2014; Kyle, Bosaeus, et al., 2004a).

For more understanding regarding the relation between R, X, and Z, Eqs. (14.22)–(14.24) (Kyle, Bosaeus, et al., 2004a).

$$Z = R + jX_C \quad (14.29)$$

$$\text{where}, |Z| = \sqrt{(R^2 + X_C^2)} \quad (14.30)$$

$$\text{then}, \varnothing = tan^{(-1)} \times (X_C/R) \quad (14.31)$$

The polar coordinates, explaining the association between (Z, R, Xc, and φ).

Phase angle (PhA)

The phase angle has two elements: resistance and reactance. On the other hand, PhA is the displacement between the alternating current (I) and the voltage (V) on the estimated impedance shown in Fig. 14.21.

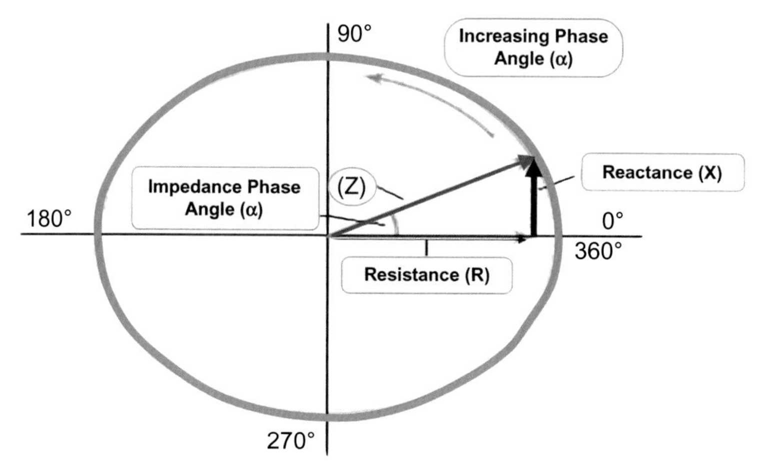

FIG. 14.20 The Polar coordinates explain the association between (Z, R, Xc, and φ). *Reprinted, by permission, from Biodynamics Corporation, Seattle, WA, U.S.A.*

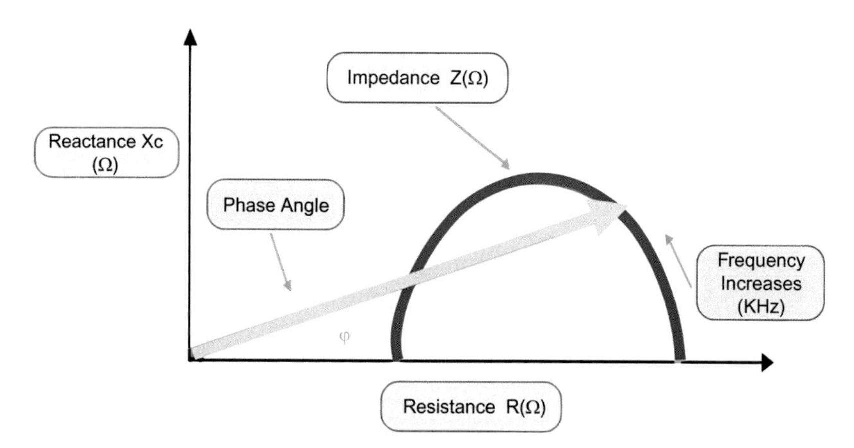

FIG. 14.21 A graphical diagram of the PhA graphic initial (PhA); Its reference to the PhA, R, Xc, and Z. *Reprinted, by permission, from Biodynamics Corporation, Seattle, WA, U.S.A.*

Impedance (Z) is applied at 50 kHz because, in humans, 50 kHz is considered the ideal frequency to improve the reaction and conclude the point at which the cells are more effective in the current resistance, thus creating the essential PhA (Lukaski, Kyle, & Kondrup, 2017; Norman, Stobäus, Pirlich, & Bosy-Westphal, 2012). PhA is an index of cell health and probity. The study in humans confirmed that the association between PhA and cellular fitness increments appears linear (Kyle et al., 2001) (Máttar, 1996). The low PhA corresponds to cells' inefficiency to store power, also a sign of a defect in the chosen porousness of the cell layers. PhA is an index of cell layer function and is generally practiced for a nutritional estimate and risk assessment of many diseases (Norman et al., 2012; Tanaka et al., 2019).

Advantages of BIA

1. BIA is a portable instrument. It is widely used in many hospitals and clinical laboratories, including epidemiological measures for chronic disease patients and rehabilitation. Therefore, BIA is noninvasive and cheap compared to other estimates of health status and BC parameters similar to "dilution techniques." TBK or DXA (González-Correa, 2018; Lingwood, Colditz, & Ward, 1999).
2. BIA depends on "age, race, gender, ethnicity, nutrition, and clinical position," demanding an equation for a specific condition (Chumlea & Guo, 1994).
3. It should be remarked that BIA (using some approved formulas) was more connected to the percentage of fat mass than the BMI and thus can generally be applied as a machine for estimating body fat in clinical practice (Coppini, Waitzberg, & Campos, 2005; Harrington et al., 2001).
4. The segmental composition of the body, comprising "left arm, right arm, trunk, left leg, and right leg lean mass," is estimated by BIA (González-Correa, 2018).
5. Bioimpedance is a simple-to-use instrument that affords a valuable dry weight assessment for patients. A lack of standards of normalcy still limits the method (Hlubik et al., 2013).
6. Bio-impedance has proven to be a unique technique for measuring BC and a patient's health. Furthermore, in the status of obesity, this way could be quite beneficial because it is suitable to detect accurate BC without invasive or ionization techniques (Barbosa-Silva & Barros, 2005; Hlubik, 2015)

Limitations of BIA

1. BIA has some limitations that concern molds and types of restrictions. "The first is related to the geometry of the human body: the human body is not a cylinder"; Alternatively, five cylinders joined in a better series could describe it (legs and trunk, arms, except for the head) (Snel, Brummer, Doerga, Zelissen, & Koppeschaar, 1995).
2. We must consider the fact that electrophysical patterns change and biological transmitters are not static. It will vary depending on the muscles "exact anatomy, hydration state, and the electrolytic atoms" concentration (Genton, Herrmann, Spörri, & Graf, 2018; Kushner et al., 1992).
3. Results can vary conditional on the device producer from <1% to approximately 20%, and the matched impedance meter cables can offer additional capacitance (C) (Genton et al., 2018; Oldham, 1996).
4. The distribution of body build (particularly in those who are obese in the abdomen) will evaluate body fat percentage. Additionally, a new equation may be necessary to verify the bioelectrical impedance evaluation in obese patients (Coppini et al., 2005).
5. The practice might require a reliable standard procedure for different methods to compare the body composition results with gold standard methods (González-Correa & Caicedo-Eraso, 2012).
6. The conditions for measuring biological resistance should be standard, and consideration must be given to each athlete's characteristics (Bertemes-Filho & Simini, 2018).

Conclusion

BIA is a noninvasive, portable, and generally inexpensive technique and may be utilized in persons with various characteristics, given that particular equations concerning age, sex, and ethnicity are validated and accepted. The sixteen parameters of body composition evaluated in this chapter were valid for populations already established in various literature; however, because of the lack of particular BIA equations in the literature for some diseases. BIA still needs to practice and validate equations in all diseases. Therefore, it is recommended to evaluate and practice BIA for body composition models in clinical assessment.

References

Aldobali, M., & Pal, K. (2021). Bioelectrical impedance analysis for evaluation of body composition: A review. In *2021 International Congress of Advanced Technology and Engineering, ICOTEN 2021* Institute of Electrical and Electronics Engineers Inc. https://doi.org/10.1109/ICOTEN52080.2021.9493494.

Barbosa-Silva, M. C. G., & Barros, A. J. D. (2005). Bioelectrical impedance analysis in clinical practice: A new perspective on its use beyond body composition equations. *Current Opinion in Clinical Nutrition and Metabolic Care, 8*(3), 311–317. https://doi.org/10.1097/01.mco.0000165011.69943.39.

Bera, T. K. (2014). Bioelectrical impedance methods for noninvasive health monitoring: A review. *Journal of Medical Engineering*, 1–28. https://doi.org/10.1155/2014/381251.

Bertemes-Filho, P., & Simini, F. (2018). Bioimpedance in biomedical applications and research. In *Bioimpedance in biomedical applications and research* (pp. 1–279). Springer International Publishing. https://doi.org/10.1007/978-3-319-74388-2.

Brožek, J., Grande, F., Anderson, J. T., & Keys, A. (1963). Densitometric analysis of body composition: Revision of some quantitative assumptions. *Annals of the New York Academy of Sciences, 110*(1), 113–140. https://doi.org/10.1111/j.1749-6632.1963.tb17079.x.

Chen, K. T., Chen, Y. Y., Wang, C. W., Chuang, C. L., Chiang, L. M., Lai, C. L., Lu, H. K., Dwyer, G. B., Chao, S. P., Shih, M. K., & Hsieh, K. C. (2016). Comparison of standing posture bioelectrical impedance analysis with DXA for body composition in a large, healthy Chinese population. *PLoS One, 11*(7). https://doi.org/10.1371/journal.pone.0160105.

Chertow, G. M., Lowrie, E. G., Wilmore, D. W., Gonzalez, J., Lew, N. L., Ling, J., Leboff, M. S., Gottlieb, M. N., Huang, W., Zebrowski, B., College, J., & Lazarus, J. M. (1995). Nutritional assessment with bioelectrical impedance analysis in maintenance hemodialysis patients. *Journal of the American Society of Nephrology, 6*(1), 75–81.

Chumlea, W. C., & Guo, S. S. (1994). Bioelectrical impedance and body composition: Present status and future directions. *Nutrition Reviews, 52*(4), 123–131. https://doi.org/10.1111/j.1753-4887.1994.tb01404.x.

Conference, N. A. (1994). Bioelectrical impedance analysis in body composition measurement: National institutes of health technology assessment conference statement. *American Journal of Clinical Nutrition, 64*, 749–762. https://doi.org/10.1093/ajcn/64.3.524s

Coppini, L. Z., Waitzberg, D. L., & Campos, A. C. L. (2005). Limitations and validation of bioelectrical impedance analysis in morbidly obese patients. *Current Opinion in Clinical Nutrition and Metabolic Care, 8*(3), 329–332. https://doi.org/10.1097/01.mco.0000165013.54696.64.

Corporation, B. (2021). *Biodynamics corporation company*. https://www.biodyncorp.com/tools/450/resistance.html.

Cox-Reijven, P., & Soeters, P. (2000). Validation of bio-impedance spectroscopy: Effects of degree of obesity and ways of calculating volumes from measured resistance values. *International Journal of Obesity, 24*(3), 271–280. https://doi.org/10.1038/sj.ijo.0801123.

De Lorenzo, A., Sorge, S. P., Iacopino, L., Andreoli, A., Petrone De Luca, P., & Sasso, G. F. (1998). Fat-free mass by bioelectrical impedance vs dual-energy X-ray absorptiometry (DXA). *Applied Radiation and Isotopes*, 739–741. https://doi.org/10.1016/s0969-8043(97)00099-7.

Dehghan, M., & Merchant, A. T. (2008). Is bioelectrical impedance accurate for use in large epidemiological studies? *Nutrition Journal, 7*(1). https://doi.org/10.1186/1475-2891-7-26.

Deurenberg, P. (1996). Limitations of the bioelectrical impedance method for the assessment of body fat in severe obesity. *American Journal of Clinical Nutrition, 64*(3). https://doi.org/10.1093/ajcn/64.3.449S. American Society for Nutrition.

Deye, N., Vincent, F., Michel, P., Ehrmann, S., Da Silva, D., Piagnerelli, M., Kimmoun, A., Hamzaoui, O., Lacherade, J. C., de Jonghe, B., Brouard, F., Audoin, C., Monnet, X., & Laterre, P. F. (2016). Changes in cardiac arrest patients' temperature management after the 2013 "TTM" trial: Results from an international survey. *Annals of Intensive Care, 6*(1). https://doi.org/10.1186/s13613-015-0104-6.

Ellis, K. J., Bell, S. J., Chertow, G. M., Chumlea, W. C., Knox, T. A., Kotler, D. P., Lukaski, H. C., & Schoeller, D. A. (1999). Bioelectrical impedance methods in clinical research: A follow-up to the NIH technology assessment conference. *Nutrition, 15*(11 – 12), 874–880. https://doi.org/10.1016/S0899-9007(99)00147-1.

Formenti, P., Bolgiaghi, L., & Chiumello, D. (2018). (pp. 275–290). Springer Nature. doi:https://doi.org/10.1007/978-3-319-73670-9_22.

Foster, K. R., & Lukaski, H. C. (1996). Whole-body impedance—What does it measure? *American Journal of Clinical Nutrition, 64*(3), 388S–396S. https://doi.org/10.1093/ajcn/64.3.388S.

Genton, L., Herrmann, F. R., Spörri, A., & Graf, C. E. (2018). Association of mortality and phase angle measured by different bioelectrical impedance analysis (BIA) devices. *Clinical Nutrition, 37*(3), 1066–1069. https://doi.org/10.1016/j.clnu.2017.03.023.

Ghassemlooy, Z. (1997). Book Review: Principles of electrical engineering materials and devices. *The International Journal of Electrical Engineering & Education, 35*, 94–95.

González-Correa, C. H. (2018). Body composition by bioelectrical impedance analysis. In *Bioimpedance in biomedical applications and research* (pp. 219–241). Springer International Publishing. https://doi.org/10.1007/978-3-319-74388-2_11.

González-Correa, C. H., & Caicedo-Eraso, J. C. (2012). Bioelectrical impedance analysis (BIA): A proposal for standardization of the classical method in adults. *Journal of Physics: Conference Series, 407*(1). https://doi.org/10.1088/1742-6596/407/1/012018. Institute of Physics Publishing.

Gudivaka, R., Schoeller, D. A., Kushner, R. F., & Bolt, M. J. G. (1999). Single- and multifrequency models for bioelectrical impedance analysis of body water compartments. *Journal of Applied Physiology, 87*(3), 1087–1096. https://doi.org/10.1152/jappl.1999.87.3.1087.

Hannan, W. J., Cowen, S. J., Freeman, C. P., & Wrate, R. M. (1993). Can bioelectrical impedance improve the prediction of body fat in patients with eating disorders? *European Journal of Clinical Nutrition, 47*(10), 741–746.

Harrington, K., Robson, P., Kiely, M., Livingstone, M., Lambe, J., & Gibney, M. (2001). The north/South Ireland food consumption survey: Survey design and methodology. *Public Health Nutrition, 4*(5 A), 1037–1042. https://doi.org/10.1079/PHN2001184.

Haslam, D., & James, W. (2005). Obesity. *Lancet*. https://doi.org/10.1016/S0140-6736(05)67483-1, 67483-1.

Hlubik, I. J. (2015). Bioimpedance measurement of spesific body resistance. *Hlubik Dissertation Bioimpedance* (p. 43). Scribd. https://www.scribd.com/document/373382407/Hlubik-disertace.

Hlubik, J., Stritecka, H., & Hlubik, P. (2013). Bioelectrical impedance analysis or basic anthropometrical parameters for evaluating weight loss success? *Central European Journal of Medicine, 8*(5), 565–570. https://doi.org/10.2478/s11536-013-0206-1.

Hoffer, E. C., Meador, C. K., & Simpson, D. C. (1969). Correlation of whole-body impedance with total body water volume. *Journal of Applied Physiology, 27*(4), 531–534. https://doi.org/10.1152/jappl.1969.27.4.531.

Hwang, S., Yu, Y. D., Park, G. C., Park, P. J., Choi, Y. I., Choi, N. K., Kim, K. W., Song, G. W., Jung, D. H., Yun, J. S., Choi, S. Y., & Lee, S. G. (2010). Bioelectrical impedance analysis for evaluation of donor hepatic steatosis in living-donor liver transplantation. *Transplantation Proceedings, 42*(5), 1492–1496. https://doi.org/10.1016/j.transproceed.2010.03.137.

Jaffrin, M. Y., & Morel, H. (2008). Body fluid volumes measurements by impedance: A review of bioimpedance spectroscopy (BIS) and bioimpedance analysis (BIA) methods. *Medical Engineering and Physics, 30*(10), 1257–1269. https://doi.org/10.1016/j.medengphy.2008.06.009.

Janaway, R. C., Percival, S. L., & Wilson, A. S. (2009). Decomposition of human remains. In *Microbiology and aging: Clinical manifestations* (pp. 313–334). Humana Press. https://doi.org/10.1007/978-1-59745-327-1_14.

Janssen, I., Heymsfield, S. B., Baumgartner, R. N., & Ross, R. (2000). Estimation of skeletal muscle mass by bioelectrical impedance analysis. *Journal of Applied Physiology, 89*(2), 465–471. https://doi.org/10.1152/jappl.2000.89.2.465.

Jeukendrup, A. E. (2017). Periodized nutrition for athletes. *Sports Medicine, 47*, 51–63. https://doi.org/10.1007/s40279-017-0694-2.

Khalil, S. F., Mohktar, M. S., & Ibrahim, F. (2014). The theory and fundamentals of bioimpedance analysis in clinical status monitoring and diagnosis of diseases. *Sensors (Switzerland)*, *14*(6), 10895–10928. https://doi.org/10.3390/s140610895.

Kremer, M. M., Latin, R. W., Berg, K. E., & Stanek, K. (1998). Validity of bioelectrical impedance analysis to measure body fat in air force members. *Military Medicine*, *163*(11), 781–785. https://doi.org/10.1093/milmed/163.11.781.

Kushner, R. F., & Schoeller, D. A. (1986). Estimation of total body water by bioelectrical impedance analysis. *American Journal of Clinical Nutrition*, *44*(3), 417–424. https://doi.org/10.1093/ajcn/44.3.417.

Kushner, R. F., Schoeller, D. A., Fjeld, C. R., & Danford, L. (1992). Is the impedance index (ht2/R) significant in predicting total body water? *American Journal of Clinical Nutrition*, *56*(5), 835–839. https://doi.org/10.1093/ajcn/56.5.835.

Kyle, U. G., Bosaeus, I., De Lorenzo, A. D., Deurenberg, P., Elia, M., Gómez, J. M., Heitmann, B. L., Kent-Smith, L., Melchior, J. C., Pirlich, M., Scharfetter, H., Schols, A. M. W. J., & Pichard, C. (2004a). Bioelectrical impedance analysis—Part I: Review of principles and methods. *Clinical Nutrition*, *23*(5), 1226–1243. https://doi.org/10.1016/j.clnu.2004.06.004.

Kyle, U. G., Bosaeus, I., De Lorenzo, A. D., Deurenberg, P., Elia, M., Gómez, J. M., Heitmann, B. L., Kent-Smith, L., Melchior, J. C., Pirlich, M., Scharfetter, H., Schols, A. M. W. J., & Pichard, C. (2004b). Bioelectrical impedance analysis—Part II: Utilization in clinical practice. *Clinical Nutrition*, *23*(6), 1430–1453. https://doi.org/10.1016/j.clnu.2004.09.012.

Kyle, U. G., Genton, L., Gremion, G., Slosman, D. O., & Pichard, C. (2004). Aging, physical activity and height-normalized body composition parameters. *Clinical Nutrition*, *23*(1), 79–88. https://doi.org/10.1016/S0261-5614(03)00092-X.

Kyle, U. G., Genton, L., Karsegard, L., Slosman, D. O., & Pichard, C. (2001). Single prediction equation for bioelectrical impedance analysis in adults aged 20-94 years. *Nutrition*, *17*(3), 248–253. https://doi.org/10.1016/S0899-9007(00)00553-0.

Kyle, U. G., Gremion, G., Genton, L., Slosman, D. O., Golay, A., & Pichard, C. (2001). Physical activity and fat-free and fat mass by bioelectrical impedance in 3853 adults. *Medicine and Science in Sports and Exercise*, *33*(4), 576–584. https://doi.org/10.1097/00005768-200104000-00011.

Lee, R. C., Wang, Z., Heo, M., Ross, R., Janssen, I., & Heymsfield, S. B. (2000). Total-body skeletal muscle mass: Development and cross-validation of anthropometric prediction models. *The American Journal of Clinical Nutrition*, 796–803. https://doi.org/10.1093/ajcn/72.3.796.

Lichtenbelt, W. (2010). The use of bioelectrical impedance analysis (BIA) for estimation of body composition. In *Body composition analysis of animals*. https://doi.org/10.1017/cbo9780511551741.008.

Lingwood, B. E., Colditz, P. B., & Ward, L. C. (1999). Biomedical applications of electrical impedance analysis. In *Vol. 1. ISSPA 1999—Proceedings of the 5th International Symposium on Signal Processing and Its Applications* (pp. 367–370). IEEE Computer Society. https://doi.org/10.1109/ISSPA.1999.818188.

Lohman, T. G., Caballero, B., Himes, J. H., Davis, C. E., Stewart, D., Houtkooper, L., Going, S. B., Hunsberger, S., Weber, J. L., Reid, R., & Stephenson, L. (2000). Estimation of body fat from anthropometry and bioelectrical impedance in native American children. *International Journal of Obesity*, *24*(8), 982–988. https://doi.org/10.1038/sj.ijo.0801318.

Lorenzo, A. D., Andreoli, A., & Deurenberg, P. (1997). Impedance ratio as a measure of water shifts. *Annals of Nutrition and Metabolism*, *41*(1), 22–28. https://doi.org/10.1159/000177974.

Lukaski, H. C. (1987). Methods for the assessment of human body composition: Traditional and new. *American Journal of Clinical Nutrition*, *46*(4), 537–556. https://doi.org/10.1093/ajcn/46.4.537.

Lukaski, H. C. (1999). Requirements for clinical use of bioelectrical impedance analysis (BIA). *Annals of the New York Academy of Sciences*, *873*, 72–76. New York Academy of Sciences https://doi.org/10.1111/j.1749-6632.1999.tb09451.x.

Lukaski, H. C., Johnson, P. E., Bolonchuk, W. W., & Lykken, G. I. (1985). Assessment of fat-free mass using bioelectrical impedance measurements of the human body. *American Journal of Clinical Nutrition*, *41*(4), 810–817. https://doi.org/10.1093/ajcn/41.4.810.

Lukaski, H. C., Kyle, U. G., & Kondrup, J. (2017). Assessment of adult malnutrition and prognosis with bioelectrical impedance analysis: Phase angle and impedance ratio. *Current Opinion in Clinical Nutrition and Metabolic Care*, *20*(5), 330–339. https://doi.org/10.1097/MCO.0000000000000387.

Martinsen, G., & Schwan, H. P. (2015). Biological tissues: Interfacial and dielectric properties. In *Encyclopedia of surface and colloid science*. https://doi.org/10.1081/E-ESCS-120000618.

Máttar, J. A. (1996). Application of total body bioimpedance to the critically ill patient. Brazilian group for bioimpedance study. *New Horizons: Science and Practice of Acute Medicine, 4*(4), 493–503.

McCarthy, H. D., Samani-Radia, D., Jebb, S. A., & Prentice, A. M. (2014). Skeletal muscle mass reference curves for children and adolescents. *Pediatric Obesity, 9*(4), 249–259. https://doi.org/10.1111/j.2047-6310.2013.00168.x.

Mikes, D. M., Cha, B. A., Dym, C. L., Baumgaertner, J., Hartzog, A. G., Tacey, A. D., & Calabria, M. R. (1999). Bioelectrical impedance analysis revisited. *Lymphology, 32*(4), 157–165.

Norman, K., Stobäus, N., Pirlich, M., & Bosy-Westphal, A. (2012). Bioelectrical phase angle and impedance vector analysis—Clinical relevance and applicability of impedance parameters. *Clinical Nutrition, 31*(6), 854–861. https://doi.org/10.1016/j.clnu.2012.05.008.

Nyboer, J. (1960). Regional pulse volume and perfusion flow measurement: Electrical impedance plethysmography. *A.M.A. Archives of Internal Medicine, 105*(2), 264–276. https://doi.org/10.1001/archinte.1960.00270140086010.

Olde Rikkert, M. G. M., Deurenberg, P., Jansen, R. W. M. M., Van't Hof, M. A., & Hoefnagels, W. H. L. (1997). Validation of multi-frequency bioelectrical impedance analysis in detecting changes in fluid balance of geriatric patients. *Journal of the American Geriatrics Society, 45*(11), 1345–1351. https://doi.org/10.1111/j.1532-5415.1997.tb02934.x.

Oldham, N. M. (1996). Overview of bioelectrical impedance analyzers. *American Journal of Clinical Nutrition, 64*(3). https://doi.org/10.1093/ajcn/64.3.405S. American Society for Nutrition.

Ott, M., Fischer, H., Polat, H., Helm, E. B., Frenz, M., Caspary, W. F., & Lembcke, B. (1995). Bioelectrical impedance analysis as a predictor of survival in patients with human immunodeficiency virus infection. *Journal of Acquired Immune Deficiency Syndromes and Human Retrovirology, 9*(1), 20–25. https://doi.org/10.1097/00042560-199505010-00003.

Ozeki, Y., Masaki, T., Yoshida, Y., Okamoto, M., Anai, M., Gotoh, K., Endo, Y., Ohta, M., Inomata, M., & Shibata, H. (2018). Bioelectrical impedance analysis results for estimating body composition are associated with glucose metabolism following laparoscopic sleeve gastrectomy in obese Japanese patients. *Nutrients, 10*(10). https://doi.org/10.3390/nu10101456.

Patel, R. V., Matthie, J. R., Withers, P. O., Peterson, E. L., & Zarowitz, B. J. (1994). Estimation of total body and extracellular water using single-and multiple-frequency bioimpedance. *Annals of Pharmacotherapy, 28*(5), 565–569. https://doi.org/10.1177/106002809402800501.

Pichler, G. P., Amouzadeh-Ghadikolai, O., Leis, A., & Skrabal, F. (2013). A critical analysis of whole body bioimpedance spectroscopy (BIS) for the estimation of body compartments in health and disease. *Medical Engineering and Physics, 35*(5), 616–625. https://doi.org/10.1016/j.medengphy.2012.07.006.

Popa, M., Sirbu, D., Curseu, D., & Ionutas, A. (2006). The measurement of body composition by bioelectrical impedance. In *IEEE international conference on automation, quality and testing, robotics*. https://doi.org/10.1109/aqtr.2006.254676.

Riyadi, M. A., Muthouwali, A. N., & Prakoso, T. (2017). Design of automatic switching bio-impedance analysis (Bia) for body fat measurement. In *Vol. 4. International conference on electrical engineering, computer science and informatics (EECSI)* (pp. 176–180). Institute of Advanced Engineering and Science. https://doi.org/10.11591/eecsi.4.1014.

Roberts, M. D., Drinkard, B., Ranzenhofer, L. M., Salaita, C. G., Sebring, N. G., Brady, S. M., Pinchbeck, C., Hoehl, J., Yanoff, L. B., Savastano, D. M., Han, J. C., & Yanovski, J. A. (2009). Prediction of maximal oxygen uptake by bioelectrical impedance analysis in overweight adolescents. *Journal of Sports Medicine and Physical Fitness, 49*(3), 240–245.

Sargent, J. A., & Gotch, F. A. (1989). *Principles and biophysics of dialysis* (pp. 87–143). Springer Science and Business Media LLC. https://doi.org/10.1007/978-94-009-1087-4_4.

Siri, W. E., & Lukaski, H. C. (1993). Body composition from fluid spaces and density: Analysis of methods … prospective overview. *Nutrition, 9*(5), 480–492.

Snel, Y. E. M., Brummer, R. J. M., Doerga, M. E., Zelissen, P. M. J., & Koppeschaar, H. P. F. (1995). Validation of extracellular water determination by bioelectrical impedance analysis in growth hormone-deficient adults. *Annals of Nutrition and Metabolism, 39*(4), 242–250. https://doi.org/10.1159/000177869.

Stolarczyk, L. M., Heyward, V. H., Hicks, V. L., & Baumgartner, R. N. (1994). Predictive accuracy of bioelectrical impedance in estimating body composition of native American women. *American Journal of Clinical Nutrition, 59*(5), 964–970. https://doi.org/10.1093/ajcn/59.5.964.

Sun, S. S., Chumlea, W. C., Heymsfield, S. B., Lukaski, H. C., Schoeller, D., Friedl, K., Kuczmarski, R. J., Flegal, K. M., Johnson, C. L., & Hubbard, V. S. (2003). Development of bioelectrical impedance analysis prediction equations for body composition with the use of a multicomponent model for use in epidemiologic surveys. *American Journal of Clinical Nutrition*, *77*(2), 331–340. https://doi.org/10.1093/ajcn/77.2.331.

Tanaka, S., Ando, K., Kobayashi, K., Seki, T., Hamada, T., Machino, M., Ota, K., Morozumi, M., Kanbara, S., Ito, S., Ishiguro, N., Hasegawa, Y., & Imagama, S. (2019). The decreasing phase angles of the entire body and trunk during bioelectrical impedance analysis are related to locomotive syndrome. *Journal of Orthopaedic Science*, *24*(4), 720–724. https://doi.org/10.1016/j.jos.2018.12.016.

Thomasset, M. A. (1962). Bioelectric properties of tissue. Impedance measurement in clinical medicine. Significance of curves obtained. *Lyon Médical*, *94*, 107–118.

Tuorkey, J. M. (2012). Bioelectrical impedance as a diagnostic factor in the clinical practice and prognostic factor for survival in cancer patients: Prediction, accuracy and reliability. *Journal of Biosensors & Bioelectronics*. https://doi.org/10.4172/2155-6210.1000121.

Wang, J., Kotler, D. P., Russell, M., Burastero, S., Mazariegos, M., Thornton, J., Dilmanian, F. A., & Pierson, R. N. (1992). Body-fat measurement in patients with acquired immunodeficiency syndrome: Which method should be used? *American Journal of Clinical Nutrition*, *56*(6), 963–967. https://doi.org/10.1093/ajcn/56.6.963.

Zingaretti, G., Nuñez, C., Gallagher, D., & Heymsfield, S. B. (2000). A new theoretical model for predicting bioelectrical impedance analysis. *Annals of the New York Academy of Sciences*, *904*, 227–228. New York Academy of Sciences https://doi.org/10.1111/j.1749-6632.2000.tb06457.x.

CHAPTER

15

Detection of cancer from histopathology medical image data using ML with CNN ResNet-50 architecture

Shadan Alam Shadab[a], *M.A. Ansari*[b], *Nidhi Singh*[a], *Aditi Verma*[a], *Pragati Tripathi*[a], *and Rajat Mehrotra*[c]

[a]Department of Electrical Engineering, Gautam Buddha University, Greater Noida, UP, India [b]Department of Electrical Engineering, School of Engineering, Gautam Buddha University, Greater Noida, UP, India [c]Department of Electrical & Electronics Engineering, GL Bajaj Institute of Technology & Management, Greater Noida, UP, India

Introduction

Nowadays in the medical field and in the healthcare system, computer-assisted diagnosis (CAD) and therapy are mostly dependent on methods that are called image processing. These images are presented to the medical experts on the screen for examine diseases. In this field, image processing helps the doctor to see the inner portion of the body part of human being very easily for examine purpose and do the keyhole operation or any surgery without opening too much part of that body (Xia, Zhang, Mu, Guan, & Wang, 2020). An endless advancement related to the study of cancer detection has been performed over the past 10 years. Researchers are applying different techniques, for example, the initial detection process, to find the types of cancer before they cause warning signs (Chtihrakkannan, Kavitha, Mangayarkarasi, & Karthikeyan, 2019). Furthermore, new techniques have been developed by researchers for the initial prediction of cancer treatment results. A very huge amount of cancer datasets have been collected with the introduction of new technologies in the medical field and are easily available to the medical research community. However, it is one of the

Computational Intelligence in Healthcare Applications
https://doi.org/10.1016/B978-0-323-99031-8.00007-7

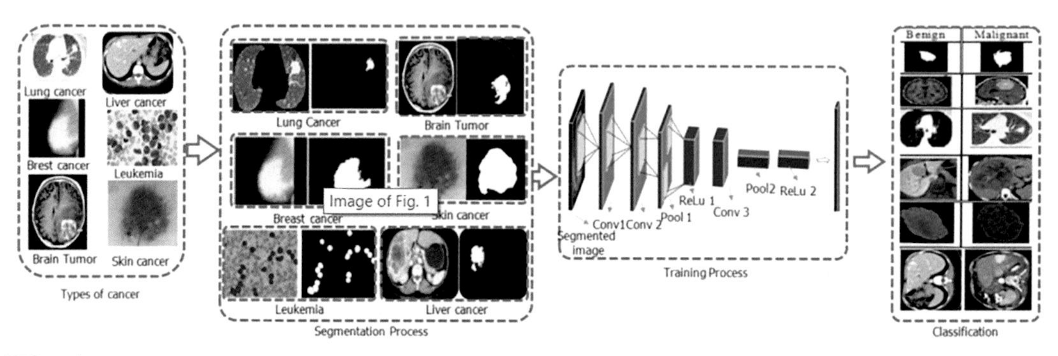

FIG. 15.1 Machine-assisted system for cancer detection. *From Saba, T. (2020). Recent advancement in cancer detection using machine learning: Systematic survey of decades, comparisons and challenges.* Journal of Infection and Public Health, *13(9), 1274–1289. https://doi.org/10.1016/j.jiph.2020.06.033.*

most interesting and challenging tasks for doctors to get more precise prediction results of a disease. As a result, machine learning (ML) techniques have become a standard tool for medical researchers. ML techniques can determine and classify patterns between them, and also see the relationship between them. ML techniques give us the hope for reaching our goal, which is to make the detection process simple and easy (Ponnada & Naga Srinivasu, 2019). In the recent study by Saba (2020), different types of cancer detection techniques and classification methods have been done by using machine assistance which gives a new way for research areas for early direction. This also shows the capacity to reduce manual system losses (Saba, 2020). In Fig. 15.1, basic structure of the machine-assisted system for cancer detection has been shown.

In the above figure, we have seen that for detection technique we may use any type of cancer image, then do a segmentation process after that we trained our model and at last we do classification and get the results.

Related work

This paper mainly focused on detection process of cancer. So, we have studied relevant paper to this topic when we have started our work and got some proposed method by the authors. In 2020, Mingyang et al. proposed a detection process of cervical cancer cells with the help of a deep learning (DL) method (Xia et al., 2020). DL is a subset of ML and with this they also try to improve the classification process network which is commonly used in DL techniques. They proposed an SPF network which is more appropriate for object detection tasks. Moreover, to design the most appropriate network structure for the detection task, five different head components are useful. It has 78.4% accuracy (Xia et al., 2020). In Fig. 15.2, the basic structure proposed by Mingyang et al. is shown.

Similarly in 2019, Chtihrakkannan et al. have proposed a breast cancer diagnosis in the early stage. The accuracy got by him is about 96% for the detection process by the use of deep neural network (DNN) algorithm (Chtihrakkannan et al., 2019). They have also tried to reduce training and computational time by using Python language. Nowadays, Python language is

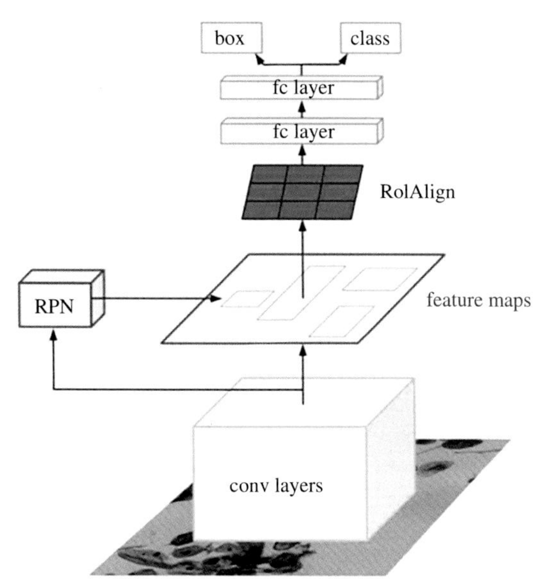

FIG. 15.2 The structure proposed by Mingyang et al. *From Xia, M., Zhang, G., Mu, C., Guan, B., & Wang, M. (2020). Cervical cancer cell detection based on deep convolutional neural network. In Chinese Control Conference, CCC (Vols. 2020-, pp. 6527–6532). IEEE Computer Society. https://doi.org/10.23919/CCC50068.2020.9188454.*

also in trend and this is easy language also. It is used in ML very commonly. In 2019 Venkata et al. has proposed an efficient convolutional neural network (EFFI-CNN) for detection of lung cancer. It consists of seven CNN layers. The first one is Convolution layer, the second one is Max-Pool layer, third one is again Convolution layer, fourth one is again Max-Pool layer, fifth one is fully connected layer, sixth one is again fully connected layer, and the last seventh one is Soft-Max layer. EFFI-CNN has a unique arrangement of CNN layers with limitations of height, breadth, depth, filter breadth, and filter height (Ponnada & Naga Srinivasu, 2019). In Fig. 15.3, convolution layer demo has been shown.

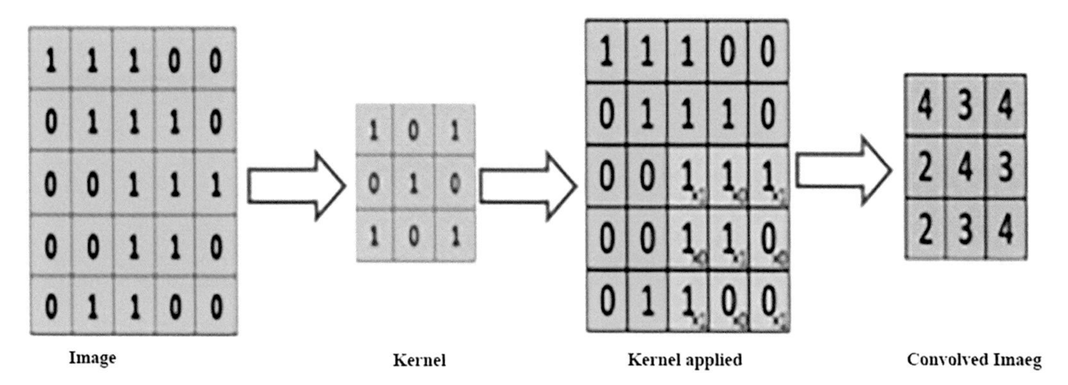

FIG. 15.3 Convolution layer demo. *From Ponnada, V. T., & Naga Srinivasu, S. V. (2019). Efficient CNN for lung cancer detection.* International Journal of Recent Technology and Engineering, *8(2), 3499–3503. https://doi.org/10.35940/ijrte.B2921.078219.*

Besides these techniques, earlier in 2017, Udhaya et al. have proposed a detection process of Brain Tumor using Medical Image Fusion based on DWT method for the examination and enhanced treatment. The process of mixing two or more than two images from a single-modality or multiple-modality is known as image fusion. In this proposed work, DWT-based image enhancement and fusion technique has been carried out. The proposed technique consists of two main processes, the first one is image enhancement and the second one is image fusion and both are dependent on DWT. For image enhancement techniques Lagrange's interpolation method is used. From the proposed method MRI and Positron Emission Tomography (PET) image were fused. From both images: MRI and PET, the fused image has matching information and by that visual information has been improved. All kinds of Brain Tumor can be detected from the proposed method (Kaur, Sharma, Dharwal, & Bakshi, 2018). In 2014, Konstantina et al. has proposed the ideas of ML while outlining its application in cancer detection (Kourou, Exarchos, Exarchos, Karamouzis, & Fotiadis, 2014). According to authors, since last year most of the studies that have been proposed are using supervised ML methods and the classification algorithm targeting to predict valid disease results. According to them which is also justified by their results, it is unmistakable that the mixing of multidimensional dissimilar data which is combined with the application of different techniques for feature selection and classification can deliver encouraging tools for conclusion in the cancer field (Kourou et al., 2014).

With these proposed methods, a survey has been done by Nikhil et al. in 2020 where they have seen that advancement in technology experienced great change in medical field when it comes (Nikhil, Srinivasa Reddy, & Dhanalaxmi, 2020). It continues to carry out, not just increase the cancer reduction methodology rate but also raise awareness among individuals. Many people are unaware of their health condition if the current technologies were not developed in cancer detection techniques. This awareness has managed to reach a huge number of cancer fighters and dropped the risk of death by a great ratio. The diseases such as Leukemia, Brain Cancer, Breast Cancer, etc. have carried out a huge beating in the death's record percentage per year in some recent years, and this record of death continues to carry out due to our modern technology and growth in the number of medical teams as well. As the development of technology has played its own role, it is also to be taken into taught that raising awareness about cancer also proves helpful to people and its effects are just as essential. Several organizations are working on it which spread awareness among people and they fight cancer. This awareness program has experienced great support and has shown enormous results by saving the lives of many individuals person and this reduces the number of deaths due to cancer every year (Nikhil et al., 2020).

Cancer detection process

The body of a human is made up of cells which are of different types and each type of cell has its own particular task. These cells in the body of a human get mature and split in an organized way and form new cells. These new cells help to possess the human body fit and make sure proper functioning. When some of these cells drop their ability to regulate their growth they grow rapidly and they are not in any specific order. The additional cells formed

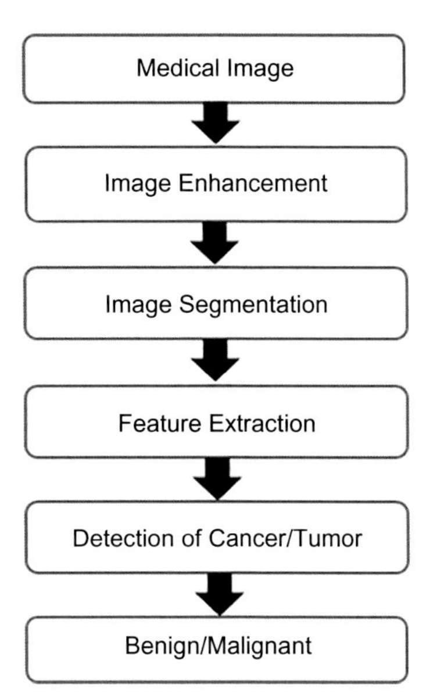

FIG. 15.4 Basic Cancer Detection Process's Block diagram. *No permission required.*

from a mass of tissue which is known as a tumor. Tumors are basically of two types one is Malignant which is a type of cancer and another one is Benign which is a non-cancer type. So nowadays, it is a challenging task to detect cancer in the human body in the initial stage. For this with the help of ML techniques, researchers are trying to make cancer detection process easier and simple with high accuracy. There are some steps for the detection process which are shown below in Fig. 15.4.

The very first one step is that one should collect medical image dataset. After that images should be enhanced and then images should be segmented to get more information from that. Feature extraction should be done to get mathematical parameter of that images. At last, detection process should be done to see whether the image is cancerous or non-cancerous. Cancerous type is known as malignant and non-cancerous type is known as Benign.

Cancer detection techniques

Different techniques are used for cancer detection which are as follows:

Biopsy

A biopsy test is a procedure for cancer detection in which the doctor remove a sample of tissue from infected area and pathologist examine that under a microscope and identify weather it is has cancer or not.

FNAC

A fine needle aspiration cytology (FNAC) is a type of biopsy test in which fine needle is inserted to examine the tissue instead of surgery.

CT

It helps the doctor to find cancer and it also shows the shape and size of the tumor. It is very often an outpatient procedure. The scanning process is not painful or we can say that it is a totally painless process and it takes only 10–30 min. It shows a slice, or cross section of the body part. It shows bones, organs, and soft tissue more clearly than any standard X-ray methods (Udhaya Suriya & Rangarajan, 2017) Fig. 15.5 shows the CT image.

MRI

MRI is a process in which powerful magnetic field radio frequency pulses are being used and it is a non-invasive technique. It takes the detailed image of organs, soft tissue, bone, and other internal body parts which are helpful for the doctor to examine the disease and give advice or medicine according to the disease. MRI scanning techniques don't produce any harmful radiation that is why it is more advisable for the patient. When we do MRI of the brain, a tumor can be detected with the help of darker or brighter appearance of normal tissue or sometimes, it is of the same intensity (Udhaya Suriya & Rangarajan, 2017) Fig. 15.6 shows the MRI image.

PET

An appropriate instrument for patient nursing and for early calculation of tumor in oncology is PET. The useful feature of the check-up is measured by PET, which is typically the amount of metabolic activity. It is also used to determine the status of a tumor (Udhaya Suriya & Rangarajan, 2017).

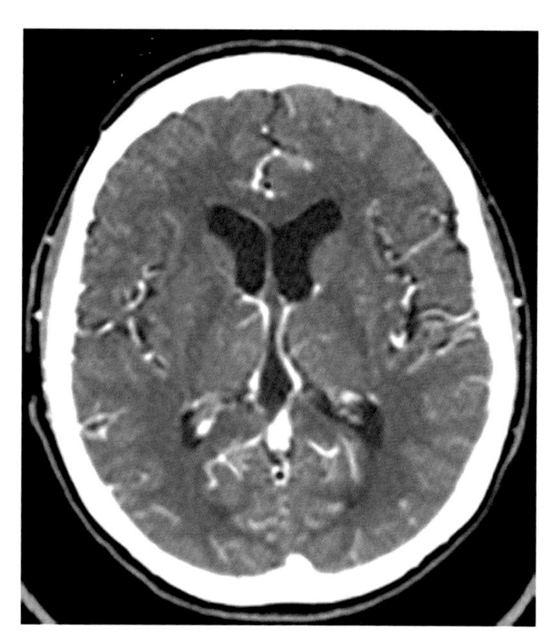

FIG. 15.5 CT image. *From MRI and CT image. (n.d.). http://16.https://4.bp.blogspot.com/-DJu6R81yTLc/V35ixNAMFXI/AAAAAAAACm4/uUWNhbXV7w8_b_DRIM5STp3Ol1BqncIswCLcB/s1600/CT_MRI.jpg.*

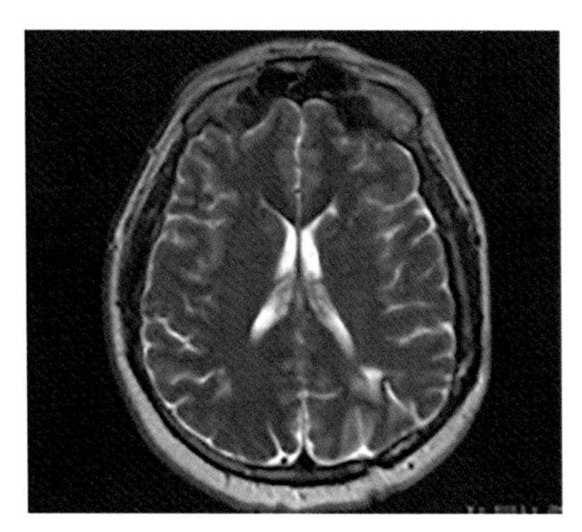

FIG. 15.6 MRI image. *From MRI and CT image. (n.d.). http://16.https://4.bp.blogspot.com/-DJu6R81yTLc/V35ixNAMFXI/AAAAAAAACm4/uUWNhbXV7w8_b_DRIM5STp3Ol1BqncIswCLcB/s1600/CT_MRI.jpg.*

Application of machine learning in cancer detection

ML is a subset of artificial intelligence (AI) which gives a system the capability of learning by itself and improving its results from past experiences without changing the executed program. ML has a basic principle of forming algorithms in which we import datasets and use mathematical analysis to calculate our output. When we calculate output, new data is available while updating outputs (Alpaydin, 2020; Nikhil et al., 2020).

In the cancer detection, process ML basically trained our model so that when we pass any random data from that trained model we got precise results. For training our model we need datasets for training our model, testing our model and for validation of our model (Alpaydin, 2020; Singh, Mahender, Singh, Kumar, & Gangwar, 2019).

The following dataset are used to train and testing ML model

Training data

From the whole dataset the large amount of data is to be used as training purpose. So training data is that data which train our model (Singh et al., 2019).

Validation data

For the cross check of our training model, we use some amount of our data to validate model. When the model is actually training this data plays its own role (Singh et al., 2019).

Testing data

When our model is finally trained then we use these data to check our results. When we pass these data from our trained model then our trained model predict some results which will further cross check by our actual data information in testing data. By this, we can check our accuracy of the results (Singh et al., 2019).

Fig. 15.7 shows how we use our dataset in different training and testing dataset.

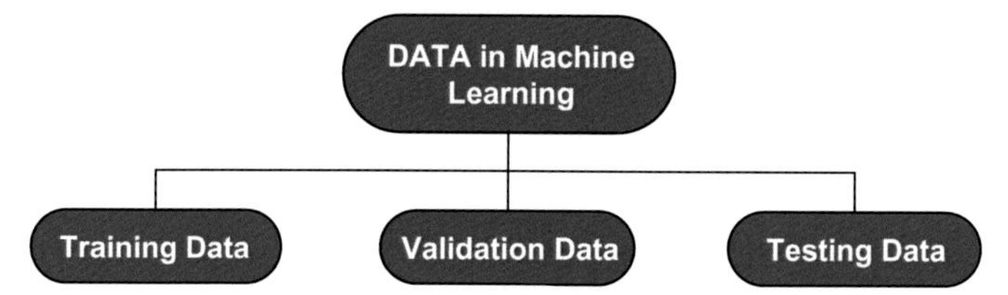

FIG. 15.7 Machine learning dataset splitting block diagram. *From Singh, A., Mahender, P., Singh, M., Kumar, A., & Gangwar. (2019). Characterization of electronic components using machine learning algorithm.*

Input dataset

We have taken Histopathology image dataset for our detection process. Histopathology images are those image which are examine under the microscope for the examination of tissue in order to see the appearances of disease by placing onto the glass slides. In medical field Histopathology refers to the examination of a biopsy test. In biopsy test, doctor takes the tissue from infected area by doing surgery and sends it to the laboratory for doing examination which shows in Fig. 15.8.

Proposed methodology

In the proposed work, we have taken a dataset to train our CNN network which has two classes, one is cancerous and another one is non-cancerous. Next, we split our dataset into two categories. One is for training my network and another one for testing our network. From the training dataset, we have trained our network and from the testing dataset we have validated our trained model. In Fig. 15.9 shows the flow chart of proposed system.

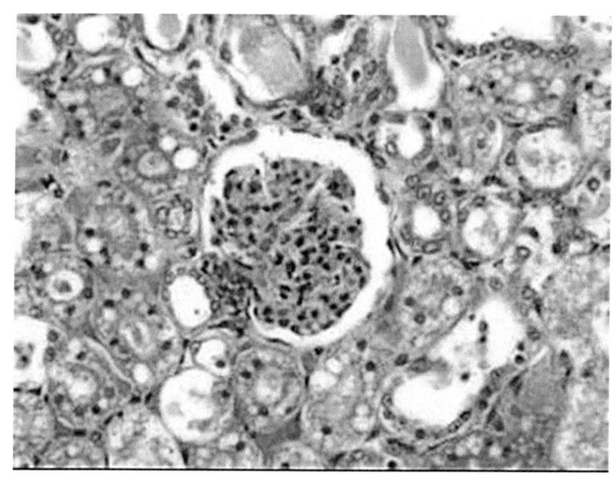

FIG. 15.8 Histopathology image. *From Breast Histopathology Images. (n.d.). https://www.kaggle.com/paultimothymooney/breasthistopathology-images.*

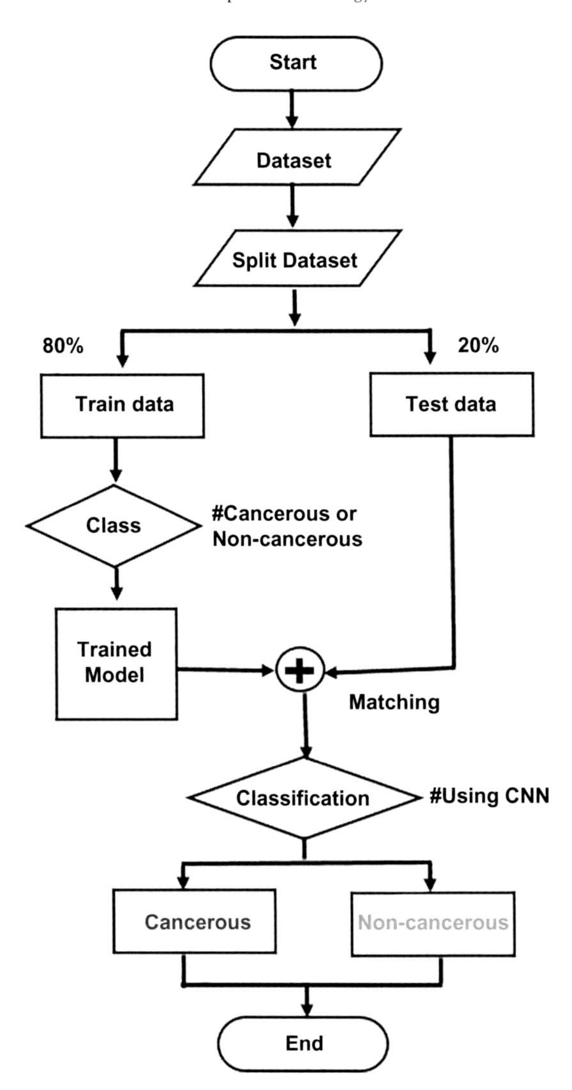

FIG. 15.9 Proposed Flow chart. *No permission required.*

These are the following proposed algorithm:

- Step-I: Start
- Step-II: Import Medical image dataset to start the program
- Step-III: Pre-process the image dataset (read image, resize image, RGB to GRAY image, enhanced image)
- Step-IV: Now split dataset into 8:2 ratio for training and testing
- Step-V: From training dataset trained the CNN model having two classes cancerous and non-cancerous
- Step-VI: Now form testing dataset check the trained model
- Step-VII: End

The various operations performed in the process of the cancer detection are as follows:

1. *Image acquisition:* Firstly, the medical images are acquired and then these images are given as input to the pre-processing stage.
2. *Pre-processing:* Image pre-processing comes with parts such as reading images, resizing as well as removing noise to enhance image quality to perform further operations. It converts image in digital form and also performs some other operations to acquire an enhanced image and to extract some advantageous data (Chtihrakkannan et al., 2019).
3. *Segmentation:* The process of separating an image into multiple segments is known as segmentation. The aim of segmentation is to make easier the analysis of image representation. It is the process of separating the tumor from normal brain tissues.
4. *Feature extraction*: It is used in image processing. It reduces the number of resources required to describe a large set of data. It builds derived features intended to be non-redundant and informative. Basically, it is the transformation of the original image to a dataset with a decreased number of variables that contains discriminated information (Chtihrakkannan et al., 2019).
5. *Classification*: Classification is a type of pattern recognition which is used in machine learning. Classification is of two types: supervised and unsupervised classification. Clustering is an unsupervised classification method. To classify an image supervised classification uses the spectral signature which is obtained from training samples (Chtihrakkannan et al., 2019).

Experimental results and discussion

We have taken our dataset from Kaggle site which has around 1000 images. These are Histopathology image datasets. This dataset is an open source dataset which can be used for any research purpose (Breast Histopathology Images, 2022).

Cancer image dataset for training network is shown in Fig. 15.10

In the above figure (Fig. 15.9), these are cancer type image dataset of Breast Histopathology image. There are approximately 700 images in this dataset.

Non-cancer image dataset for training network shows in Fig. 15.11

In the above figure (Fig. 15.11), these are non-cancer type image dataset of Breast Histopathology image. There are approximately 300 images in this dataset.

Loaded dataset shows in Fig. 15.12

After importing training dataset in my model, the random images shown here in Fig. 15.12 which are further enhanced by changing it into grayscale image, i.e., RGB to Grayscale image.

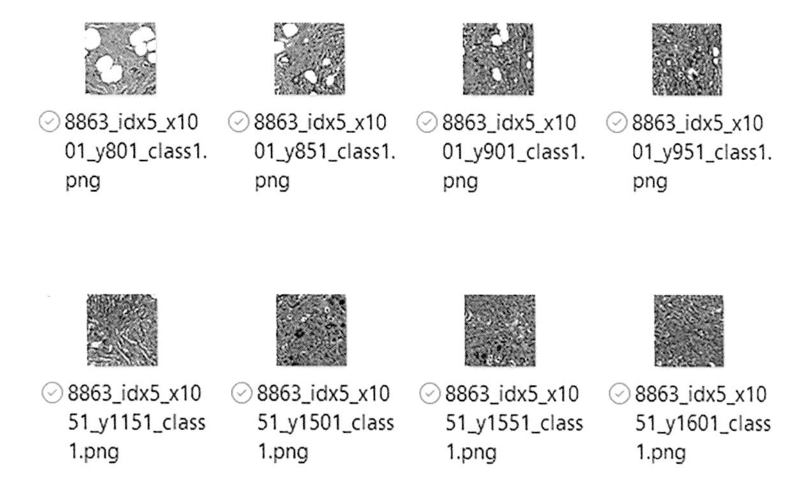

FIG. 15.10 Dataset of Cancer image. *No permission required.*

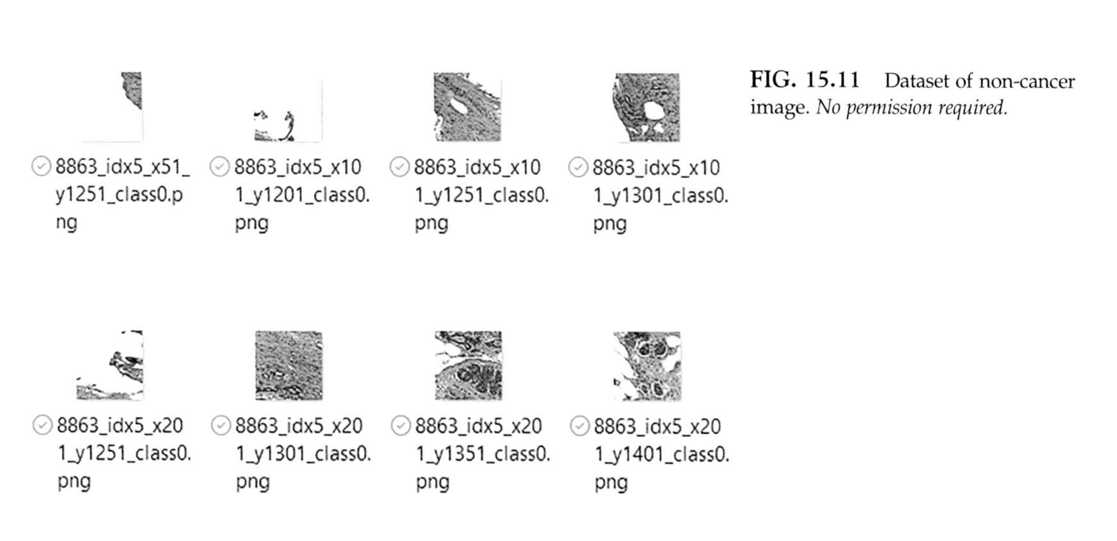

FIG. 15.11 Dataset of non-cancer image. *No permission required.*

Enhanced image shows in Fig. 15.13

Dataset we have used here is RGB image which is converted into grayscale image and enhanced it to get more clear information from that image.

Architecture of ResNet-50 for classification shows in Fig. 15.14

The Architecture of ResNet-50 is shown in above Fig. 15.14 and it comprises the following layers convolution layers, max pooling layers, activation layer, and a fully connected layer. A ResNet-50 model was pre-trained on a million images from the ImageNet database and can classify images into 1000 object categories.

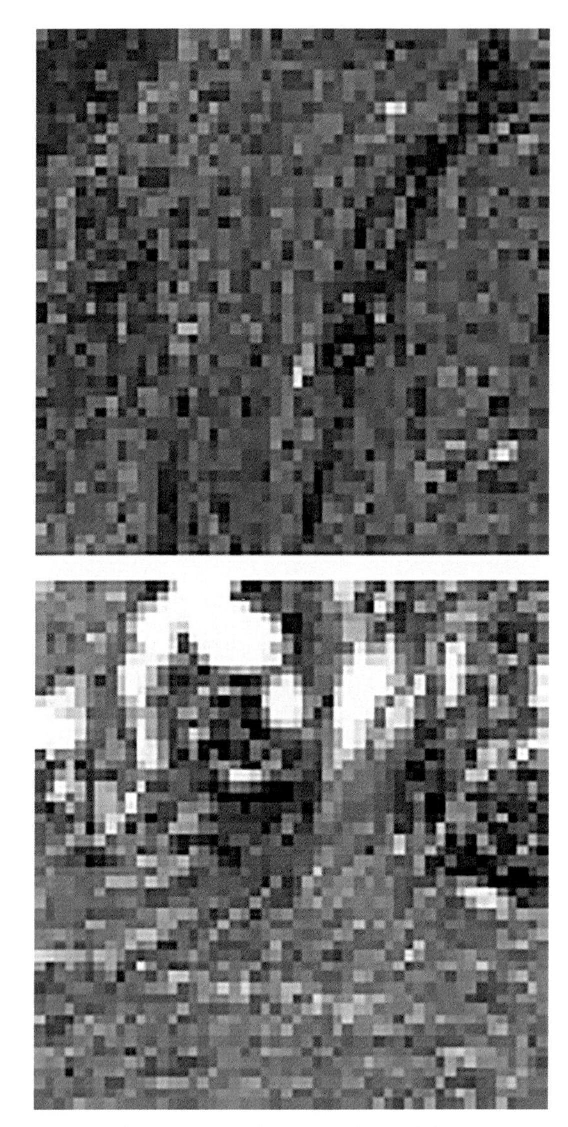

FIG. 15.12 Random images shown from training dataset in the model. *No permission required.*

Training features

See Table Table 15.1.

Testing features

See Table Table 15.2.

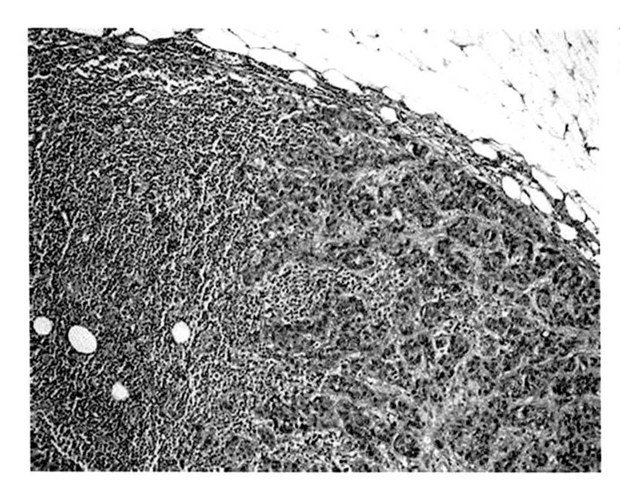

FIG. 15.13 Enhanced image. *No permission required.*

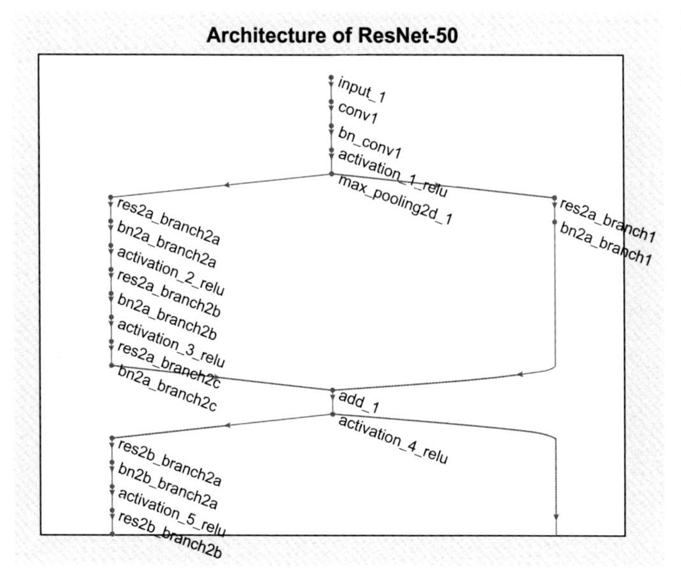

FIG. 15.14 flow cart of layer wise architecture of ResNet-50. *No permission required.*

In the above Tables Table 15.1 and Table 15.2, we have seen the features of our dataset. When we import large datasets, it is important to find out features of our dataset so that it reduces the complexity of our network. Here we have seen in Tables Table 15.1, training features of given dataset which has 1000 classes and 124 types of features. Similarly in Table Table 15.2, we have seen testing features of given dataset which has 1000 classes and 290 types of feature. These features come from ResNet-50 which has 1000 classification (fc1000).

TABLE 15.1 Training features of the given dataset.

	Feature 1	Feature 2	Feature 3	Feature 4	Feature 5	Feature 6	Feature 7	Feature 8	Feature 9	....	Feature 124
Class 1	−3.8033	−1.6159	−3.6644	−1.3083	−1.9590	−2.4765	−2.2145	−2.0836	−3.0319	....	−2.9101
Class 2	−0.8733	0.0402	−0.9730	1.9400	0.2312	−0.3775	−0.6918	0.8421	−0.1512	....	1.2342
Class 3	−4.3398	−3.3191	−4.2911	−1.7989	−3.1751	−2.9942	−3.1928	−3.7826	−3.5578	....	−4.9326
Class 4	−3.5396	−2.6236	−3.4329	−1.7003	−3.1398	−2.4441	−3.5359	−2.9557	−3.8253	....	−3.9523
Class 5	−3.2040	−2.4668	−3.2761	−1.5216	−2.2301	−1.5880	−3.3011	−2.2640	−2.8496	....	−2.5553
Class 6	4.0452	4.2917	2.9944	5.6937	5.7963	4.1637	2.7501	5.0693	4.1630	....	6.3178
Class 7	−3.9514	−2.6286	−3.5547	−1.6795	−2.4045	−2.4753	−4.1127	−2.8250	−3.8014	....	−2.6574
....	....	....	....	....	....	....	....	....	....	....	....
Class 1000	−1.0557	−0.3511	−0.8054	−0.4774	−0.0304	−0.2058	−1.4728	−1.2312	−0.9134	....	1.2775

No permission required.

TABLE 15.2 Testing features of the given dataset.

	Feature 1	Feature 2	Feature 3	Feature 4	Feature 5	Feature 6	Feature 7	Feature 8	Feature 9	.....	Feature 290
Class 1	−1.4770	−2.0640	−2.2768	−1.9618	−3.4294	−4.6654	−2.5695	−2.2375	−2.9888	.....	−1.2804
Class 2	−0.0221	−2.1866	−0.1816	1.8034	0.1869	−0.5529	−0.6125	−0.6717	−1.2139	.....	2.4297
Class 3	−2.3989	−2.7329	−3.3958	−5.8195	−5.9121	−6.2539	−4.1496	−3.3188	−4.2411	.....	−2.3889
Class 4	−2.6582	−2.6059	−3.4744	−3.8575	−4.3942	−4.7985	−3.0982	−2.9303	−3.2356	......	−2.4975
Class 5	−2.9858	−3.1009	−2.7074	−2.7761	−3.5867	−3.2883	−2.7584	−3.0508	−2.3533	......	0.0331
Class 6	3.5006	3.0556	4.3166	4.5639	6.7711	3.2998	2.9166	2.9788	4.5482	......	2.0625
Class 7	−3.0633	−5.4788	−3.5667	−2.6916	−1.6562	−4.3180	−2.7321	−3.4179	−3.3103	....	−2.7699
.....	......	......	....	......	......	......	....	....	....	....	....
Class 1000	−1.4611	0.4113	0.0919	2.2749	1.0091	−0.2348	0.3313	−0.5806	−0.1769	......	3.5440

No permission required.

Confusion matrix

Confusion matrix is a useful ML method which helps to measure accuracy, error, precision, etc. of our model. It is basically a two by two matrix in which there are mainly four parameters: True positive, true negative, false positive, and false negative (National Cancer Institute, 2020).

See Table Table 15.3.

$$\begin{aligned}\text{Training accuracy} &= (\text{TP} + \text{TN}/\text{TP} + \text{TN} + \text{FN} + \text{FP}) \; * \; 100 \\ &= 86.36\%\end{aligned} \quad (15.1)$$

Loaded image for testing shows in Fig. 15.15.

In the figure (Fig. 15.15), when I have tested it from my trained network I got 86.5% testing accuracy and it detected as non-cancer image.

In the Fig. 15.16, when I have tested it from my trained network I got 91.7% testing accuracy and it detected as Cancer image.

TABLE 15.3 Confusion matrix.

	Positive	Negative
Positive	0.9586	0.0414
Negative	0.2345	0.7655

No permission required.

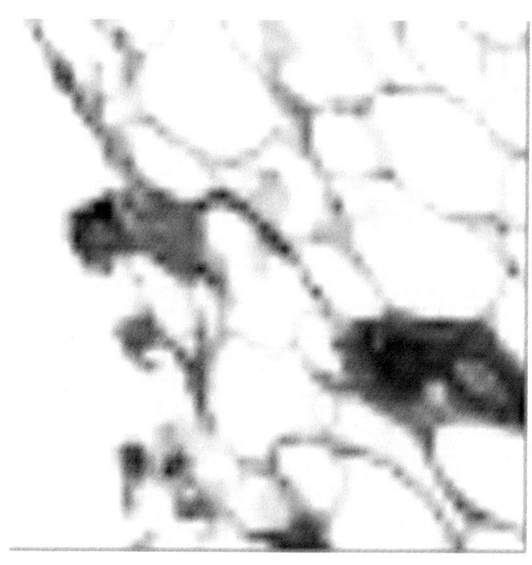

FIG. 15.15 First image for testing (test1.png). *No permission required.*

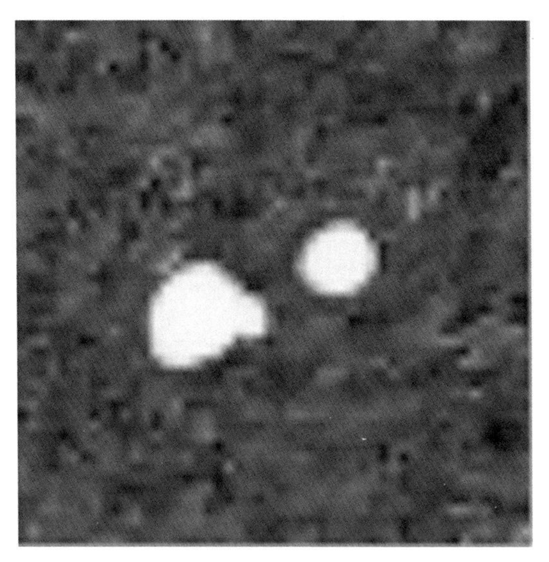

FIG. 15.16 Second image for testing (test2.png). *No permission required.*

Conclusion

In the above work we have tried to enhance the detection process of cancer in a human body. By doing this we increase curing percentage of patients in the initial stages. As we can see the rapid growth in size and number of medical databases, but instead of this most of these data are not analyzed for finding valuable and out of sight information. Advanced technology can be used to find out unknown patterns and relationships. Models developed from these techniques are useful for doctors to make right decisions. I have tried to do this by using MLtechniques. I have taken breast histopathology images dataset and trained a CNN using ResNet-50 architecture with the help of coding. In Tables Table 15.1 and Table 15.2, there is a training feature and testing features of our dataset are presented, respectively. We got different accuracy for different testing image. In Fig. 15.15, we got 86.5% accuracy. In Fig. 15.16, we got 91.7% accuracy. Our overall average accuracy comes out to 86.36% which shows in the equation no. 1 and the error comes out to be 13.63%.

References

Alpaydin, E. (2020). *Introduction to machine learning*. MIT Press.

Breast Histopathology Images. (2022). https://www.kaggle.com/paultimothymooney/breasthistopathology-images.

Chtihrakkannan, R., Kavitha, P., Mangayarkarasi, T., & Karthikeyan, R. (2019). Breast cancer detection using machine learning. *International Journal of Innovative Technology and Exploring Engineering*, *8*(11), 3123–3126. https://doi.org/10.35940/ijitee.K2498.0981119.

Kaur, L., Sharma, M., Dharwal, R., & Bakshi, A. (2018). Lung cancer detection using CT scan with artificial neural netwok. In *2018 international conference on recent innovations in electrical, electronics and communication engineering, ICRIEECE 2018* (pp. 1624–1629). Institute of Electrical and Electronics Engineers Inc. https://doi.org/10.1109/ICRIEECE44171.2018.9009244.

Kourou, K., Exarchos, T. P., Exarchos, K. P., Karamouzis, M. V., & Fotiadis, D. I. (2014). Machine learning applications in cancer prognosis and prediction. *Computational and Structural Biotechnology Journal, 13*, 8–17.

National Cancer Institute. (2020). https://www.cancer.gov/research/areas/diagnosis/artificial-intelligence.

Nikhil, D., Srinivasa Reddy, K., & Dhanalaxmi, B. (2020). Image processing based cancer detection techniques using modern technology—A survey. In *Proceedings of the 5th international conference on communication and electronics systems, ICCES 2020* (pp. 1279–1282). Institute of Electrical and Electronics Engineers Inc. https://doi.org/10.1109/ICCES48766.2020.09137997.

Ponnada, V. T., & Naga Srinivasu, S. V. (2019). Efficient CNN for lung cancer detection. *International Journal of Recent Technology and Engineering, 8*(2), 3499–3503. https://doi.org/10.35940/ijrte.B2921.078219.

Saba, T. (2020). Recent advancement in cancer detection using machine learning: Systematic survey of decades, comparisons and challenges. *Journal of Infection and Public Health, 13*(9), 1274–1289. https://doi.org/10.1016/j.jiph.2020.06.033.

Singh, A., Mahender, P., Singh, M., Kumar, A., & Gangwar. (2019). *Characterization of electronic components using machine learning algorithm*.

Udhaya Suriya, T. S., & Rangarajan, P. (2017). Brain tumour detection using discrete wavelet transform based medical image fusion. *Biomedical Research, 28*(2), 684–688 (Unpublished paper).

Xia, M., Zhang, G., Mu, C., Guan, B., & Wang, M. (2020). Cervical cancer cell detection based on deep convolutional neural network. In (Vols. 2020). *Chinese control conference, CCC* (pp. 6527–6532). IEEE Computer Society. https://doi.org/10.23919/CCC50068.2020.9188454.

CHAPTER

16

Performance analysis of augmented data for enhanced brain tumor image classification using transfer learning

Preet Sanghavi[a], *Shrey Dedhia*[a], *Siddharth Salvi*[a], *Pankaj Sonawane*[a], *and Sonali Jadhav*[b]

[a]Dwarkadas J. Sanghvi College of Engineering, Computer Engineering, Mumbai, India

[b]Computer Department, Thadomal Shahani Engineering College, Mumbai, India

Introduction

The brain is one of the most essential as well as intricate organs of the human body. A healthy brain efficiently and effectively optimizes its cognitive as well as psychological functioning (Brain health, n.d.). A brain tumor is the escalated growth of mass or abnormal cells in or around the brain. It can be either harmful (malign), containing active cells, or harmless (benign), not containing active cells. Although the principal sources/causes of brain tumor are yet to be determined, many factors meteorically increase its risk, a few of which are—exposure to hazardous radiation or noxious chemicals, eclectic history of past diseases like cancer, and obesity. The World Health Organization has classified more than 120 types of brain tumors. The brain tumor was ranked as the 10th most common disease in India. The overall process of identifying, diagnosing, managing, and treating brain tumors is quite a tortuous one. It requires the highly coordinated work of various healthcare professionals, including medical and radiation oncologists, neurologists, neurosurgeons, as well as various hospital staff (McFaline-Figueroa & Lee, 2018). This calls for a dire need of a system that can swiftly diagnose the disease, list down the exact type as well as severity, which would prompt the entire team to take timely as well as appropriate steps in alleviating the tumor,

https://doi.org/10.1016/B978-0-323-99031-8.00010-7

thus greatly increasing the chances of survival. This is where one must utilize the benefits of the advances made in the field of artificial intelligence, machine learning, and deep learning.

Related work

Colossal amounts of medical imaging data in the form of X-rays, magnetic resonance imaging (MRI), computed tomography scans, and positron emission tomography pave the way for computer vision techniques to assist healthcare professionals in taking faster steps toward patient's treatment. Appreciable results in diagnosis have been achieved in fields such as radiology, pathology, ophthalmology, and dermatology (Esteva et al., 2019). Significant advances have been made in the diagnosis of brain tumors. Convolutional neural networks, which consist of a rectified linear unit (ReLu), a convolutional layer, and a pooling layer (Wu, 2017), are widely used for tumor classification. Abiwinanda, Hanif, Hesaputra, Handayani, and Mengko (2019) proposed an architecture that consisted of two layers of convolutional, ReLu activation, and max pool each. A single hidden "dense" layer comprising 64 neurons was located. The mentioned model achieved training and validation accuracies of 98.51% and 84.19%. Pereira, Pinto, Alves, and Silva (2016) proposed the implementation of deep convolutional neural networks with 3×3 filters, precluding overfitting. Li, Li, and Wang (2019) introduced a model utilizing the U-Net architecture, which was modified through inception blocks and a cross-layer architecture. In this study, the adaptive moment estimation optimizer was used to fine-tune the network's trainable parameters. Building and training a convolutional neural network from scratch can be a tedious task due to multiple challenges, some of which are—diminutive quantity of images in medical datasets, risk of overfitting, and hiring healthcare specialists for parameter tuning, which can prove to be expensive (Swati et al., 2019). To overcome the mentioned challenges, the use of transfer learning via pretrained models is highly encouraged. Yang et al. (2018) employed both GoogLeNet and AlexNet architecture for glioma grading before operation. GoogLeNet displayed better accuracy. Deepak and Ameer (2019) made use of transfer learning through the GoogLeNet model along with classifiers such as support vector machine and k-nearest neighbors. Despite the high accuracy rates, several drawbacks persisted, such as misclassification of meningioma class samples and overfitting engendered due to the small size of the dataset. Lu et al. (Swati et al., 2019) employed the VGG-19 model for the classification of brain tumor images, which was assisted by blockwise fine-tuning. The work achieved an accuracy of 94.82%. Kaur and Gandhi (2020) performed classification using multiple pretrained models including GoogLeNet, Alexnet, Resnet50, VGG-16, VGG-19, and Resnet101. The last three layers of the models were replaced for customization to suit the classes of the images in Harvard, Figshare, and clinical datasets. Excluding ResNet, Arbane, Benlamri, Brik, and Djerioui (2021) also explored the possibility of applying MobileNetV2 and Xception models for brain tumor classification. MobileNetV2 showed superior performance with 98.24% accuracy and 98.42% F-score. Razzak et al. (Rehman, Naz, Razzak, Akram, & Imran, 2020) utilized VGG-16, AlexNet, and GoogLeNet architectures for classification on the figshare dataset. To overcome the limitations of limited dataset size, image augmentation was done through rotation and flipping. The VGG16 model performed the best on the augmented dataset with a classification accuracy of 98.69%. Pravitasari et al. (2020) introduced a hybrid architecture consisting of U-Net and VGG16 models for brain tumor segmentation. An accuracy of 96.1% was achieved on the training set. Apart from classification, brain tumor detection (whether it is benign or malign) was also performed using pretrained models and transfer learning. Soumik and Hossain (2020) made use of the InceptionV3 pretrained model for three-class classification on the Figshare CE-MRI dataset. Grampurohit, Shalavadi, Dhotargavi, Kudari, and Jolad (2020) applied both convolutional neural networks and VGG-16 for brain tumor detection, in which the latter approach outperformed the former. A 97.16% training accuracy was reached using the pretrained model. The mentioned works demonstrate the advantages of augmenting the sparse dataset of medical images and classifying using pretrained models, which would have otherwise taken hours or days and utilized immense computational resources.

Pipeline and implementation

Pipeline

As illustrated in Fig. 16.1, the approach followed for brain tumor classification first involved preprocessing of the images in the Figshare dataset. The preprocessing involves

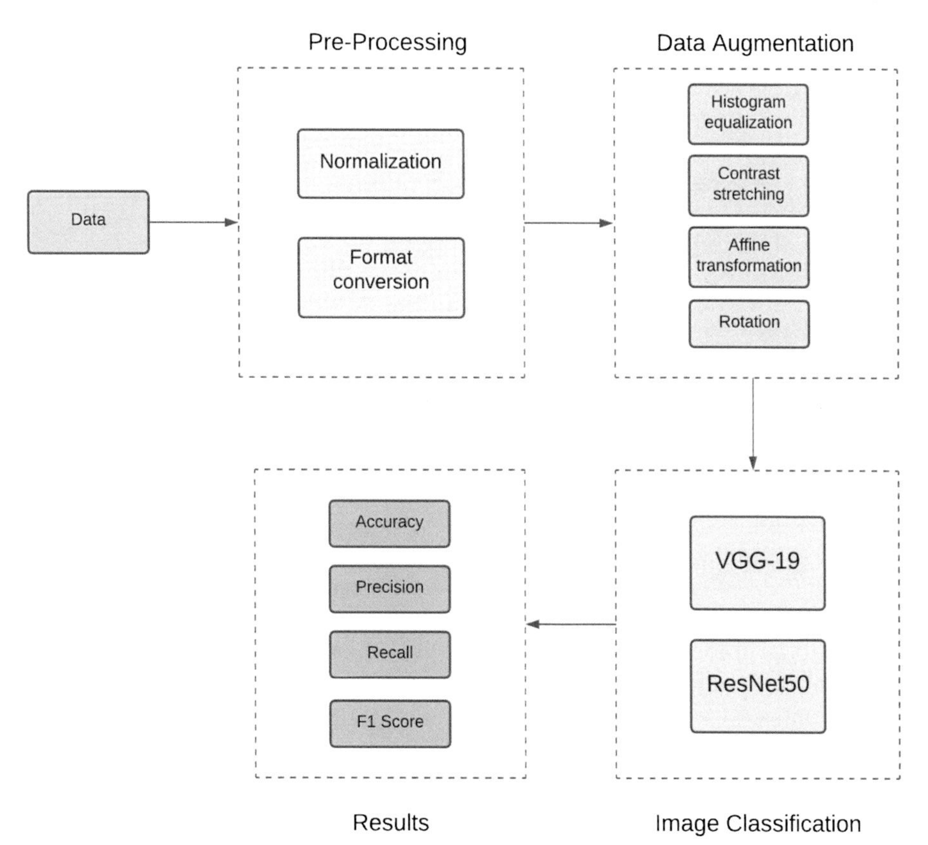

FIG. 16.1 Pipeline of the proposed architecture. *No permission required.*

converting the format of the images and normalizing them. After all the images are preprocessed, augmentation is carried out due to the lack of availability of abundant data for appropriate training. The techniques utilized for augmentation included histogram equalization, contrast stretching, affine transformation, and rotation. Image augmentation hampers overfitting and amplifies the generalization ability of the model trained. Potential expenses incurred due to the labeling of large amounts of data are also warded off. The original images from the dataset, along with augmented data are then sent for classification training. The transfer learning approach is applied for classification, which is inclined toward more accurate results. Oxford's VGG19 and Microsoft's ResNet50 are selected as the pretrained models, which are then modified and customized to suit for classification of the dataset in hand. All except a few of the last layers of the model architecture were frozen to preserve the weights of parameters calculated during pretraining. The last layers were built per the target classes in the dataset, namely—glioma, meningioma, and pituitary tumor. The model was trained for a variety of epochs to gauge the general trend in performance and accuracy. The performance parameters including precision, recall, f1-score, accuracy, macro average, and weighted average were calculated on nonaugmented as well as augmented images to track the impact and demonstrate the effectiveness of dataset augmentation.

Dataset description

For our approach, we will be using the figshare brain tumor dataset (Brima, Tushar, Kabir, & Islam, 2021). This data comes from mainly 230 patients who have three different types of tumor. These types are glioma tumor that has approximately 1400 slices, meningioma that has approximately 700 slices, and lastly, a pituitary tumor that has approximately 900 slices. This dataset is processed as it is divided into four separate folders in order to meet certain limitations. Each of the aforementioned folders possesses around 760 slices of images, where the cross-validation sets are separately mentioned.

Dataset preprocessing

In our implementation, the process of data preprocessing is mainly divided into two separate parts.

Part1: Conversion of files from mat to png.

The primary concern with respect to the dataset is that the images are provided in *.mat* format. Every mat file essentially has a structure with predefined features for a particular photograph. For each image, there are mainly five fields. These are *label*, *PID*, *image*, *tumorBorder*, and *TumorMask*. For each image, we retrieve the image data and the corresponding data label associated with it, thereby converting the image from mat to png by keeping the cmap as gray. The final images are then passed to the next step of processing.

Part2: Normalization of images.

We have normalized the data after preparing relevant datasets from the figshare data files. Data normalization is generally done in order to bring all the datasets together with respect to their image size and value of pixel intensities. There are multiple image normalization techniques, like pixel normalization, pixel centering, and pixel standardization. In our approach, we have implemented the pixel normalization technique. We have normalized the pixel values to be from 0 to 1. This scaling of data between 0 and 1 is generally called normalization.

Image augmentation for brain tumor images

In our implementation, we aim to use data augmentation in order to increase the number of images for training, testing as well as validation. As we know, the lack of training data usually leads the model to underfit, thereby providing poor testing results. In our study, we tried to get rid of this problem by using data augmentation techniques for medical imaging. From all of the aforementioned techniques, we aim to use data augmentation, in particular three techniques. These involve affining image rotation, adding Gaussian noise and cropping the image to a certain extent, and perspective transformation. While implementing data augmentation, we mainly made use of 80 preprocessed images for each of the 3 types of tumor, namely, glioma, meningioma, and pituitary. The following are the images as shown in Fig. 16.2 as compared with before and after image data augmentation along with their techniques. However, for training, augmentation is implemented fairly among all three different types of tumor.

Training and testing

Table 16.1 shows the number of images in the original dataset and the number of images after augmentation.

The images in both the original dataset and the augmented dataset were loaded and read using the imread() function of the OpenCV library, which returned the images in the Numpy array format. These images were further processed, to resize them to the (224, 224, 3) shape, which is the optimum shape of images for input to a VGG19 model or the ResNet50 transfer learning

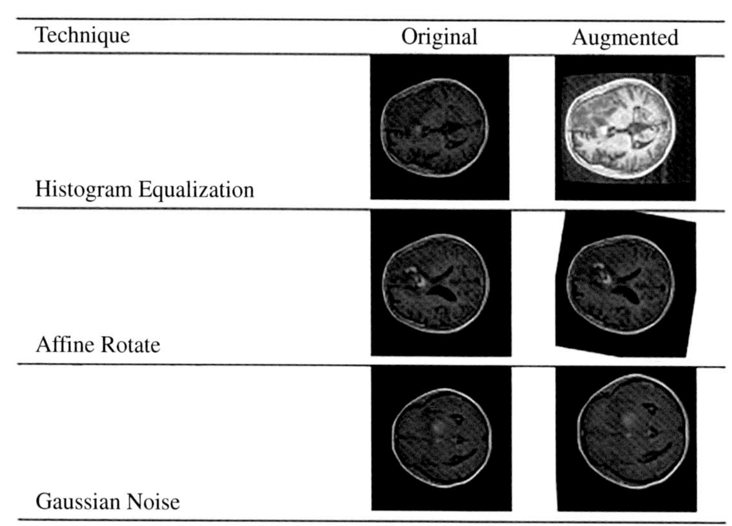

FIG. 16.2 Original vs augmented comparison. *No permission required.*

TABLE 16.1 Dataset description.

	Original dataset	Augmented dataset
Glioma	200	500
Meningioma	183	513
Pituitary tumor	178	490
Total	561	1503

model. Subsequently, the datasets were randomly split into training and validation sets, with 25% of the images being utilized for validation purposes and 75% being utilized for training purposes.

The original dataset composed of 421 images and the corresponding testing set consisted of 140 images. Similarly, the training set of the augmented dataset consisted of 1127 images, and the testing set comprised 376 images. We used 2 transfer learning models, namely, VGG19 and ResNet50, for the implementation of automatic multiclass brain tumor segmentation for both the original and the augmented image dataset. The datasets as well as training and testing set comparison are shown in Table 16.1.

b.KerasLayer function. The *input shape* parameter of this function was set to (224, 224, 3), which was the shape of the input images after resizing, and the *trainable* parameter was set to False, *rendering* this layer nontrainable. The *Sequential* class was subsequently defined, to append layers to the model one by one. The first dense layer added had three nodes and an output shape like (None, 7, 7, 3). This neural network layer was flattened to convert the entire pooled feature map matrix into a single column (Stephen, Sain, Maduh, & Jeong, 2019). The feature map procured after the application of the flatten layer was fed to a dense neural network layer consisting of an input layer, a completely connected hidden layer, and the output layer with the number of nodes equal to the number of classes that the images need to be classified into, which was the *activation* parameter for the neural network layer was set to, *softmax*, as the image was to be classified into multiple classes. The softmax activation function gives the probability for each of the classes for the given input and hence estimates the most probable output (Srinivasan & Muthu, 2018). Subsequently, all the layers added were combined together to form a complete deep neural network architecture using the *compile()* method, where the optimizer parameter of the method was set to "adam," which is based on the stochastic gradient descent. Since the labels of the training dataset were not one-hot encoded, the *loss* parameter was set to *sparse categorical cross-entropy*. This was followed by the training of the data, where the number of epochs was set to 25. The model was then evaluated on the testing data of the original dataset. Subsequently, the augmented dataset was trained and tested using steps similar to those mentioned previously for fair comparison purposes.

Results and accuracy

The original and the augmented dataset were tested on both VGG19 and ResNet50 transfer learning models. The metrics used for the purpose of comparison between the original and the augmented datasets were accuracy, recall, precision, and f1-score.

VGG19 result analysis: The accuracy obtained for the original dataset after training for 25 epochs was approximately 78% with a loss of about 1.032. The accuracy improved when validating the test data for the augmented dataset to approximately 86%, which is an increase of around 8%, while the loss diminished to about 0.4515, which was a decrease of about 0.5. The loss being calculated here is the cross-entropy loss, which primarily compares the probability of each of the predicted classes to the actual output of the test data and essentially penalizes the probability based on the distance from the actual value. It is given by

$$Loss = \sum_{i=1}^{N} y_{o,c} * \log(p_{o,c})$$

where N is the number of classes, y is the binary indicator (0 or 1) if the label assigned to the observation is the correct classification, and p is the predicted probability that the observation o belongs to class c. The precision, recall, and f1-score of the augmented dataset also increased by about 6% as compared to the original dataset.

Fig. 16.3 demonstrates the comparison between the training loss for the 25 epochs between the original and the augmented dataset. Table 16.2 shows the performance comparison between the original and the augmented dataset based on the previously mentioned metrics. Fig. 16.4 shows the comparison between the training accuracy for the original and the augmented dataset for the 25 epochs.

ResNet50 result analysis: The accuracy obtained for the original dataset after training for 25 epochs was approximately 78%, while the loss was found to be around 0.823. The accuracy procured for the augmented dataset was around 91%, which was about 14% greater than the accuracy obtained for the original dataset. The loss obtained from the augmented dataset also diminished as compared to the original dataset by about 0.6. Moreover, the precision, recall, and the F1-score obtained for the augmented dataset were greater than the original dataset by about 15%. Table 16.3 shows the performance comparison between the original and the augmented dataset based on the previously mentioned metrics. Fig. 16.5 shows the comparison of

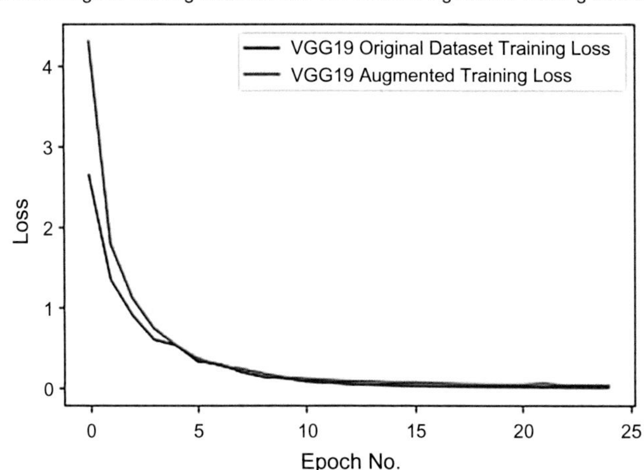

FIG. 16.3 VGG19 training loss comparison for original and augmented dataset. *No permission required.*

TABLE 16.2 VGG19 result comparison.

	Original dataset	Augmented dataset
No. of epochs	25	25
Loss	1.0326714515686035	0.4516
Accuracy (%)	77.99999713897705	86.01
Precision	0.7833	0.8717
Recall	0.7967	0.86
F1-score	0.77	0.86

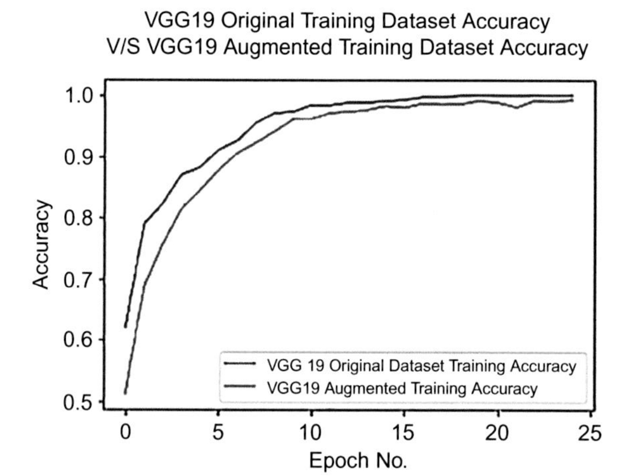

FIG. 16.4 VGG19 training accuracy comparison for original and augmented dataset. *No permission required.*

TABLE 16.3 ResNet50 result comparison.

	Original dataset	Augmented dataset
No. of epochs	25	25
Loss	0.823776125907898	0.24960991740226746
Accuracy (%)	77.999997139	91.19170904159546
Precision	0.7866	0.91
Recall	0.76667	0.9133
F1-score	0.77	0.91

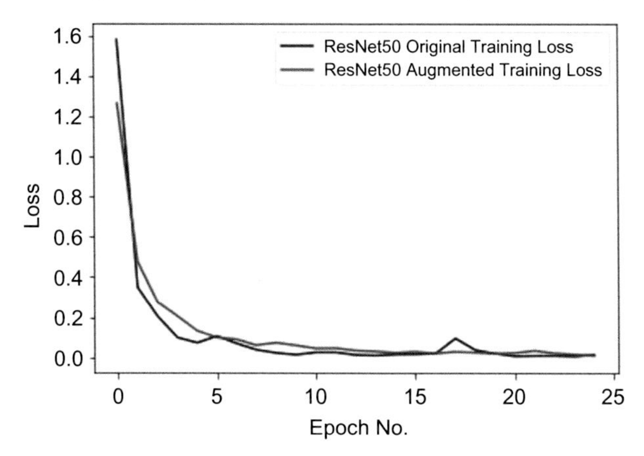

FIG. 16.5 ResNet50 training loss comparison for original and augmented dataset. *No permission required.*

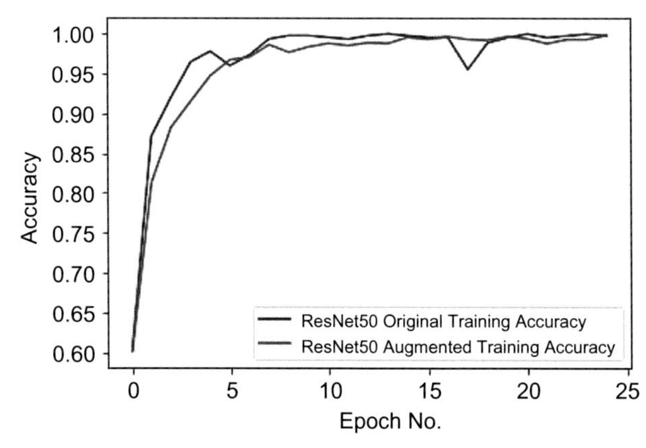

FIG. 16.6 ResNet50 training accuracy comparison for original and augmented dataset. *No permission required.*

the training loss when using the ResNet50 model, on the original and the augmented datasets. It was observed that the loss for the original dataset started at a high value of around 2.2 and decreased to around 0.096, while for the augmented dataset, the value of the loss started at about 1.6 and reduced to 0.082. Fig. 16.6 demonstrates the comparison between the accuracy obtained for the training set for 25 epochs between the original and the augmented datasets.

Conclusion and future scope

The chapter presented an approach to classifying brain tumors. Our work demonstrates the propitious results of applying data augmentation and utilizing transfer learning via pretrained models, which gave considerable results in a relatively lower number of epochs, thereby saving computational time and resources. The following system presented can be further enhanced by the availability of more authentic data and images, utilizing generative adversarial networks (GANs) for image augmentation and customizing more unfrozen layers on the pretrained model. The results show the potential of the discussed pipeline to be employed in healthcare systems early, as well as accurate analysis, thereby saving many lives.

References

Abiwinanda, N., Hanif, M., Hesaputra, S. T., Handayani, A., & Mengko, T. R. (2019). Brain tumor classification using convolutional neural network. *IFMBE Proceedings, 68*(1), 183–189. Springer Verlag https://doi.org/10.1007/978-981-10-9035-6_33.

Arbane, M., Benlamri, R., Brik, Y., & Djerioui, M. (2021). Transfer learning for automatic brain tumor classification using MRI images. In *2020 2nd international workshop on human-centric smart environments for health and well-being, IHSH 2020* (pp. 210–214). Institute of Electrical and Electronics Engineers Inc. https://doi.org/10.1109/IHSH51661.2021.9378739.

Brain health. (n.d.). Retrieved May 24, 2021, from https://www.who.int/health-topics/brain-health.

Brima, Y., Tushar, M. H. K., Kabir, U., & Islam, T. (2021). *Brain MRI Dataset*. https://doi.org/10.6084/m9.figshare.14778750.v2.

Deepak, S., & Ameer, P. M. (2019). Brain tumor classification using deep CNN features via transfer learning. *Computers in Biology and Medicine, 111*. https://doi.org/10.1016/j.compbiomed.2019.103345, 103345.

Esteva, A., Robicquet, A., Ramsundar, B., Kuleshov, V., DePristo, M., Chou, K., et al. (2019). A guide to deep learning in healthcare. *Nature Medicine, 25*(1), 24–29. https://doi.org/10.1038/s41591-018-0316-z.

Grampurohit, S., Shalavadi, V., Dhotargavi, V. R., Kudari, M., & Jolad, S. (2020). Brain tumor detection using deep learning models. In *Proceedings—2020 IEEE India council international subsections conference, INDISCON 2020* (pp. 129–134). Institute of Electrical and Electronics Engineers Inc. https://doi.org/10.1109/INDISCON50162.2020.00037.

Kaur, T., & Gandhi, T. K. (2020). Deep convolutional neural networks with transfer learning for automated brain image classification. *Machine Vision and Applications, 31*(3). https://doi.org/10.1007/s00138-020-01069-2. Springer.

Li, H., Li, A., & Wang, M. (2019). A novel end-to-end brain tumor segmentation method using improved fully convolutional networks. *Computers in Biology and Medicine, 108*, 150–160. https://doi.org/10.1016/j.compbiomed.2019.03.014.

McFaline-Figueroa, J. R., & Lee, E. Q. (2018). Brain tumors. *American Journal of Medicine, 131*(8), 874–882. https://doi.org/10.1016/j.amjmed.2017.12.039.

Pereira, S., Pinto, A., Alves, V., & Silva, C. A. (2016). Brain tumor segmentation using convolutional neural networks in MRI images. *IEEE Transactions on Medical Imaging, 35*(5), 1240–1251. https://doi.org/10.1109/TMI.2016.2538465.

Pravitasari, A. A., Iriawan, N., Almuhayar, M., Azmi, T., Irhamah, Fithriasari, K., et al. (2020). UNet-VGG16 with transfer learning for MRI-based brain tumor segmentation. *Telkomnika (Telecommunication Computing Electronics and Control), 18*(3), 1310–1318. https://doi.org/10.12928/TELKOMNIKA.v18i3.14753.

Rehman, A., Naz, S., Razzak, M. I., Akram, F., & Imran, M. (2020). A deep learning-based framework for automatic brain tumors classification using transfer learning. *Circuits, Systems, and Signal Processing, 39*(2), 757–775. https://doi.org/10.1007/s00034-019-01246-3.

Soumik, M. F. I., & Hossain, M. A. (2020). Brain tumor classification with inception network based deep learning model using transfer learning. In *2020 IEEE region 10 symposium, TENSYMP 2020* (pp. 1018–1021). Institute of Electrical and Electronics Engineers Inc. https://doi.org/10.1109/TENSYMP50017.2020.9230618.

Srinivasan, K., & Muthu, N. (2018). A comparative study and analysis of contrast enhancement algorithms for MRI brain image sequences. In *2018 9th international conference on computing, communication and networking technologies, ICCCNT 2018* Institute of Electrical and Electronics Engineers Inc. https://doi.org/10.1109/ICCCNT.2018.8494068.

Stephen, O., Sain, M., Maduh, U. J., & Jeong, D.-U. (2019). An efficient deep learning\napproach to pneumonia classification in healthcare. *Journal of Healthcare Engineering, 2019*.

Swati, Z. N. K., Zhao, Q., Kabir, M., Ali, F., Ali, Z., Ahmed, S., et al. (2019). Brain tumor classification for MR images using transfer learning and fine-tuning. *Computerized Medical Imaging and Graphics*, *75*, 34–46. https://doi.org/10.1016/j.compmedimag.2019.05.001.

Wu, J. (2017). *Introduction to convolutional neural networks*. China (p. 5). China: National Key Lab for Novel Software Technology, Nanjing University.

Yang, Y., Yan, L. F., Zhang, X., Han, Y., Nan, H. Y., Hu, Y. C., et al. (2018). Glioma grading on conventional MR images: A deep learning study with transfer learning. *Frontiers in Neuroscience*, *12*. https://doi.org/10.3389/fnins.2018.00804.

CHAPTER

17

Brain tumor detection through MRI using image thresholding, k-means, and watershed segmentation

Aditi Verma[a], M.A. *Ansari*[b], *Pragati Tripathi*[a], *Rajat Mehrotra*[c], *and Shadan Alam Shadab*[a]

[a]Department of Electrical Engineering, Gautam Buddha University, Greater Noida, UP, India [b]Department of Electrical Engineering, School of Engineering, Gautam Buddha University, Greater Noida, UP, India [c]Department of Electrical & Electronics Engineering, GL Bajaj Institute of Technology & Management, Greater Noida, UP, India

Introduction

The brain is described as a critical part of the human body. It is not easy to study the brain because of its complex structure. The brain controls almost every part and action of our body; even during sleep, it is in operating mode. The brain is responsible for controlling the actions of other body parts; it also performs the function of learning, memorizing, and understanding. The abnormal growth of brain cells or brain tissues is generally considered the occurrence of a brain tumor. A brain tumor is considered a very dangerous disease; several researchers are conducting studies for the easy detection and cure of the brain tumor. As in the early stage, the tumor can easily be treated; however, if it is left undetected, it will develop into brain cancer, and then it will be more difficult to cure it properly. Thus, the detection of a tumor at an early stage is highly important. In recent years, the presence of the brain is mainly found among the younger generation and older groups of society. Brain tumor spreads very rapidly, making the disease more fatal and is also one of the basic issues in tumor detection. In our body, the new cells are formed when the old cells are damaged; because of some disorder, these new cells grow at a rapid speed even when the old cells have not yet been replaced. Due to this rapid growth of cells, the tumor is formed, which is a lump of rapidly developing cells. The growth of a tumor can be benign or malignant. Benign brain tumors do not contain

https://doi.org/10.1016/B978-0-323-99031-8.00006-5

cancer cells, whereas malignant brain tumors contain cancer cells; the growth of malignant brain tumors is considerably greater than that of benign brain tumors. For the detection of tumors, a brain computed tomography (CT) scan or brain MRI is mainly used. In this study, the MR images of the brain were mainly used to detect the tumor. In the medical field, MR images are widely used for diagnosis and study as it is a noninvasive technique; it uses some powerful magnetic and radio waves to obtain the image of the body. Tumors are treated using various techniques such as surgery, radiotherapy, and chemotherapy. Before the treatment of a tumor, its detection is an important task. For accurate detection of tumors, the MR images of the brain must be filtered because they may contain certain kinds of unwanted noise that may affect the accuracy of tumor detection. Filtration techniques used mainly make the brain MR images more clear. Filtering removes the blur and preserves the edges of the image. The filters used mainly help to detect the tumor more precisely and accurately. For tumor detection, there are three major steps, namely, preprocessing, segmentation, and feature extraction. First, we aim to make the image clear and remove the noises; then we segment the image to obtain the tumor, and lastly, we extract the features of the image. In this study, four filtration techniques were applied to remove noises from MR images. These are anisotropic filtering, adaptive filtering, unsharp mask filtering, and median filtering. These filtration techniques are used for preprocessing the image, and then their results are discussed by comparing the filtered images. These filters mainly reduce and sometimes remove the impulse noises from the brain MRI. After completing preprocessing, an image is obtained that contains less noise; small details of the image are also preserved. Preprocessing of MR images is necessary for the accurate detection of the tumor. After the completion of the filtration process, the MR image is segmented. Here, we used three different segmentation algorithms to study three different kinds of segmentation techniques, namely, k-means segmentation, image thresholding, and watershed segmentation. After these segmentation techniques were applied, the features were extracted using the GLCM method. These features provide some detailed information regarding the detected tumor. Some of the features used in this work are contrast, energy, entropy, correlation, autocorrelation, variance, maximum probability, homogeneity, cluster shade, and cluster prominence. Extracted features are mainly utilized for diagnosis or research purposes in the medical field.

Literature review

The brain is one of the most complex organs in the human body that works with billions of cells. A cerebral tumor occurs when there is an uncontrolled division of cells that forms an abnormal group of cells around or within the brain. A tumor is a mass of abnormal cells that accumulate forming a tissue. These abnormal cells feed on the normal body cells and destroy them and keep growing. One of these tumors is a brain tumor. A tumor can be diagnosed via imaging and pathology. A brain tumor is imaged with MRI, giving cross-section images of the brain. In recent days, medical image processing plays a vital role in medical research. Image classification, filtration, and segmentation are the major processing techniques. This image is resized and converted into a grayscale image. The preprocessing technique is used to remove noise from images. Processing of MR images is one of the components of image processing in the medical field, which is the most emerging field these days. Tumor detection is often the preliminary phase (Phillips et al., 2020). Electronic modalities are used to diagnose brain tumors. Filter operation was used to remove unwanted noises as much as possible to assist in better segmentation. Brain tumor detection is considered a challenging mission in medical image processing. Automatic detection of a brain tumor is a difficult task due to variations in type, size, location, and shape of tumors (Erickson, Korfiatis, Akkus, & Kline, 2017). Sonar and Bhosle (2016) suggested many nature-inspired algorithms such as the Ant Colony Optimization (ACO) algorithm, Cuckoo Search (CS), Firefly Algorithm (FA), and Particle Swarm Optimization (PSO) for the classification of brain tumors from an MRI image (Sonar & Bhosle, 2016). In a similar work related to brain tumor classification, Taie and Ghonaim applied CS for feature reduction (Taie & Ghonaim, 2017). Ozic, Ozbay, and Baykan (2014), in their contemporary work, implemented PSO with OTSU's thresholding algorithm to detect a tumor in the brain. Gopal and Karnan (2010) compared the classification accuracy level of two methods, namely, PSO and GA along with fuzzy C-mean to detect brain tumors (Gopal & Karnan, 2010). Akhil, Aashish, and Manikantan (2016) extended the artificial bee colony (ABC) algorithm with a binary input mechanism to obtain more precise results.

Methodology

Benign tumors can easily be ejected and they rarely grow again, whereas malignant brain tumors are more serious. If benign tumors are not detected at an early stage, they might convert into malignant brain tumors causing a threat to life. Tumor detection using MR images involves various steps including preprocessing, segmentation, and feature extraction. These steps are performed to detect the tumor precisely. Most of the MRI images contain some unwanted intensity variations due to the imperfections of MRI scanners. For reducing these intensity variations, different filters are used. Removing or reducing different forms of noises or intensity variations from the MR images is known as preprocessing of the image.

Image preprocessing

MRI scans taken as input were resized to 256×256 pixels for processing. The preprocessing of a brain MRI improves the accuracy to detect the tumor region. This step is done before segmentation. It removes different kinds of noises from the MR image. Noise is an unwanted signal that restricts the efficiency of a system. Therefore, the noise of an MR image should be concealed before forwarding the image to segmentation and classification algorithms. The preprocessing of MR images is used to enhance and make it more suitable to analyze reduction and removal of noise, image sharpening, and contrast enhancement. The noise removal or reduction process should not produce any kind of fake data. Along with different types of filters, two morphological operations are used for image enhancement, namely, dilation and erosion. Erosion is a mathematical operation used for erasing the foreground pixels from the image and dilation is the opposite of erosion; it causes the addition of background pixels to the input image. Medical images might include some kind of unwanted signals or noise, poor contrast, or weak edges and boundaries. These problems are mainly solved by using a filter. The objective of the filter is to increase or reduce the number of pixels of the dataset and improve the visualization by brightening the dataset. In this work, four filters were used to improve the condition of an MR image. Median filter, unsharp mask filter, adaptive filter, and anisotropic filter are four filters used for tumor detection.

Image segmentation

Depending on the interaction of humans in the segmentation process, this process is categorized into three different types. The manual segmentation technique has some problems with MR images, making it impossible to reproduce the image; in addition, it deteriorates the result of segmentation. During the segmentation of an image, the image was just divided into subgroups of pixels. There are mainly three segmentation methods used with MR imaging, which are classification, region, and contour-based segmentation. The purpose of this division is to represent the medical image in a more purposeful and significant form, making it easier to detect the tumor. The segmentation process is very commonly used because of the advantages of being robust and very fast; as it is generally automated, it reduces time consumption and provides an accurate result. In this work, three segmentation techniques were used, namely, k-means, image thresholding, and watershed segmentation techniques. The input image to this segmentation process is a well-defined and filtered image that consists of little noise or disturbance. Using these segmentation techniques, we detected the tumor shape, the boundary of the tumor, and the area affected by the tumor. Ahammed Muneer, Rajendran, and Paul Joseph (2019) utilized the genuine dataset from facilities in the United States. They utilized a redid classification calculation dependent on Wndchrm. The device has the ideas of neighborhood distance measure utilizing morphology and profound learning for CNN. The creators with this proposed arrangement accomplished an exactness of 92.86% notwithstanding 98.25% that was accomplished through the Wndchrm classification. Banerjee, Mitra, Masulli, and Rovetta (2019) examined the convolutional neural organization to upgrade MRI image classification using a succession of various MRI images. They proposed ConvNet models that were redone to be created from the stomach muscle initio, in view of ideas of cutting and fixing MRI images. Kabir Anaraki, Ayati, and Kazemi (2019) proposed a procedure dependent on CNN and GA (hereditary calculation) to arrange different kinds of Glioma images using MRI information. The prop framework utilized GA for a programmed choice of the CNN structure. They obtained 90.9% precision foreseeing Glioma images of three sorts. Additionally, the investigation revealed an exactness of 94.2% in the classification of glioma, meningioma, and pituitary.

Feature extraction

After the process of tumor segmentation, feature extraction is performed. GLCM method was used for feature extraction. We represented input data in the form of a set of features and this technique is known as feature extraction. Feature extraction techniques are mainly applied for the recognition of images and classification purposes. This process is used to connect the high-level information of the given MRI. Feature extraction facilitates the learning process and is extracted from the initial data set. It is mainly extracting the subset from the set of primary features. The given features include the information regarding the dataset so that the reduced representation of the data can be done instead of the complete dataset. This step is used to increase the accuracy of tumor detection. Using the feature extraction method, we might get some more information regarding the tumor, which is used for the diagnosis as well as for research purposes. Features can be texture-based, edge-based, pixel intensity-based, etc.

Ten features were extracted in this work. Contrast, correlation, cluster prominence, cluster shade, dissimilarity, energy, entropy, maximum probability, variance, and autocorrelation are 10 features extracted. Feature extraction is also considered a special form of reduction of dimensionality. Feature extraction is when we reconstruct the input data into the set of appearances.

Filtration techniques

Detecting brain tumor works more accurately and precisely when the MR image of the brain does not contain any kind of unwanted signals. Before proceeding to the segmentation technique, all noises and disturbances must be removed or reduced for more accuracy. Filtering is the process used for reducing the noise as it expects certain undesirable segments. Filtering preserves the edges of the image, smoothens the noisy pixels, and removes the blur. Filtration of an MR image is important because it finally determines the accuracy of the system. The accuracy of tumor detection largely depends on filtration techniques. Fig. 17.1 shows the chart of different filtration techniques discussed. A more precise filtration leads to a more accurate detection. In this work, four different filtration techniques are used and their performance compared, namely, adaptive filter, anisotropic filter, unsharp mask filter, and median filter.

Adaptive filter

The adaptive filter is performed on degraded images and depends on the two statistical measures, which are mean and variance. This is a linear filter. An optimization algorithm is used to adjust those parameters. The use of adaptive filters has become more common due to the advancement of digital signal processors. Adaptive filters for commonly used in communication devices, digital cameras, medical monitoring equipment, etc. Adaptive

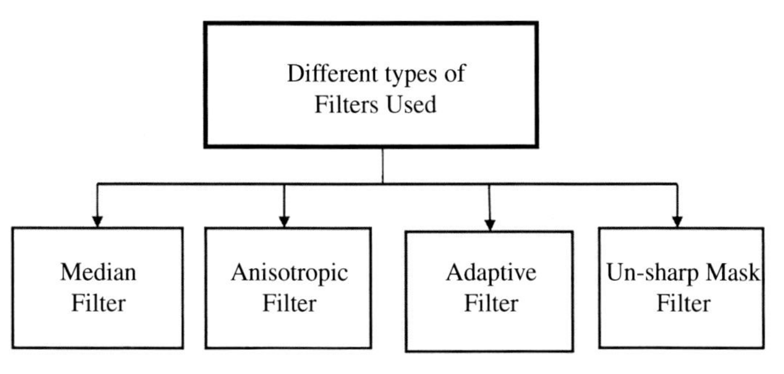

FIG. 17.1 Different filtration techniques. *No permission required.*

filters generally do not blur the image structure and restore or enhance the data with the removal of noises. An adaptive filter does not blur the structures of the image. Adaptive filters are used to enhance the quality of the image, which contains the noise and the two statistical measures, that is, the mean and variance of a defined window, on which an adaptive filter depends. For adaptive filtering of local noise, the adaptive expression is given as:

$$F(x,y) = g(x,y) - (\sigma 2n/\sigma 2L)\,[g(x,y) - mL]$$

$\sigma 2n$, variance of overall noise; $\sigma 2L$, local variance; mL, local mean; $g(x, y)$, pixel value at position (x, y).

It also adapts itself to the local surroundings of the image. It works well for images that have abrupt intensity changes. An adaptive filter is self-adjusting in nature. This filter adapts itself to the surroundings primarily to the mean and variance of the pixels. It removes the impulse noise and speckle noise.

Median filter

The median filter is generally used to remove the noise from an image or signal. It is a nonlinear digital filter. This filter is widely preferred in digital image processing because it preserves the edges of an image while removing the noise. In addition, it does not make the image blur or deteriorate its quality. Median filter shorts the neighborhood values of the pixel from low to high. The center pixel is considered as the test pixel, which is replaced by taking the average or the median of the surrounding pixels. Its surrounding pixels or neighborhood pixels are known as a window. If the size of the window is more, the smoothening of the image is also more. There are shrinking windows near the boundaries; thus, this process is avoided at the boundary. Median filtering can also be called a smooth-running technique that is used for removing impulse noises. It is not good at dealing with a large amount of Gaussian noises.

Anisotropic filter

The anisotropic filter is required to enhance the quality of textures of an image on the surface of computer graphics. This filter eliminates the aliasing effect; it is also common in modern graphics. It removes the Gaussian noise from the image. There are four name-value pair arguments used in this filter. The gradient threshold controls the conduction process. In this filter, the value of the gradient threshold is increased and the default value of the gradient threshold is 10% to make the image more smooth. The number of iterations is defined by a positive integer to provide an idea about how many times the process has to be done. In this work, the number of iterations used was 10. Connectivity is defined as the relation of pixels to their neighbors and it can be maximal or minimal in nature. The maximal connectivity considers the eighth nearest neighbors of the 2D image; the minimal connectivity considered the four nearest neighbors of the 2D image. The conduction method was used to determine the contrast of images; it can be exponential or quadratic. Exponential diffusion favors high-contrast edges, whereas quadratic diffusion favors wide regions.

Unsharp mask filter

Unsharp mask filter is simply a sharpening operator used as an image-sharpening technique; it is mainly available in digital image-processing software. This filter acts as a sharpening operator that enhances the edges of the image. However, when a certain high amount of noise is present in the image, this filter produces poor results. This filter highlights the edges of the image but lowers the quality of the image. This filter subtracts the unsharp input image from the original one. In addition, it enhances the high-frequency components present in the image. The filtered image has clear and sharp edges and the high-frequency components are highlighted. However, in the case of noise removal, it yields a poor result. Different settings are available for the unsharp mask. For soft-aged subjects, the threshold must be 6–10, a radius around 1–1.5, and an amount of 100%–150%, whereas, for portraits, the threshold must be 4–6, a radius around 1–2, and an amount of 100%–120%. Unsharp mask filter only performs the sharpening of the image; however, it does not correct a severely blurred image. The operation of this filter is considered a convolution operation on the input image with the kernel mask.

Segmentation techniques

Segmentation is the procedure, in which the image is divided into different segments or different regions. Using this, we represent the image in the most significant manner, making it easier to detect the tumor. Segmentation techniques can be manual or automatic. These are divided into methods including surface-based approaches or region-based approaches. Image thresholding and k-mean segmentation fall under the category of region-based segmentation methods. In this work, three segmentation techniques were used—image thresholding segmentation, k-means clustering segmentation, and watershed segmentation. These segmentation techniques are discussed and their results are also displayed in this work.

k-means segmentation

k-means segmentation is the simplest unsupervised learning technique. It divides the image into different sections, making it easy to detect the tumor location. In this segmentation technique, the image is divided into different segments or clusters. These segments were created based on some kind of similarities between them. The number of segments or clusters was initialized in the beginning. First, we input the value of k, which gives the number of clusters and their centers. The number of clusters was given by k, which was initialized at the beginning. The points k from the data set were randomly selected. For every pixel of the image, the Euclidean distance was computed. Euclidean distance is the distance between the center and every pixel of the image.

We represent this distance by d. Depending on the Euclidean distance, each pixel of the image was allocated to the closest center. The algorithm is also shown in Fig. 17.2. After each pixel was located, the position of the center was recalculated and this process was repeated until it satisfied the error value. Cluster pixels were then reshaped into the image form. k-means segmentation is very popular in the clinical field. In this technique, we do not know

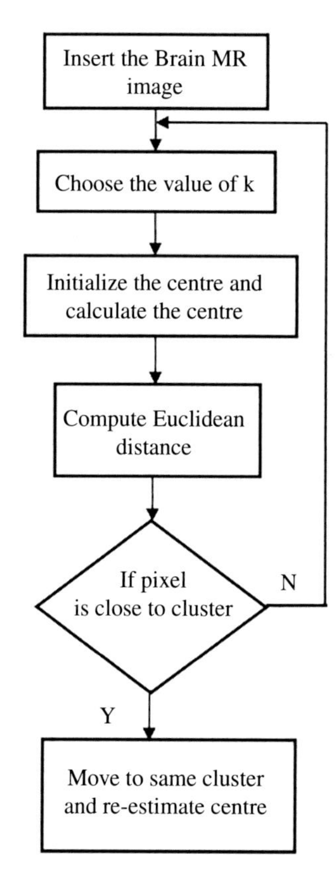

FIG. 17.2 k-means segmentation algorithm. *No permission required.*

what we are looking for but by dividing it into segments, we can easily obtain the structures or groups of pixels having different properties. There are two steps, one of which is the E-step, which is used to assign the data points to the cluster, which is the closest and another step is the M-step, which is to identify the centroid of each cluster.

k-means segmentation is the simplest unsupervised learning technique. It divides the image into different sections making it easy to detect the tumor location. It is also used to segment the interest area from the background. This segmentation technique is utilized to detect the tumor from the brain MRI using MATLAB software.

Image thresholding segmentation

In digital image processing, image thresholding is the simplest segmentation method. The input is a grayscale image and it is converted into a binary image. This segmentation technique is used to separate the desired object from the background of the input image. In the image thresholding method, we considered the intensity of the pixel to analyze the image or detect the tumor. This method separates the region with different intensities as it separates the light and dark regions of the image.

A threshold value was set and every pixel of the image was compared with the set threshold value. By comparing every pixel of the image with the threshold value, the images were divided into two different regions. One region contains the part of the image that had the intensity of pixels greater than the threshold value and another region contains the part of the image that had the intensity of pixels less than the threshold value.

Thresholding value can be global or local. If the threshold value remains constant throughout the process, it is known as local thresholding. The threshold value is dependent on the position of the image pixels in the case of global thresholding (Celenk, 1991). In the thresholding method, the pixel values are taken as $I_{i,j}$, and the threshold value is taken as T; hence, according to the simple thresholding technique, the system follows the equation given below:

$$I_{i,j} = 0, n(i,j) < T$$

$$I_{i,j} = 1, n(i,j) > T$$

The threshold value was set and every pixel of the image was compared with that of the set threshold value.

Here, we used two arguments, namely, solidity and area to detect the tumor. Solidity means the intensity of the pixel; it is used to determine the presence of the tumor if the solidity or intensity of the pixel is greater than the defined threshold value. Fig. 17.3 shows the algorithm of image thresholding segmentation.

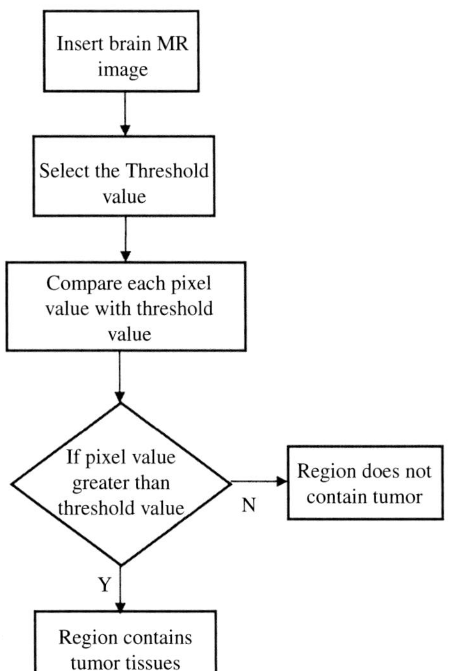

FIG. 17.3 Image thresholding segmentation algorithm. *No permission required.*

Watershed segmentation

This segmentation technique is region-based and is defined on a grayscale image. In the watershed segmentation technique, the input is a two-dimensional (2D) grayscale image and it uses the pixel values for gradient magnitude of the grayscale image to evaluate the input. In this technique, the image is regarded as a topographic landscape with ridges and valleys. The elevation values are gradient magnitudes or values of pixels. Watershed segmentation utilizes image morphology. In this technique, at least one marker interior or seed point to each object was selected by an operator. After the objects were marked, they can be grown using a morphological watershed transformation. Fig. 17.4 shows the algorithm of watershed segmentation.

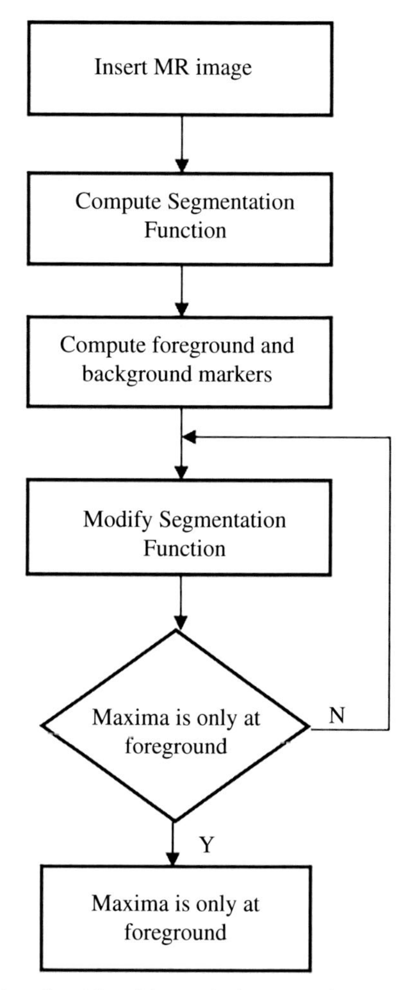

FIG. 17.4 Watershed segmentation algorithm. *No permission required.*

Feature extraction

After preprocessing and tumor segmentation and detection, feature extraction was performed. To reduce the dimensions of an image and represent it in the numeric form, we performed feature extraction. There are broadly two types of linear algorithms (Kailath et al., 1974) and nonlinear algorithms (Addabbo, Biondi, Clemente, Orlando, & Pallotta, 2019). If the size of input data was excessively large to be processed for a particular algorithm, then its dimensional had to be reduced. The feature extraction technique is used to obtain high-level information on tumor detection and detect tumor location. GLCM was introduced by Haralick (Kharat & Kulkarni, 2012). This approach describes the spatial relation between the pixels of various gray-level values (Haralick, Shanmugam, & Dinstein, 1973).

Gray-level co-occurrence matrix (GLCM) is a 2D histogram, in which (p, q)th elements refer to the frequency of event p that occurs with q. It is a function of distance $S=1$, angle (at 0 (horizontal), 45 degrees (with the positive diagonal), 90 degrees (vertical) and 135 degrees (negative diagonal) and gray scales p and q, and calculates how often a pixel with intensity p occurs in relation to another pixel q at a certain distance S and orientation. In this method, GLCM was initiated and the textural features, such as contrast, correlation, energy, homogeneity, entropy, and variance, were obtained from LL and HL subbands of the first four levels of wavelet decomposition (Study of different brain tumor MRI image segmentation techniques, 2014). To obtain the difference between normal and abnormal tissues of the brain for visual perception, we mainly used texture analysis. In this work, statical features were extracted using GLCM. In feature extraction, we calculated the numeric value of various features, and 10 features were extracted. Features can be intensity-based, texture-based, edge-based, etc. Different data sets were used to observe the features of different MR images. We took seven different images and their results are displayed in a tabular format. The features extracted in this work are contrast, correlation, cluster prominence, cluster shade, dissimilarity, energy, entropy, maximum probability, variance, and autocorrelation. The result of these 10 features of seven different MR images was extracted and displayed for obtaining high-level information for research purposes.

Implementation

In the era of developing technologies, the detection of tumors and classification of normal or abnormal MR images have become easier and more accurate. The presents study discusses major steps, that is, preprocessing, segmentation, and feature extraction. For preprocessing, four filtration techniques were used. The aim of these techniques was to remove the unwanted disturbances from the MR image of the brain. Some of the morphological operations were also used. Preprocessing is essential before tumor detection because the accuracy of detection can be less due to the presence of disturbances in the image. Thus, for more accurate tumor detection, preprocessing is necessary. The main purpose of the study was to detect the tumor accurately. The input was taken from the user in jpg or png format, following which it was resized to 256×256 pixels. The algorithm of different filtration techniques was used to obtain and compare the output. Four different filtration techniques as discussed earlier have their merits and setbacks. Anisotropic filter, adaptive filter, median filter, and unsharp mask filter are four different filters used in this work. These filters were used to remove the unwanted signals from the image. These filters work on highlighting the edges and reducing the blur of the image. These filters work on the texture of the image to make it smooth and also work on the sharpness of the image. These filtration techniques were used to preserve the detail of the image. After preprocessing, the segmentation of the image is done, in which the image is segmented into different regions to make the detection of the tumor easy. The differences in segmentation algorithms used in this work are also discussed along with their merits and methodology. Image thresholding is the simplest technique of segmentation of the image; k-means segmentation is an unsupervised learning technique, whereas watershed segmentation is a gradient-based segmentation technique. After the implementation of four different filters and checking their results, these three segmentation techniques were applied to the images and their results are discussed. These three segmentation techniques work on different principles and are used to detect the tumor from an MR image in different ways. After filtration and segmentation of the MRI image, the next step is feature extraction, in which certain selected features are extracted in numeric values.

These features mainly provide better information regarding the tumor detected. Feature extraction is a helpful step to study more about the detected tumor. In total, 10 different features of seven MR images were extracted and displayed in a tabulated format. The processes mentioned above have to be performed in the same given order to obtain better results in tumor detection.

Results and discussion

The necessity of automated brain tumor segmentation and detection is high. Obtaining an accurate MRI image of the brain tumor is challenging. An MRI image has high contrast, indicating regular and irregular tissues that help in differentiating the overlap margins (Mehrotra, Ansari, Agrawal, & Anand, 2020; Samantaray, Panigrahi, Patra, Panda, & Mahakud, 2016). In this work, four filtration techniques were used for preprocessing the image; using MATLAB GUI, they were displayed adjacent to each other so that the comparison between the four filtration techniques can be done easily. The program for filtration techniques was modified according to the GUI application. Different push buttons were used to obtain the result of different filtered images.

All four filtration images along with the input image are shown in Fig. 17.5. Anisotropic filter, unsharp mask filter, median filter, and adaptive filter are four filters displayed in Fig. 17.5. It is clearly seen in the unsharp mask filter image that the resulting image was a blur and its quality was reduced. In the median filter image, the edges of the image were preserved and the image quality was also better. In an adaptive filter image, the edges were preserved and the image quality was also good. For anisotropic filter images, the images were clear but edges were not preserved.

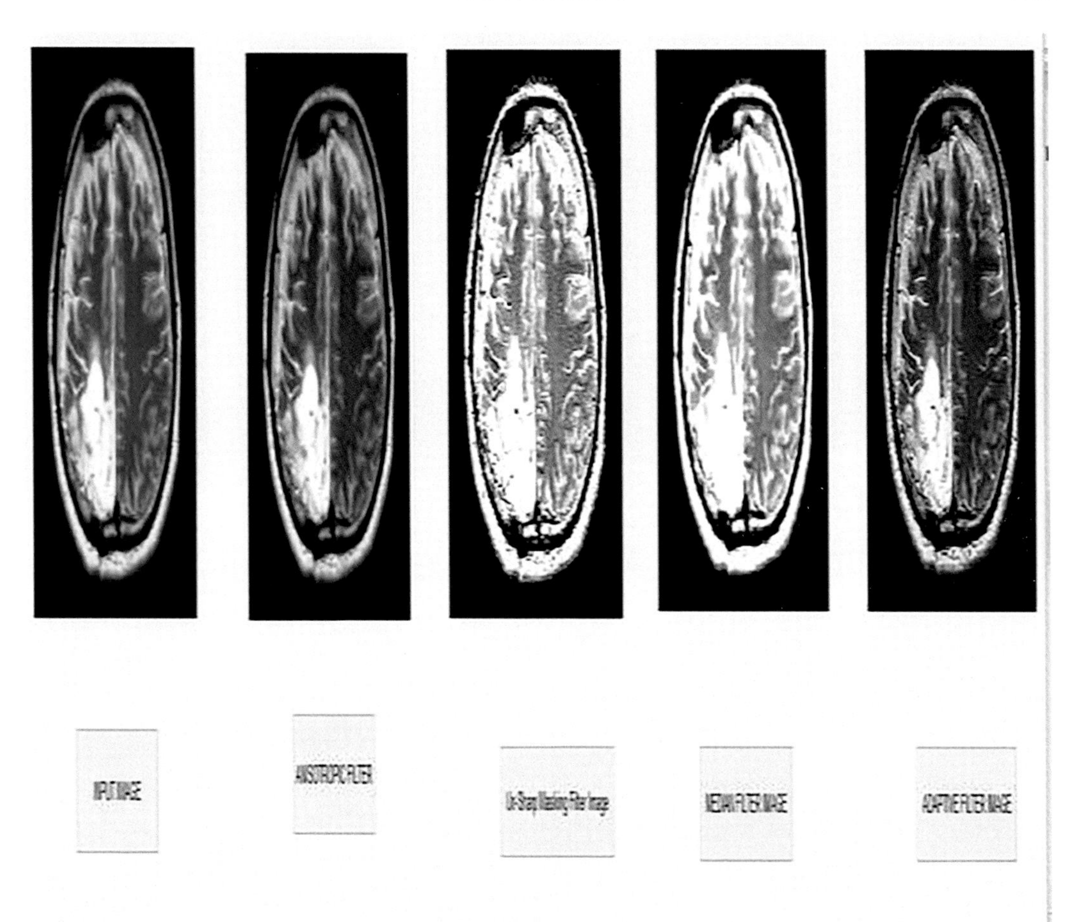

FIG. 17.5 GUI result of different filtration techniques. *No permission required.*

Using MATLAB GUI, we can easily determine the difference between the four filtration techniques, as shown in Fig. 17.5. The comparison study can also be done with ease if all resulting images are placed adjacent to each other.

After preprocessing using four different filtration techniques, the segmentation step was performed. The k-means segmentation technique was done by filtering the image through a median filter as, among all four filtration techniques, the median filter had better results.

After obtaining the resulting image of the median filter, the foreground image of input was also obtained. Next, the final image in the result displays the segmented image. The k-means segmentation technique was an unsupervised learning technique; in this technique, the images were divided into the number of clusters of data introduced by the user. Using MATLAB GUI, the input image, processed image, median filter image, foreground image, and k-means segmented image were displayed in the same frame, as shown in Fig. 17.6.

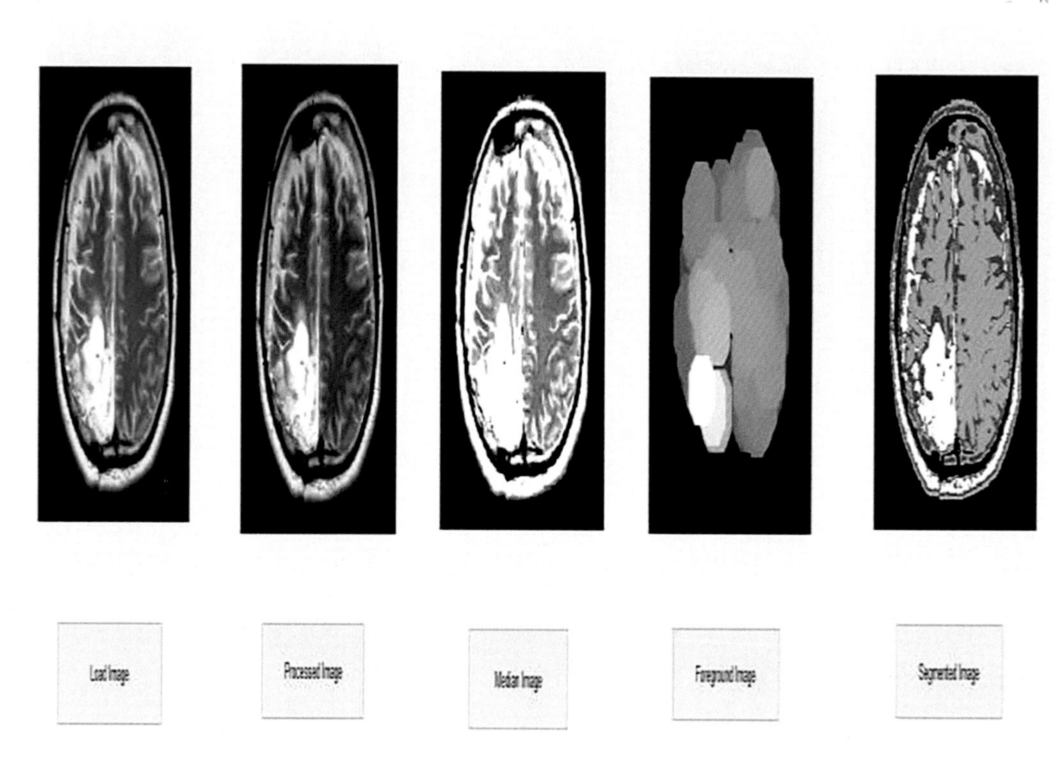

FIG. 17.6 GUI result of k-means segmentation. *No permission required.*

After observing the result of the k-means segmentation technique, two other segmentation techniques, that is, image thresholding and watershed segmentation were displayed in the same frame. The input image was a grayscale image, which was then preprocessed and filtered using a median filter. Afterward, the images were segmented by both techniques to obtain the threshold image and watershed segmented image, as shown in Fig. 17.7.

Finally, the tumor-alone image is displayed. Watershed segmentation is suitable for images with a high value of intensity. The image thresholding segmentation technique uses the intensity of the image to determine the presence of the tumor.

Both these techniques use different principles to detect the tumor. Using MATLAB GUI, the results of both segmentation techniques are displayed along with the processed image, median filter image, and tumor alone image. The tumor alone image provides the shape of the tumor. The last step is feature extraction. After tumor segmentation and detection, feature extraction is used to obtain more information regarding the detected tumor, mainly in the form of numeric values. It transforms the input data into a set of features; we obtained five features and their values are displayed for six different images, as shown in Fig. 17.8. The extracted features can be seen in Table 17.1. GLCM method was used to determine the features of the detected tumor. The feature extraction table provides high-level information regarding the detector tumor. All input images showed the tumor.

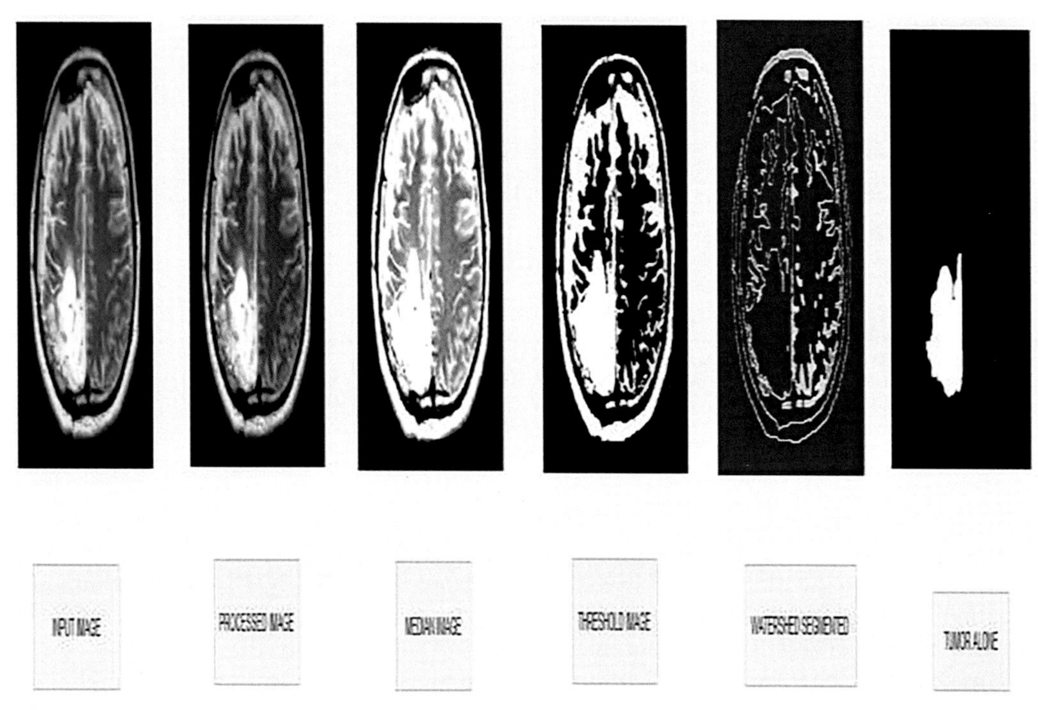

FIG. 17.7 GUI results of image thresholding and watershed segmentation technique. *No permission required.*

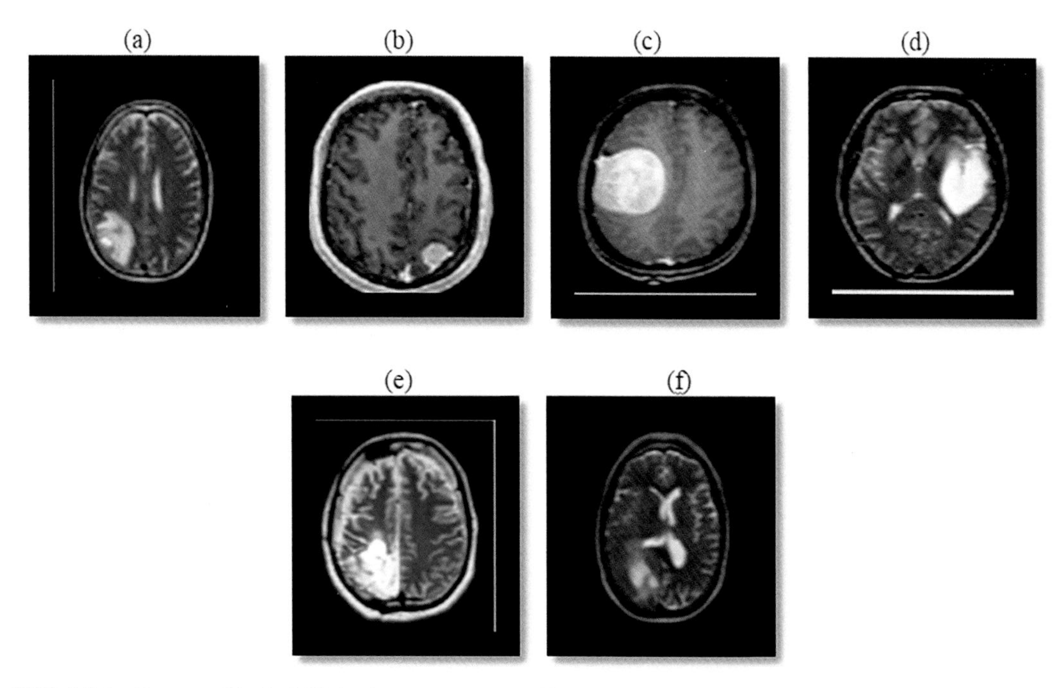

FIG. 17.8 Dataset of brain MRI for feature extraction. *No permission required.*

TABLE 17.1 Feature extraction for different MR images.

Images	Contrast	Correlation	Cluster prominence	Cluster shade	Dissimilarity
(a)	0.3711	0.8676	112.4961	17.5972	0.1854
(b)	0.7191	0.8750	151.0537	4.0617	0.2747
(c)	0.3226	0.9253	120.4991	13.6395	0.1753
(d)	1.2360	0.6968	132.1170	17.9983	0.3495
(e)	1.0212	0.7861	113.1955	10.0868	0.4004
(f)	0.6361	0.6727	58.6371	8.7795	0.2418

This table provides the features of the tumor detected.
No permission required.

Conclusion

Based on the technique of image processing, this study constructed a model for monitoring different filtration techniques and compared their results. Along with the filtration techniques, different segmentation algorithms were applied. Both image enhancement and segmentation are important pillars of this work. Accurate results of both of these steps are necessary. For better representation of the results, MATLAB GUI was created and the results are represented in the form of GUI. Different filtration techniques and all three segmentation algorithms are displayed along with their input images so that a better study can be conducted to detect tumors.

A median filter is used with these three segmentation techniques. After the preprocessing and segmentation of the image, some of the features are extracted for better study purposes (Table 17.1). These features are displayed in the form of a table and different data sets are used so that their results can be analyzed.

Limitations and future scope

This work aims to detect the tumor only for the 2D grayscale image; however, nowadays, many three-dimensional (3D) images can be used along with real-time monitoring of data. Many undetected noises can also be present in this MRI used for analysis. In this work, different segmentation and filtration techniques are used, whose advanced and better versions are being developed.

In the future, tumors can be detected using real-time data and this GUI application results can be accessed from the phone of the patient.

References

Addabbo, P., Biondi, F., Clemente, C., Orlando, D., & Pallotta, L. (2019). Classification of covariance matrix eigenvalues in polarimetric SAR for environmental monitoring applications. *IEEE Aerospace and Electronic Systems Magazine*, *34*(6), 28–43. https://doi.org/10.1109/maes.2019.2905924.

Ahammed Muneer, K. V., Rajendran, V. R., & Paul Joseph, K. (2019). Glioma tumor grade identification using artificial intelligent techniques. *Journal of Medical Systems*, *43*(5). https://doi.org/10.1007/s10916-019-1228-2.

Akhil, M. B. S. S., Aashish, P., & Manikantan, K. (2016). Feature selection using binary-ABC algorithm for DWT-based face recognition. In *2015 IEEE international conference on computational intelligence and computing research, ICCIC 2015*Institute of Electrical and Electronics Engineers Inc. https://doi.org/10.1109/ICCIC.2015.7435632.

Banerjee, S., Mitra, S., Masulli, F., & Rovetta, S. (2019). *Deep radiomics for brain tumor detection and classification from multi-sequence mri. arXiv*. https://arxiv.org.

Celenk, M. (1991). Colour image segmenatation by clustering. *IEEE Proceedings E Computers and Digital Techniques, 138*.

Erickson, B. J., Korfiatis, P., Akkus, Z., & Kline, T. L. (2017). Machine learning for medical imaging. *Radiographics*, *37*(2), 505–515. https://doi.org/10.1148/rg.2017160130.

Gopal, N. N., & Karnan, M. (2010). Diagnose brain tumor through MRI using image processing clustering algorithms such as fuzzy C means along with intelligent optimization techniques. In *2010 IEEE international conference on computational intelligence and computing research, ICCIC 2010* (pp. 694–697). https://doi.org/10.1109/ICCIC.2010.5705890.

Haralick, R. M., Shanmugam, K., & Dinstein, I. (1973). Textural features for image classification. *IEEE Transactions on Systems, Man, and Cybernetics*, 610–621. https://doi.org/10.1109/TSMC.1973.4309314.

Kabir Anaraki, A., Ayati, M., & Kazemi, F. (2019). Magnetic resonance imaging-based brain tumor grades classification and grading via convolutional neural networks and genetic algorithms. *Biocybernetics and Biomedical Engineering*, *39*(1), 63–74. https://doi.org/10.1016/j.bbe.2018.10.004.

Kailath, T., Vie, A., Biondi, F., Clemente, C., Orlando, D., & Pallotta, L. (1974). Classification of covariance matrix eigenvalues in polarimetric SAR for environmental monitoring applications. *IEEE Transactions on Information Theory*, *20*(2), 146–181.

Kharat, K. D., & Kulkarni, P. P. (2012). Brain tumor classification using neural network based methods. *International Journal of Computer Science and Informatics*, 112–117. https://doi.org/10.47893/ijcsi.2012.1075.

Mehrotra, R., Ansari, M. A., Agrawal, R., & Anand, R. S. (2020). A transfer learning approach for AI-based classification of brain tumors. *Machine Learning with Applications*, *2*. https://doi.org/10.1016/j.mlwa.2020.100003, 100003.

Ozic, M. U., Ozbay, Y., & Baykan, O. K. (2014). Beyin mr görüntüsünde otsu-pso yöntemi ile tümör tespiti. In *2014 22nd signal processing and communications applications conference, SIU 2014—Proceedings* (pp. 1999–2002). IEEE Computer Society. https://doi.org/10.1109/SIU.2014.6830650.

Phillips, S. M., Welch, W. A., Fanning, J., Santa-Maria, C. A., Gavin, K. L., Auster-Gussman, L. A., et al. (2020). Daily physical activity and symptom reporting in breast cancer patients undergoing chemotherapy: An intensive longitudinal examination. *Cancer Epidemiology, Biomarkers & Prevention*, *29*(12), 2608–2616. https://doi.org/10.1158/1055-9965.EPI-20-0659.

Samantaray, M., Panigrahi, M., Patra, K. C., Panda, A. S., & Mahakud, R. (2016). An adaptive filtering technique for brain tumor analysis and detection. In *10th international conference on intelligent systems and control (ISCO)*.

Sonar, P., & Bhosle, U. (2016). Optimized association rules for MRI brain tumor classification. In *Proceedings of the 10th INDIACom; 2016 3rd international conference on computing for sustainable global development, INDIACom 2016* (pp. 2644–2649). Institute of Electrical and Electronics Engineers Inc.

Study of different brain tumor MRI image segmentation techniques. (2014). *International Journal of Computer Science Engineering and Technology (IJCSET)*, *4*, 133–136.

Taie, S. A., & Ghonaim, W. (2017). In *Title CSO-based algorithm with support vector machine for brain tumor's disease diagnosis 2017 IEEE international conference on pervasive computing and communications workshops, PerCom workshops 2017* (pp. 183–187). Institute of Electrical and Electronics Engineers Inc. https://doi.org/10.1109/PERCOMW.2017.7917554.

CHAPTER

18

An intelligent diagnostic technique using deep convolutional neural network

Shrabana Saha, Rajarshi Bhadra, and Subhajit Kar

Electrical Engineering, Future Institute of Engineering and Management, Kolkata, West Bengal, India

Introduction

In the era of advanced pattern recognition and deep learning, the application of AI in the health care industry is very evident. An AI-aided diagnostic tool is proposed to help doctors in their diagnostic predictions. Lack of rapid testing causes premature deaths of COVID-19-affected patients. Antigen test is used for rapid COVID-19 diagnosis, but this might result in false-negative reports (Mak et al., 2020). To overcome this issue, RT-PCR plays a major role as a standard pathogenic tool for screening COVID-19 disease (Tahamtan & Ardebili, 2020). Due to several mutations of SARS-CoV-2 in recent times, RT-PCR tests are also resulting in false-negative reports (Gupta-Wright et al., 2021; Xiao, Tong, & Zhang, 2020). To regulate the spread of COVID-19, several lockdowns were declared which also affected the regular monitoring of patients with diabetic retinopathy (Chatziralli et al., 2021).

Chest X-rays are used to detect pulmonary diseases more efficiently. Besides chest X-ray analysis, remote monitoring of patients suffering from diabetic retinopathy is also important in the current scenario. In these circumstances, an automated diagnostic tool has been proposed to aid doctors in their diagnosis. A very less complex deep multilayered CNN using depth wise separable and pointwise convolution has been put forward for diagnosis of pulmonary and retinal diseases accurately from CXR and OCT scans, respectively. The propounded approach has been evaluated on various open-source datasets. Experimental results exemplify the robustness of the system in the diagnosis of pulmonary and retinal diseases.

Computational Intelligence in Healthcare Applications
https://doi.org/10.1016/B978-0-323-99031-8.00021-1

Related works

Computed tomographical scans and chest X-rays have major contributions in examining patients who are clinically suspected to be affected by the virus, especially those whose initial RT-PCR test results are negative (Chen et al., 2020). Qaqos et al. proposed a simple convolution network for diagnosis of chest X-ray scans for detection of COVID-19, pneumonia, tuberculosis, and healthy patients. In total, 6587 CXR scans have been used for this research, out of which 70% of data is used for training and the rest is used for validation and testing (Qaqos & Kareem, 2020). Loey et al. put forward a generative adversarial network for data preprocessing and augmentation (Loey, Smarandache, Khalifa, & N. E., 2020). GoogleNet, ResNet18, and Alexnet were implemented by them, where Alexnet resulted in good test accuracy for three-class classification. Sethi et al. propounded a technique using InceptionV3, ResNet50, MobileNet, and Xception models for the detection of COVID-19 cases (Sethi, Mehrotra, & Sethi, 2020). Their dataset consists of 320 CXR scans of COVID-19 cases and 5298 non-COVID cases. Bhadra et al. came up with a less complex CNN model with 0.9 million parameters, which resulted in 99.1% blind test accuracy (on 399 samples) in the identification of COVID-19 and pneumonia from chest X-ray scans (Bhadra & Kar, 2020a).

For retinal disease detection, an intelligent method has been proposed by Girard et al. by putting together deep learning and graph propagation for the identification of artery diseases from fundus images (Girard and Cheriet, 2017). In total, 93.3% test accuracy has been achieved by implementing their proposed system on the DRIVE database. Sambaturu et al. proposed a region-based convolution neural network to trace the exudates and hemorrhages for examining diabetic retinopathy, which resulted in a recall of 90% (Sambaturu et al., 2017). Calimeri et al. put forward a CNN model to diagnose retinal fundus images (Calimeri, Marzullo, Stamile and Terracina, 2016). Athar et al. proposed a CNN model to locate and identify fluid filled regions from OCT scans (Athar, Vahadane, Joshi, & Dastidar, 2018). Their CNN model consists of Dense Blocks & Scaled Exponential Unit (SeLU) activations, which resulted in a classification accuracy of 94.8%. Bhadra et al. proposed deep multilayered CNN consisting of five layers and two fully connected layers. The proposed model was trained on 59,142 retinal images and tested on 8446 retinal images with good classification accuracy (Bhadra & Kar, 2020b).

Proposed approach

Fig. 18.1 represents the block diagram of propounded approach. The CXR and OCT imaging dataset consists of images of nonuniform sizes. So to maintain uniformity, the shape of the images has been normalized by rescaling them to 128×128 pixels. These resized images have been fed to the propounded CNN architecture. In the end, the output layer having a softmax classifier classifies the images to their respective classes.

A deep CNN architecture is necessary for improved feature extraction for the classification of complex medical images. The propounded CNN model consists of 17 layers having depthwise separable convolution layers and pointwise convolution layers. The enactment of these convolution layers brings down the complexity and cost of operation of the model.

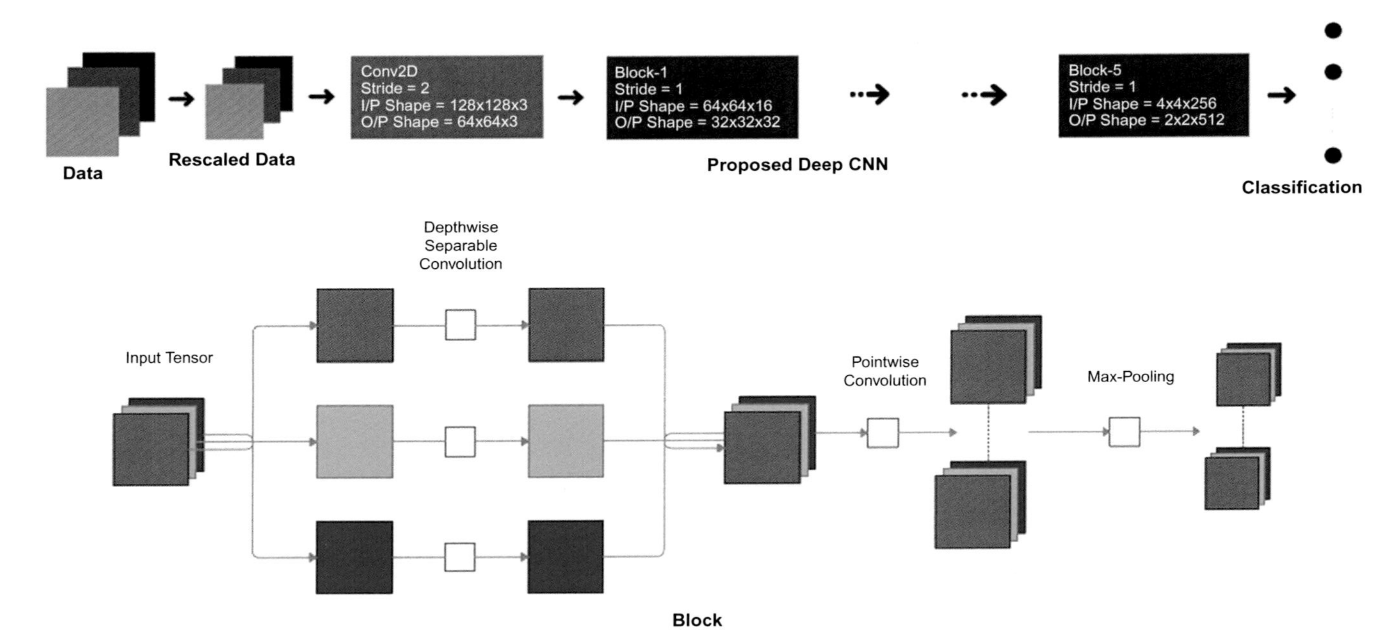

FIG. 18.1 Block diagram of the proposed approach. The workflow of the proposed system has been described here. *No Permission Required.*

The operation cost of depthwise separable and pointwise convolution is shown in Eq. (18.1).

$$(n_H \cdot n_W \cdot M) + (f_H \cdot f_W) + M \cdot N \cdot f_H \cdot f_W \quad (18.1)$$

whereas the total operation cost of standard convolution operation is shown in Eq. (18.2).

$$n_H \cdot n_W \cdot M \cdot N \cdot f_H \cdot f_W \quad (18.2)$$

where n_H and n_W are the height and width of input image, respectively, f_H and f_W are the height and width of kernel, M is the number of input channels, and N is the number of output channels.

The reduction of computation using depthwise separable and pointwise convolution is shown in Eq. (18.3).

$$\frac{(n_H \cdot n_W \cdot M) \cdot (f_H \cdot f_W) + (M \cdot N \cdot f_H \cdot f_W)}{n_H \cdot n_W \cdot M \cdot N \cdot f_H \cdot f_W} = \frac{1}{N} + \frac{1}{n_H \cdot n_W} \quad (18.3)$$

This not only results in less computation but also uses relatively very few parameters, which bring down overfitting issues. The dimension of the output of the convolution layers has been reduced using 2D Max-Pooling layers. After the feature extraction, a three-dimensional tensor has been converted to a one-dimensional feature vector using the global average-pooling having a kernel of size 2×2. Thereafter an output layer with SoftMax activation has been used for classification. The proposed architecture has been elaborated in Table 18.1.

LeakyReLU is defined as in Eq. (18.4).

$$g\left(Z^{[l]}\right) = \max\left(\alpha Z^{[l]}, Z^{[l]}\right) \quad (18.4)$$

$$\text{where } Z^{[l]} = W^{T[l]} \cdot A^{[l-1]} + b^{[l]}$$

Here, $W^{[l]}$ is the weight matrix and $b^{[l]}$ is bias vector assigned to the lth layer. $A^{[l-1]}$ is the activation of the $(l-1)$th layer. $g(Z^{[l]})$ is the output of the activation and $\alpha = 0.3$, that is, gradient assigned to $Z^{[l]}$, which averts dying ReLU while performing backpropagation (Lu, Shin, Su, & Karniadakis, 2020). SoftMax is described as

$$\hat{y}_i = \frac{e^{z_i^{[l]}}}{\sum_{j=1}^{k} e^{z_j^{[l]}}} \quad (18.5)$$

where $z^{[l]}$ is the preactivation of the ith node of output layer (l), k is total nodes in the output layer (l), and $\hat{y}_i$ symbolizes predicted probability of a neuron, which also corresponds to a class. Categorical cross-entropy loss is used to evaluate cost while training. The overall cost $J(w,b)$ can be described as

$$J(w,b) = \sum_{i=1}^{M} L(\hat{y}_i, y_i) \quad (18.6)$$

TABLE 18.1 Summary of the proposed CNN architecture.

Block	Layer type	Stride	Kernel-size	Output size	Parameters
–	Input Layer	2	–	128 × 128 × 3	0
	Conv2D	2	3 × 3	64 × 64 × 16	448
	Batch-Normalization	–	–	–	64
Block-1	Depthwise-Conv2D	1	3 × 3	64 × 64 × 16	160
	Batch-Normalization	–	–	–	64
	Pointwise-Conv2D	1	1 × 1	64 × 64 × 32	544
	Batch-Normalization	–	–	–	128
	Max-pooling2D	2	2 × 2	32 × 32 × 32	0
Block-2	Depthwise-Conv2D	1	3 × 3	32 × 32 × 32	320
	Batch-Normalization	–	–	–	128
	Pointwise-Conv2D	1	1 × 1	32 × 32 × 64	2112
	Batch-Normalization	–	–	–	256
	Max-pooling2D	2	2 × 2	16 × 16 × 64	0
Block-3	Depthwise-Conv2D	1	3 × 3	16 × 16 × 64	640
	Batch-Normalization	–	–	–	256
	Pointwise-Conv2D	1	1 × 1	16 × 16 × 128	8320
	Batch-Normalization	–	–	–	512
	Max-pooling2D	2	2 × 2	8 × 8 × 128	0
Block-4	Depthwise-Conv2D	1	3 × 3	8 × 8 × 128	1280
	Batch-Normalization	–	–	–	512
	Pointwise-Conv2D	1	1 × 1	8 × 8 × 256	33,024
	Batch-Normalization	–	–	–	1024
	Max-pooling2D	2	2 × 2	4 × 4 × 256	0
	Dropout (20%)	–	–	–	0
Block-5	Depthwise-Conv2D	1	3 × 3	4 × 4 × 256	2560
	Batch-Normalization	–	–	–	1024
	Pointwise-Conv2D	1	1 × 1	4 × 4 × 512	131,584
	Batch-Normalization	–	–	–	2048
	Max-pooling2D	2	2 × 2	2 × 2 × 512	0
–	Global-Average-Pooling2D	–	2 × 2	512	0
	Output Layer	–	–	No. of classes	–
Total Parameters:	**1.8 lakhs**				

No Permission Required.

where cross-entropy loss is defined as

$$L(\hat{y}_i, y_i) = -y_i \log(\hat{y}_i) \tag{18.7}$$

where M symbolizes number of training examples, y_i denotes true output, and $\hat{y}_i$ denotes predicted output. The loss function has been optimized using Adam optimizer (Kingma & Ba, 2015).

Dataset used

Case-1: Analysis of chest X-ray scans

For the diagnosis of COVID-19 and pneumonia, CXR scans of COVID-affected patients, patients suffering from pneumonia and healthy cases have been used for this research. The dataset is created using multiple open-sourced datasets (Cohen, Morrison, & Dao, 2020; Haque & Rahman, 2020; Kermany et al., 2018). In total, 1330 images from each case (i.e., COVID-19, pneumonia, healthy) have been collected to form a balanced dataset. Fig. 18.2 shows the CXR scans from the dataset.

Case-2: Analysis of retinal OCT scans

The database consisting of retinal OCT scans is described in Kermany et al. (2018). The dataset consists of OCT scans of patients suffering from CNV, DRUSEN, DME, and healthy patients, which comprises a total of 84,484 OCT scans. These four classes have been depicted in Fig. 18.3.

Experimental results

The proposed CNN model has been trained on a system having hexacore CPU, NVIDIA RTX 2060 GPU having 6GB of DDR6 VRAM, 8 GBX2 DDR4 memory clocked at 3200 mHz and

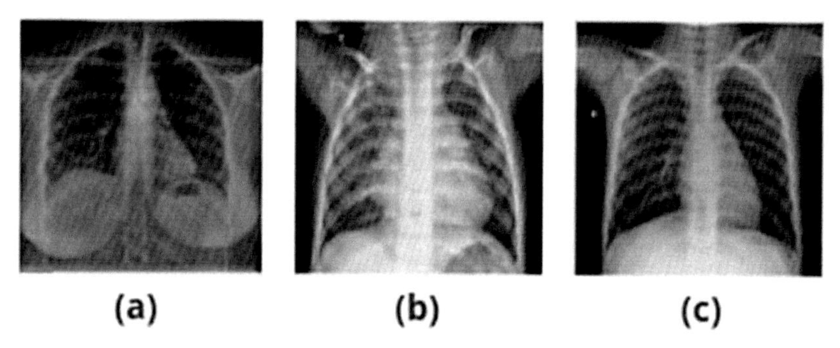

FIG. 18.2 Chest X-ray scans: (A) COVID-19, (B) pneumonia, and (C) healthy. The images of chest X-ray scans have been depicted here. *No Permission Required.*

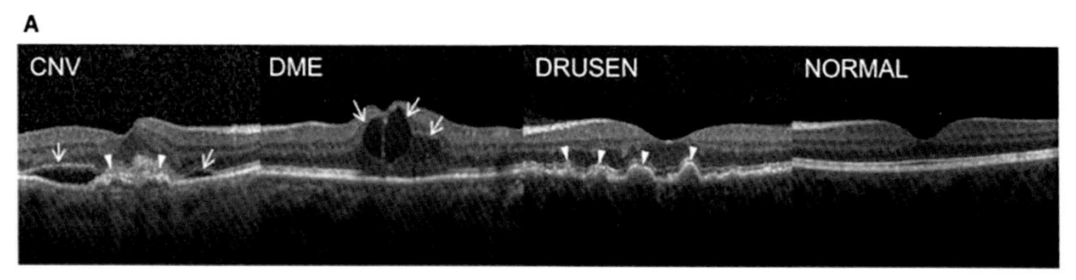

FIG. 18.3 Optical coherence tomographical scans. The images of retinal OCT scans have been depicted here. *No Permission Required.*

240GB NVMe SSD. Tensorflow-2 has been used for model creation purposes. The learning rate has been selected as 0.001 in the first occurrence. However, staircase learning rate decay has been applied, that is, learning rate is decreased after every 10 epochs by a component of 10. Early stopping has been used where the training stops if the training loss does not show any improvement over 10 consecutive epochs. The mini-batch size for this experiment has been selected as eight empirically.

Case-1: Analysis of chest X-ray scans

The proposed CNN model has been trained on 8591 CXR scans, validated on 399 CXR scans, and tested on 1995 CXR scans, respectively. Figs. 18.4 and 18.5 show the training and validation performance of the proposed system. The propounded model achieved 98.7% training accuracy, 97.8% validation accuracy, and 98.0% blind test accuracy. In total, 99.8% sensitivity and 99.9% specificity have been achieved over the test set by this model. The exploratory results have been elaborated in Table 18.2, and the confusion matrix is shown in Fig. 18.6.

Case-2: Analysis of retinal OCT scans

The proposed model resulted in 99.1% training accuracy, which is evaluated on 59,152 images, validation accuracy of 95.1%, which is evaluated on 12,666 images, and blind test

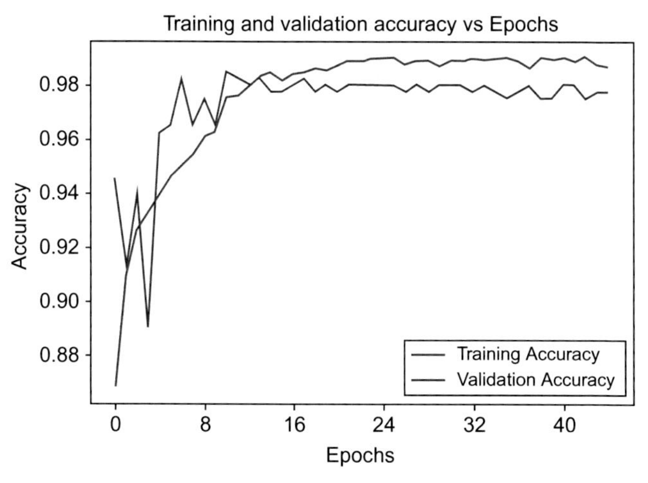

FIG. 18.4 Accuracy vs epochs (CXR analysis). The accuracy vs epochs curve of CXR analysis has been depicted here. *No Permission Required.*

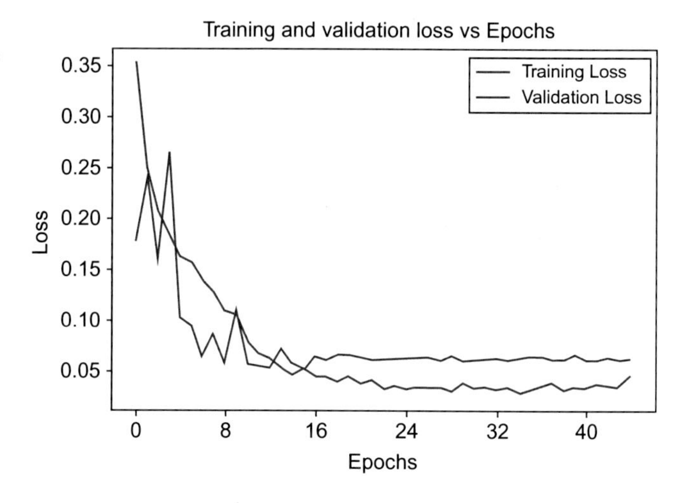

FIG. 18.5 Loss vs epochs (CXR analysis). The loss vs epochs curve of CXR analysis has been depicted here. *No Permission Required.*

TABLE 18.2 Classification result of CXR analysis.

Class	Precision	Recall	F1-score
COVID-19	0.99	1.00	1.00
Normal	0.97	0.97	0.97
Pneumonia	0.97	0.97	0.97

No Permission Required.

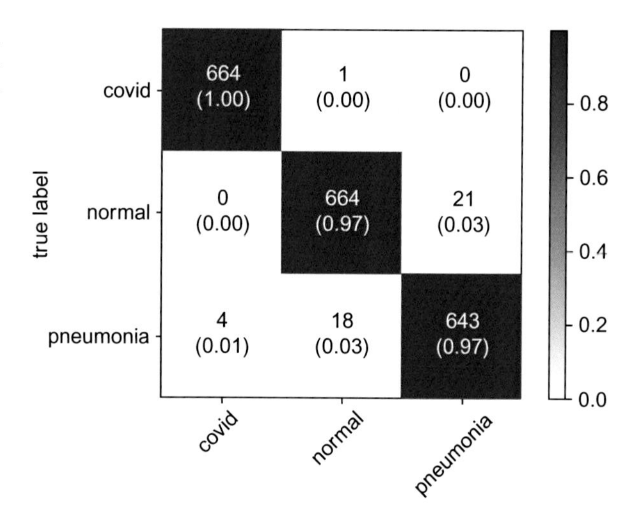

FIG. 18.6 Confusion matrix of CXR analysis. The confusion matrix shows the number of perfectly classified pulmonary diseases. *No Permission Required.*

accuracy of 95% over 12,666 images. The proposed model resulted in 99.1% sensitivity and 97.4% specificity over the test set. Training and validation performance has been shown in Figs. 18.7 and 18.8. The classification performance over the blind test set consisting of 12,666 images has been described in Table 18.3 and in the confusing matrix shown described in Fig. 18.9.

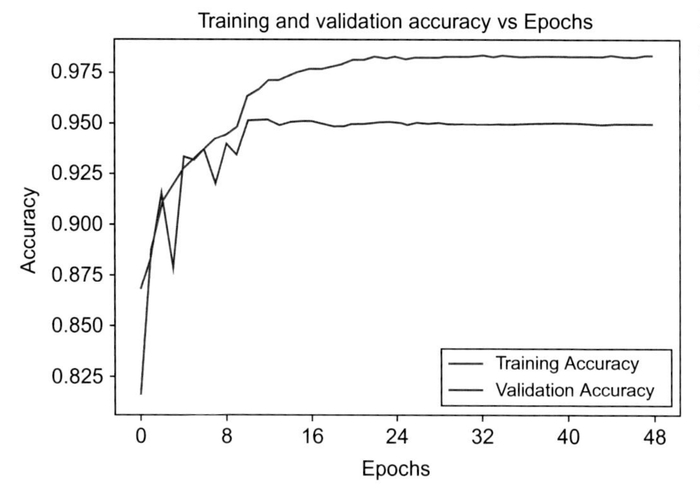

FIG. 18.7 Accuracy vs epochs (OCT analysis). The accuracy vs epochs curve of OCT analysis has been depicted here. *No Permission Required.*

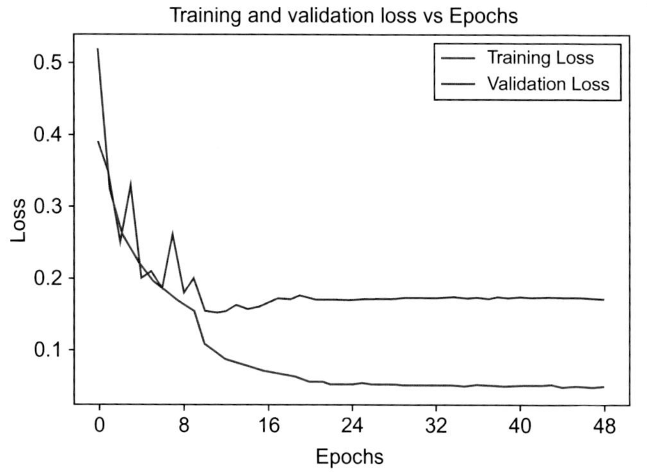

FIG. 18.8 Loss vs epochs (OCT analysis). The loss vs epochs curve of OCT analysis has been depicted here. *No Permission Required.*

TABLE 18.3 Classification result of OCT analysis.

Class	Precision	Recall	F1-score
CNV	0.97	0.97	0.97
DME	0.93	0.92	0.92
Drusen	0.87	0.82	0.84
Normal	0.95	0.98	0.96

No Permission Required.

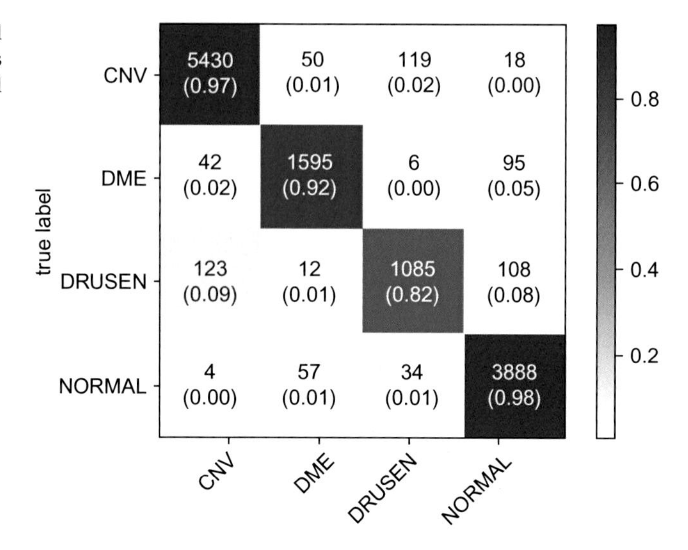

FIG. 18.9 Confusion matrix of retinal OCT analysis. The confusion matrix shows the number of perfectly classified retinal diseases. *No Permission Required.*

Discussion

The proposed model has been compared with other state-of-the-art CNN architectures. In spite of having a significantly lower number of parameters which leads to faster computation, the proposed model resulted in good classification accuracy. The detailed comparison has been described in Table 18.4.

TABLE 18.4 Model comparison.

		Accuracy	
Model	**No. of parameters**	**CXR analysis**	**OCT analysis**
VGG16	14 million	96.6%	89.8%
VGG19	20 million	96.5%	90.3%
MobileNet	3 million	97.1%	91.1%
InceptionV3	22 million	90.2%	90.7%
ResNet50	24 million	96.2%	92.4%
DenseNet121	7 million	97.8%	94.1%
Xception	21 million	97.0%	91.7%
Proposed	*0.18 million*	*98.0%*	*95.0%*

No Permission Required.

Conclusion

This system aims to deliver accurate disease detection and classification using deep learning. The proposed system resulted in 98% and 95% blind test accuracies on chest X-ray scans and on retinal OCT scans respectively. Lack of data might be a challenge which might cause negative effects in classification accuracies so in that case data will be augmented using generative adversarial networks. In the future, many more disease detection systems based on images, numeric data, and voice samples are going to be implemented in this platform.

References

Athar, S., Vahadane, A., Joshi, A., & Dastidar, T. R. (2018). Weakly supervised fluid filled region localization in retinal OCT scans. In *2018 IEEE 15th international symposium on biomedical imaging (ISBI 2018)* (pp. 1467–1470). https://doi.org/10.1109/ISBI.2018.8363849.

Bhadra, R., & Kar, S. (2020a). Covid detection from CXR scans using deep multi-layered CNN. In *2020 IEEE Bombay section signature conference (IBSSC)* (pp. 214–218). https://doi.org/10.1109/IBSSC51096.2020.9332210.

Bhadra, R., & Kar, S. (2020b). Retinal disease classification from optical coherence tomographical scans using multilayered convolution neural network. In *2020 IEEE applied signal processing conference (ASPCON)* (pp. 212–216). https://doi.org/10.1109/ASPCON49795.2020.9276708.

Calimeri, F., Marzullo, A., Stamile, C., & Terracina, G. (2016). Optic disc detection using fine tuned convolutional neural networks. In *2016 12th international conference on signal-image technology & internet-based systems (SITIS)* (pp. 69–75). https://doi.org/10.1109/SITIS.2016.20.

Chatziralli, I., Dimitriou, E., Kazantzis, D., Machairoudia, G., Theodossiadis, G., & Theodossiadis, P. (2021). Effect of COVID-19-associated lockdown on patients with diabetic retinopathy. *Cureus*, *13*(5), e14831. https://doi.org/10.7759/cureus.14831.

Chen, D., Jiang, X., Hong, Y., Wen, Z., Wei, S., Peng, G., et al. (2020). Can chest CT features distinguish patients with negative from those with positive initial RT-PCR results for coronavirus disease (COVID-19)? *American Journal of Roentgenology*, *216*(1), 66–70. https://doi.org/10.2214/AJR.20.23012.

Cohen, J., Morrison, P., & Dao, L. (2020). *COVID-19 image data collection*. arXiv:2003.11597v1.

Girard, F., & Cheriet, F. (2017). Artery/vein classification in fundus images using CNN and likelihood score propagation. In *2017 IEEE global conference on signal and information processing (GlobalSIP)* (pp. 720–724). https://doi.org/10.1109/GlobalSIP.2017.8309054.

Gupta-Wright, A., Macleod, C. K., Barrett, J., Filson, S. A., Corrah, T., Parris, V., et al. (2021). False-negative RT-PCR for COVID-19 and a diagnostic risk score: A retrospective cohort study among patients admitted to hospital. *BMJ Open*, *11*(2), e047110. https://doi.org/10.1136/bmjopen-2020-047110.

Haque, A. K. M. B., & Rahman, M. (2020). *Augmented COVID-19 X-ray images dataset (Mendely) analysis using convolutional neural network and transfer learning*. https://doi.org/10.13140/RG.2.2.20474.24003.

Kermany, D. S., Goldbaum, M., Cai, W., Valentim, C. C. S., Liang, H., Baxter, S. L., et al. (2018). Identifying medical diagnoses and treatable diseases by image-based deep learning. *Cell*, *172*(5), 1122–1131.e9. https://doi.org/10.1016/j.cell.2018.02.010.

Kingma, D. P., & Ba, J. (2015). *Adam: A method for stochastic optimization*. http://arxiv.org/abs/1412.6980.

Loey, M., Smarandache, F., Khalifa, M., & N. E. (2020). Within the lack of chest COVID-19 X-ray dataset: A novel detection model based on GAN and deep transfer learning. *Symmetry*, *12*(4). https://doi.org/10.3390/sym12040651.

Lu, L., Shin, Y., Su, Y., & Karniadakis, G. (2020). Dying ReLU and initialization: Theory and numerical examples. *Communications in Computational Physics*, *28*, 1671–1706. https://doi.org/10.4208/cicp.OA-2020-0165.

Mak, G. C., Cheng, P. K., Lau, S. S., Wong, K. K., Lau, C. S., Lam, E. T., et al. (2020). Evaluation of rapid antigen test for detection of SARS-CoV-2 virus. *Journal of Clinical Virology: The Official Publication of the Pan American Society for Clinical Virology*, *129*, 104500. https://doi.org/10.1016/j.jcv.2020.104500.

Qaqos, N. N., & Kareem, O. S. (2020). COVID-19 diagnosis from chest X-ray images using deep learning approach. In *2020 international conference on advanced science and engineering (ICOASE)* (pp. 110–116). https://doi.org/10.1109/ICOASE51841.2020.9436614.

Sambaturu, B., Srinivasan, B., Prabhu, S. M., Rajamani, K. T., Palanisamy, T., Haritz, G., et al. (2017). A novel deep learning based method for retinal lesion detection. In *2017 international conference on advances in computing, communications and informatics (ICACCI)* (pp. 33–37). https://doi.org/10.1109/ICACCI.2017.8125812.

Sethi, R., Mehrotra, M., & Sethi, D. (2020). Deep learning based diagnosis recommendation for COVID-19 using chest X-rays images. In *2020 second international conference on inventive research in computing applications (ICIRCA)* (pp. 1–4). https://doi.org/10.1109/ICIRCA48905.2020.9183278.

Tahamtan, A., & Ardebili, A. (2020). Real-time RT-PCR in COVID-19 detection: Issues affecting the results. *Expert Review of Molecular Diagnostics*, *20*(5), 453–454. https://doi.org/10.1080/14737159.2020.1757437.

Xiao, A. T., Tong, Y. X., & Zhang, S. (2020). False negative of RT-PCR and prolonged nucleic acid conversion in COVID-19: Rather than recurrence. *Journal of Medical Virology*, *92*(10), 1755–1756. https://doi.org/10.1002/jmv.25855.

CHAPTER

19

Design of a biosensor for the detection of glucose concentration in urine using 2D photonic crystals

Sanjeev Sharma[a], *Arvind Kumar*[a], *and Rajeev Agrawal*[b]

[a]Department of Physics, GL Bajaj Institute of Technology and Management, Greater Noida, UP, India [b]Department of Electronics and Communications, Llyod Institute of Engineering and Technology, Greater Noida, UP, India

Introduction

Diabetes is currently a very serious health problem, affecting a large portion of the population. Diabetes is caused by the pancreas not producing enough insulin in the body, insulin being a hormone that regulates blood glucose. According to a WHO report (2014), 8.5% of adults aged 18 years and older had diabetes. Based on 2019 data, diabetes was the direct cause of 1.5 million deaths worldwide. WHO announced that there will be 366 million diabetics by 2030 (Wild et al., 2004). Generally, blood glucose monitoring is necessary to control diabetes. However, most of the conventional commercialized clinical glucose analyzers give accurate results, but are bulky and expensive. Due to the cost, the poorest/middle-income families cannot afford the equipment. Glucose in the urine indicates high blood sugar, which can be caused by diabetes.

Recently, a photonic crystal-based biosensor has been successful in the analysis and detection of different diseases. These photonic crystals play an important role in detection of various types of cancers, diagnosis of different types of fever, high glucose in the blood, urine infection problems, etc. The next section discusses the different types of photonic crystals and their applications.

https://doi.org/10.1016/B978-0-323-99031-8.00015-6

Photonic crystals and their applications

Originally photonic crystals were a new branch of science whose aim was to control the optical properties of various types of materials. In the last century, researchers had more success in controlling the optical properties of materials using semiconductors. Photonic crystals also control electromagnetic properties of materials, giving more useful results. In 1987, using the formulation of Schrödinger's wave equations, Eli Yablonovitch introduced an idea to build artificial periodic structures manipulating the permittivity in order to totally inhibit light propagation. Thus, using these concepts, a photonic bandgap (PBG) material was discovered. Generally, photonic crystals are categorized into one-dimensional (1D), two-dimensional (2D), and three-dimensional (3D) forms, according to the dimensionality of the building stacks (Fig. 19.1). A 1D PC consists of alternating layers of the dielectric and semiconductor materials having low and high refractive indices and the dielectric constant is modulated along one direction only. In 2D photonic crystals, the dielectric constant of the crystal is periodic in one plane. These include quadratic, hexagonal, and honeycomb types of lattices. The 2D PCs are less difficult to fabricate in comparison to 3D dielectric arrays. In 3D PBG structures, refractive index modulation is periodic along all the three directions. Fig. 19.1 shows the directionality of all three types of photonic crystals.

Let us discuss a brief history of photonic crystals. In 1987, many researchers showed great interest in the field of photonic crystals, mainly one-dimensional photonic crystals. In that year, E. Yablonovitch published a research paper that presented the possibility of inhibiting spontaneous emission of electromagnetic radiation using three-dimensional photonic crystals. Then, John (1987) discussed the strong localization of photons in disordered dielectric superlattices. In 1989, John published another work showing that the face centered cubic (fcc) structure of photonic crystals generated a complete photonic bandgap at the range of the second and third transmission window (S. John & Rangarajan, 1988). At the end of 1994, the Yablonovite structure had been fabricated, which shows the complete band structures and optical properties such as omnidirectional reflection, filter, WDM, etc., of the photonic crystals. Regarding the fabrication of a structure, four alternate layer-based photonic crystals were presented at the end of 1998. These crystals created a photonic bandgap effect in the range of the mid-infrared region of the proposed spectrum (Lin et al., 1998; Yamamoto et al., 1998). Lin et al. published research showing that the photons with high speed comparable to the speed of light in free space were strongly dispersed in 2D crystals when they approached the bandgap edges. Fleming and Lin (1999) presented the first three-dimensional

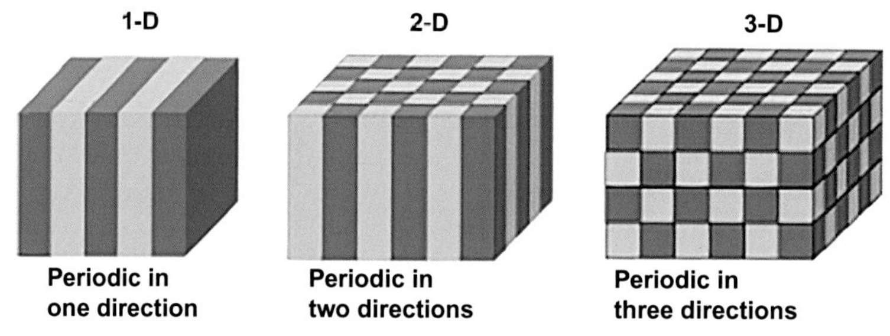

FIG. 19.1 Different types of photonic crystals.

photonic crystal working with a stop band in the range of wavelengths between 1330 nm to 1950 nm. Noda et al. (2000) fabricated an eight-layer photonic crystal using the wafer fusion method.

In 2000, the research work on photonic crystals shifted towards negative index metamaterials. The negative refraction of a composite dielectric material was experimentally verified by Shelby et al. (2001). Pendry (2000) proposed that a negative index medium could be used to make a perfect lens. Li et al. (2003) proposed a new type of structure by taking alternating layers of ordinary (positive-refractive index) and extraordinary (negative-refractive index) materials in a periodic manner. Panoiu Jr et al. (2006) demonstrated photonic superlattices consisting of alternate layers of materials (dielectric/semiconductor) with positive index of refraction and negative effective index of refraction in a photonic gap. Sang and Li (2007) demonstrated the properties of a material with a defect layer of graded material in one-dimensional photonic crystals. Srivastava and Ojha (2009) theoretically calculated large forbidden bands in one-dimensional exponentially graded structures and also designed a broadband optical reflector in a simple graded structure. Wu et al. (2011) studied the omnidirectional photonic bandgap and other properties of one-dimensional ternary plasma photonic crystals (Wu et al., 2012).

Kumar et al. (2012, 2013) and Suthar et al. (2012) theoretically designed a wavelength division multiplexer using Si- material with a defect layer in the periodic structure of a one-dimensional photonic crystal. Sharma, Kumar, et al. (2013), Sharma et al. (2014a, 2014b) demonstrated a narrow filter and dense-wavelength division multiplexing (DWDM) using a 1D periodic photonic crystal structure with single defects in 2014. In 2015–16, an omnidirectional mirror was proposed by many researchers (Chen, 2015; Qian et al., 2016; Sharma et al., 2015) and found out that they have cent percent omnidirectional reflection at a particular transmission window. Sharma et al. (2017) theoretically designed a temperature-dependent ODR reflection at the 1550-nm wavelength range. Sharma and Vimal (2019) theoretically designed an enhanced ODR reflection by using SiO_2 exponentially graded materials in one-dimensional photonic crystals. Currently a number of researchers are designing DWDM, tunable filters and ODR mirrors focused at the third transmission window using one-dimensional photonic crystals (Balaji et al., 2021; Kalhan et al., 2018; Sharma, 2021; Sharma et al., 2016; Sharma, Gupts, et al., 2020; Sharma & Kumar, 2021a; Sharma, Kumar, & Gupta, 2021; Sharma, Kumar, & Singh, 2020; Sharma & Tewari, 2020; Vimal et al., 2022). Recently, a biosensor has been designed using 1D and 2D photonic crystals. These biosensors detect various types of diseases in human blood and urine.

Photonic crystal-based biosensors

Photonic crystals can be used in the design of biosensors for the detection of multiple biological targets, such as various type of proteins in urine, nucleic acids, and glucose concentration in blood/urine samples, for the diagnosis of a broad range of diseases, including diabetes and cancer. This is a new technique for human disease detection, with the new technology of photonic sensing technology proving effective in biosensing applications for cancer, fever, urine issues, blood infection problems, and others. The schematic diagram of a biosensor is shown in Fig. 19.2. Several researchers have designed photonic sensors using

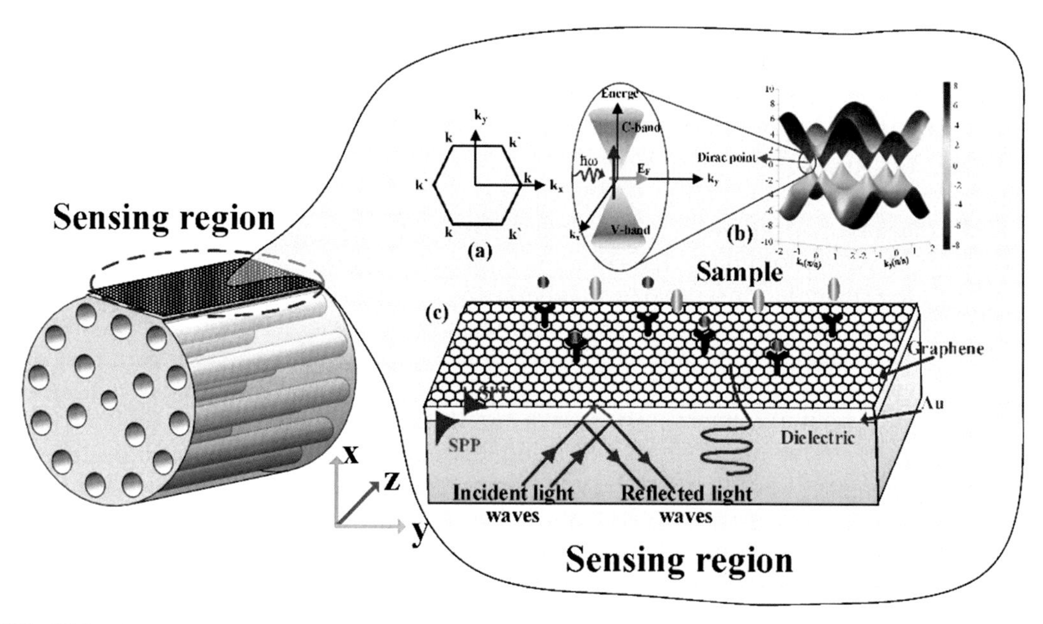

FIG. 19.2 Schematic diagram of a biosensor.

two-dimensional photonic crystals in the shape of a ring resonator that are successful in detection of glucose concentration in urine over the range of 0–15g/dL (Robinson & Dhanlaksmi, 2017; Sharma, Kumar, Singh, & Tyagi, 2021). Robinson and Dhanlaksmi (2017) proposed a biosensor consisting of two inverted L-shaped waveguides in the arrangement of a ring resonator based on 2D photonic crystals. They observed that the refractive index peak (glucose concentration in urine) shifted within a bandgap that caused sensitivity. In this work, a sample of urine filled a silicon rod and an electromagnetic wave passing through it showed various information about it. The measurement of the proposed biosensor is about 11.4μm × 11.4μm, and the proposed sensor gives an accurate analytical result. The shifting of the transmittance peak shows the sensitivity of the biosensor.

Sharma, Deshmukh, and Sharan (2013) designed a photonic crystal-based biosensor for the analysis of blood components using optical sensor technology. According to them, a biosensor is a powerful detection device with enormous application potential in medical research and healthcare. In this work, a biosensor was designed using 2D photonic crystals because they give more accurate results in comparison to 1D photonic crystals. The defect layer is used to detect different components of blood in a human. In this work, all simulation work was done using finite difference time domain (FDTD) software. The simulation results provided a sharp image of different blood components, such as blood serum albumin, blood plasma, glucose, white blood cells, and red blood cells, according to their refractive indices. The transmittance peak of these components was observed, and the shifting of the transmission peak indicated that the proposed structure works as a sensor. F.D. Gudagunti et al. successfully detected breast cancer in his research work using hybrid photonic crystals in the shape of a ring resonator. Sharma and Sharan (2015a, 2015b) in 2015 proposed a 2D optical sensor for detection of different concentrations of glucose in the urine of diabetic patients. The proposed work was done at wavelengths of 1530–1565 nm and indicated the presence of high glucose levels through a narrow peak of glucose concentration in the urine responsible for causing diabetes. Chopra et al. (2016) proposed a biosensor that is used to detect the molecules in a sample very accurately. In this work, a diamond-shaped ring resonator was used for designing a biosensor using a 2D photonic crystal. Radhouene et al. (2017) demonstrated a temperature-dependent sensor using 2D photonic crystals.

In another study, a biosensor was proposed using a 2D photonic crystal to detect glucose/hemoglobin concentration as well as refractive index in urine or blood samples (Aidinis et al., 2020; Bendib & Bendib, 2018; Islam et al., 2020; Sharma, Kumar, Singh, & Tyagi, 2021). Bendib and Bendib (2018) theoretically designed a biosensor for the detection and analysis of malaria fever in blood samples using 2D photonic crystals. Aidinis et al. (2020) theoretically designed a sensor for measuring glucose in urine using 2D photonic crystals in 2020. Sharma, Kumar, Singh, and Tyagi (2021) also theoretically designed a biosensor that detects the chikungunya virus in blood samples using 2D photonic crystal materials. Thus biosensors have been created using 2D photonic crystals for sensing various parameters such as quality factor, resonant peak wavelength, and power transmission efficiency, among others, in blood samples.

In the present research work, a double ring resonator 2D photonic crystal-based biosensor was theoretically designed to detect glucose concentration in urine samples using the FDTD method. A biosensor is theoretically proposed using 2D photonic crystals to detect glucose concentration in urine samples within a range of 0–15 g/dL. The popular FDTD method is used to analyze the transmission characteristics of glucose concentration as well as refractive index in urine. The refractive indices of the different components of the sample with their concentrations were recorded as 1.335, 1.336, 1.337, 1.338, 1.341 for 0–15 mg/dL, 0.625 g/dL, 1.25 g/dL, 2.25 g/dL, and 5 g/dL glucose in urea (Robinson & Dhanlaksmi, 2017). This new type of biosensor is proving more effective in various sensing applications. The theoretical results provide good quality factor, power transmission efficiency, and high sensitivity for better biosensor design. For both commercial and medical purposes, the proposed biosensor will be a useful tool.

Design of a biosensor

In the present research work, a double ring resonator photonic crystal-based biosensor has been theoretically designed using the FDTD method. The biosensor dimensions are taken as 21 μm × 19 μm. A 3D picture of this biosensor with its refractive index shows a large number of Ge rods along the X and Z axis, as illustrated in Fig. 19.3. The radius, lattice constant, and refractive index of the Ge rod is given as 265 nm, 550 nm, and 4.2, respectively.

Simulation and result

In this research work, the transmission efficiency and photonic bandgap of the proposed biosensor have been done by using the plane wave expansion (PWE) and FDTD methods. Fig. 19.4 shows the three photonic bandgaps of the proposed structure. In these photonic

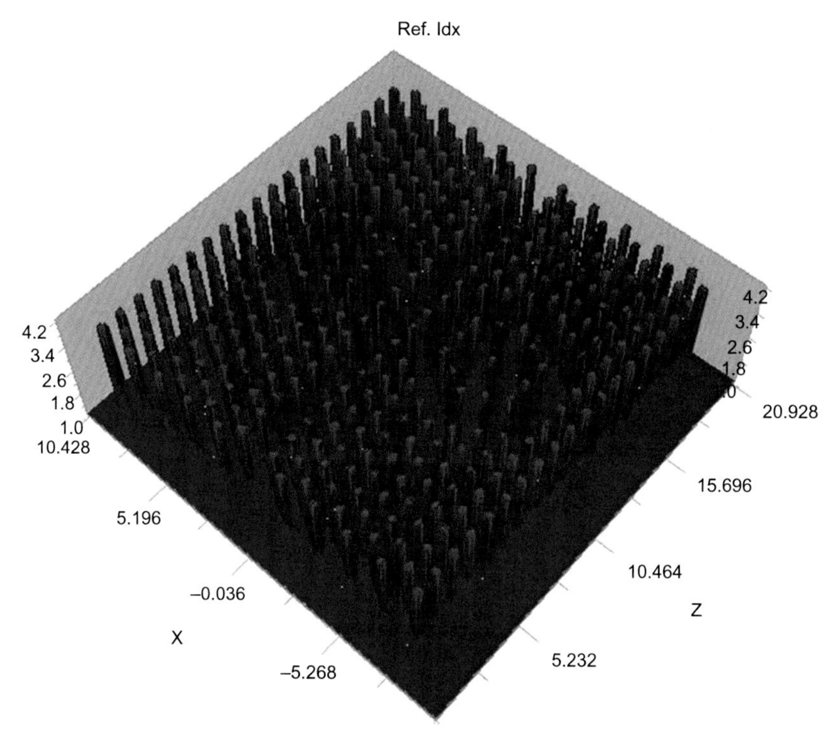

FIG. 19.3 3D structure of a biosensor with refractive index.

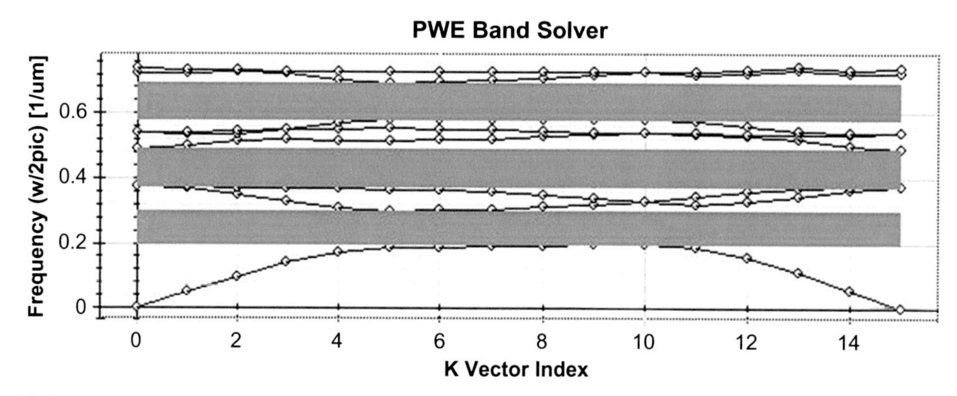

FIG. 19.4 Photonic bandgap structure of a biosensor.

bandgaps, we have chosen a bandgap range of 1445–1730nm, which gives us valuable results. The continuous pulse of this biosensor photonic crystal is central at a wavelength of 1505nm for our area of interest.

The electric field with the transmission peak of the proposed biosensor is indicated in Fig. 19.5. The DFT simulation is set at a wavelength of 1505nm, at which the input signal

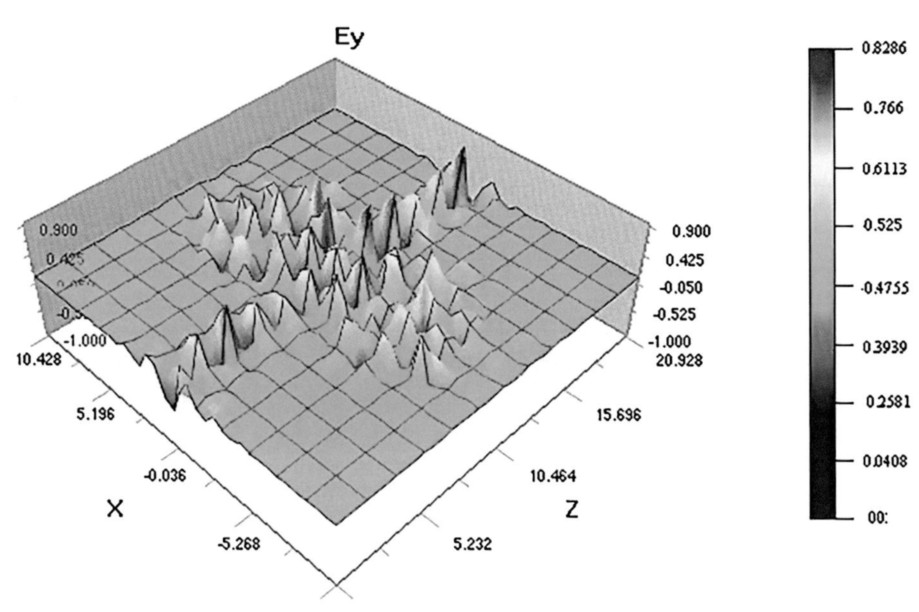

FIG. 19.5 2D Electric field distribution of the biosensor.

transmits and emerges from the ring resonator arrangement. It shows the maximum output, which is more useful in our study.

The transmission efficiency of glucose in urine has been obtained by using the FDTD method. In the case of glucose in urine, the refractive indices of the sample correspond to their concentrations, recorded as 1.335, 1.336, 1.337, 1.338, and 1.341. The transmission spectra peak indicates the glucose concentration at wavelengths 1503.2 nm, 1503.8 nm, 1504.4 nm, 1505.1 nm, and 1505.9 nm with the refractive indices of 1.335, 1.336, 1.337, 1.338, and 1.341, respectively (Fig. 19.6). The transmission peaks of glucose concentration in urine centered at wavelength 1.550 μm and the output efficiency of glucose at a concentration of 0–15 mg/dL,

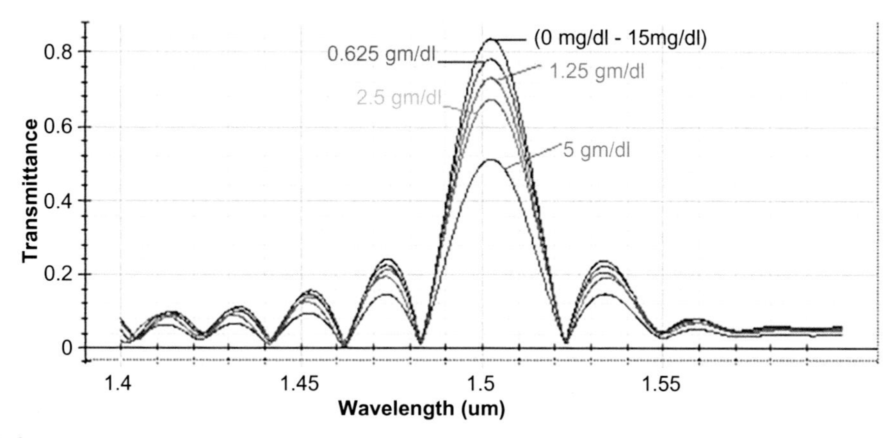

FIG. 19.6 Transmittance spectra of glucose concentration in urine.

TABLE 19.1 Transmission efficiency, refractive indices and quality factor of glucose in urine.

S. no.	Glucose concentration in urine	Refractive index	Central wavelength (nm)	Quality factor	Transmission efficiency (%)
1	Normal (0–15 mg/dL)	1.335	1503.2	72	84
2	0.625 g/dL	1.336	1503.8	70	78
3	1.25 g/dL	1.337	1504.4	68	72
4	2.5 g/dL	1.338	1505.1	67	65
5	5 g/dL	1.341	1505.9	64	52

0.625 g/dL, 1.25 g/dL, 2.25 g/dL, and 5 g/dL are 84%, 78%, 72%, 65%, and 52%, respectively. The transmission peaks of glucose in urine shifted from wavelength 1.503.2 μm to 1.505.9 μm. Fig. 19.6 shows the graph between transmission efficiency and refractive index of glucose in urine with their concentrations; from this graph it can also be observed that the transmittance of this biosensor is increased with the value of refractive indices. So, the theoretical analysis shows that the proposed structures worked with good sensitivity in comparison to the previous research work (Balaji et al., 2021; Sharma & Kumar, 2021b; Sharma, Kumar, Singh, & Tyagi, 2021). The sensitivity of our proposed biosensor is recorded as 541.75 nm/RIU, which is more than the 46.02 nm/RIU in a previously published work (Suthar & Bhargava, 2021).

Table 19.1 shows the transmission efficiency and quality factor of the proposed device with the variation of the central peak of glucose in the sample. In this case study, the output efficiency of glucose in the urine is recorded as 84% for the normal range of glucose in a patient sample. It is also observed that the transmittance of the structure increased with a decrease in the amount of glucose in the urine. At concentrations of 0–15 g/dL the transmission efficiency and quality factor are recorded as 84% and 72. Similarly, at concentrations of 0.625 g/dL, 1.25 g/dL, 2.5 g/dL, and 5 g/dL, the quality factor and efficiency are recorded as 70, 68, 67, 64 and 78%, 72%, 65%, and 52%, respectively. It can also be observed that the central peak of glucose shifted towards the higher order of the wavelength region with respect to the refractive indices. Hence, these new types of biosensor are more useful in various sensing applications. The theoretical results also provide a good quality factor, transmission efficiency, and a high sensitivity of 541.75 nm/RIU in the case of glucose in urine.

Conclusion

In this research work, a biosensor is proposed that is more efficient than existing technology and has good sensitivity and quality factor. The proposed photonic biosensor detects the glucose level in urine with the glucose concentration/refractive indices and is more useful in various sensing applications than existing methods. The quality factor, transmission efficiency, and sensitivity of this biosensor is recorded as 68, 84%, and 541.75 nm/RIU in the case of glucose in urine. Thus it has good sensitivity and better efficiency than in the previously published work. The importance of this biosensor is that it has novel and good measurement quality. For both commercial and medical purposes the proposed biosensor will be very useful.

References

Aidinis, K., Goudarzi, K., & Esmaeili, A. H. (2020). Optical sensor based on two-dimensional photonic crystals for measuring glucose in urine. *Optical Engineering*, *59*(5). https://doi.org/10.1117/1.OE.59.5.057104.

Balaji, V. R., Murugan, M., Robinson, S., & Hegde, G. (2021). A novel hybrid channel DWDM demultiplexer using two dimensional photonic crystals meeting ITU standards. *SILICON*. https://doi.org/10.1007/s12633-020-00902-7.

Bendib, S., & Bendib, C. (2018). Photonic crystals for malaria detection. *Journal of Biosensors & Bioelectronics*. https://doi.org/10.4172/2155-6210.1000257.

Chen, Z.-J. (2015). Design and optimization of omnidirectional band gap for one-dimensional periodic and quasiperiodic phononic heterostructures. *Chinese Physics Letters*, *32*(1). https://doi.org/10.1088/0256-307X/32/1/014301, 014301.

Chopra, H., Kaler, R. S., & Painam, B. (2016). Photonic crystal waveguide-based biosensor for detection of diseases. *Journal of Nanophotonics*, *10*(3). https://doi.org/10.1117/1.JNP.10.036011.

Fleming, J. G., & Lin, S. Y. (1999). Three-dimensional photonic crystal with a stop hand from 1.35 to 1.95 μm. *Optics Letters*, *24*(1), 49–51. https://doi.org/10.1364/OL.24.000049.

Islam, R., Hossain, B., & Mehedi, I. M. (2020). Modeling of highly improved SPR sensor for formalin detection. *Results in Physics*, *16*.

John, S. (1987). Strong localization of photons in certain disordered dielectric superlattices. *Physical Review Letters*, 2486–2489. https://doi.org/10.1103/physrevlett.58.2486.

John, S., & Rangarajan, R. (1988). Optimal structures for classical wave localization: An alternative to the ioffe-regel criterion. *Physical Review B*, *38*(14), 10101–10104. https://doi.org/10.1103/PhysRevB.38.10101.

Kalhan, A., Sharma, S., & Kumar, A. (2018). 16-channel DWDM based on 1D defect mode nonlinear photonic crystal. *AIP Conference Proceedings*, *1953*. https://doi.org/10.1063/1.5032777. American Institute of Physics Inc.

Kumar, A., Suthar, B., Kumar, V., Singh, K. S., & Bhargava, A. (2012). Tunable wavelength demultiplexer for DWDM application using 1-D photonic crystal. *Progress in Electromagnetics Research Letters*, *33*, 27–35. https://doi.org/10.2528/PIERL12042009.

Kumar, V., Suthar, B., Kumar, A., Singh, K. S., & Bhargava, A. (2013). Design of a wavelength division demultiplexer using Si-based one-dimensional photonic crystal with a defect. *Optik*, *124*(16), 2527–2530. https://doi.org/10.1016/j.ijleo.2012.07.025.

Li, J., Zhou, L., Chan, C. T., & Sheng, P. (2003). Photonic band gap from a stack of positive and negative index materials. *Physical Review Letters*, *90*(8), 083901/4. https://journals.aps.org/prl/issues.

Lin, S. Y., Fleming, J. G., Hetherington, D. L., Smith, B. K., Biswas, R., Ho, K. M., et al. (1998). A three-dimensional photonic crystal operating at infrared wavelengths. *Nature*, *394*(6690), 251–253. https://doi.org/10.1038/28343.

Noda, S., Tomoda, K., Yamamoto, N., & Chutinan, A. (2000). Full three- dimensional photonic band gap crystals at near- infrared wavelengths. *Science*, *289*.

Panoiu, N. C., Jr., Osgood, R. M., Zhang, S., & Brueck, S. R. J. (2006). Zero- n bandgap in photonic crystal superlattices. *Journal of the Optical Society of America B*, *23*.

Pendry, J. B. (2000). Negative refraction makes a perfect lens. *Physical Review Letters*, *85*(18), 3966–3969. https://doi.org/10.1103/PhysRevLett.85.3966.

Qian, E., Fu, Y., Xu, Y., & Chen, H. (2016). Total omnidirectional reflection by sub-wavelength gradient metallic gratings. *EPL*, *114*(3). https://doi.org/10.1209/0295-5075/114/34003.

Radhouene, M., Chhipa, M. K., Najjar, M., Robinson, S., & Suthar, B. (2017). Novel design of ring resonator based temperature sensor using photonics technology. *Photonic Sensors*, *7*(4), 311–316. https://doi.org/10.1007/s13320-017-0443-z.

Robinson, S., & Dhanlaksmi, N. (2017). Photonic crystal based biosensor for the detection of glucose concentration in urine. *Photonic Sensors*, *7*(1), 11–19. https://doi.org/10.1007/s13320-016-0347-3.

Sang, Z. F., & Li, Z. Y. (2007). Properties of defect modes in one-dimensional photonic crystals containing a graded defect layer. *Optics Communications*, *273*(1), 162–166. https://doi.org/10.1016/j.optcom.2006.12.008.

Sharma, S. (2021). Design of a tunable 8-channel DWDM covers all transmission windows using defect mode nonlinear photonic crystals. *Optical Engineering*, *60*(1). https://doi.org/10.1117/1.OE.60.1.017105.

Sharma, P., Deshmukh, P., & Sharan, P. (2013). Design and analysis of blood components by using optical sensor. *International Journal on Current Research*, 2225–2228.

Sharma, S., Dwivedi, D., Yadav, A., & Sengar, A. (2016). Temperature dependence ZnS based 1D photonic crystal. *International Journal of Physics and Research (IJPR)*, *6*(3), 2319–4499.

Sharma, S., Gupts, S., Suther, B., & Singh, K. S. (2020). Design of a tunable transmission mode filter using 1D Ge based nonlinear photonic crystal. *AIP Conference Proceedings, 2220*. https://doi.org/10.1063/5.0001128. American Institute of Physics Inc.

Sharma, S., & Kumar, A. (2021a). Analysis of silica based single-mode fiber doped with germanium at different transmission window. *SILICON*. https://doi.org/10.1007/s12633-020-00884-6.

Sharma, S., & Kumar, A. (2021b). Design of a biosensor for the detection of dengue virus using 1D photonic crystals. *Plasmonics*. https://doi.org/10.1007/s11468-021-01555-x.

Sharma, S., Kumar, V., & Gupta, S. (2021). Tunable transmittance using temperature dependence ZnS-based ID photonic crystals. *Lecture Notes in Electrical Engineering: Vol. 721* (pp. 325–330). Springer Science and Business Media Deutschland GmbH. https://doi.org/10.1007/978-981-15-9938-5_31.

Sharma, S., Kumar, A., & Singh, K. S. (2020). Design of a tunable DWDM multiplexer using four defect layers of GaAs nonlinear photonic crystals. *Optik, 212*. https://doi.org/10.1016/j.ijleo.2020.164652, 164652.

Sharma, S., Kumar, R., Singh, K. S., & Kumar, A. (2017). Temperature dependence of ODR range at 1550nm using 1D binary and ternary photonic crystals. *Journal of Optoelectronics and Advanced Materials, 19*(5–6), 319–324. https://joam.inoe.ro/index.php?option=magazine&op=view&idu=4106&catid=103.

Sharma, S., Kumar, R., Singh, K. S., Kumar, V., & Jain, D. (2013). Design of a transmission TM mode filter using one-dimensional ternary photonic crystal. *AIP Conference Proceedings, 1536*, 705–706. https://doi.org/10.1063/1.4810422.

Sharma, S., Kumar, R., Singh, K. S., Kumar, V., & Kumar, A. (2014a). Performance analysis of dense wavelength division demultiplexer based on one dimensional defect mode nonlinear photonic crystal. *Journal of Optoelectronics and Advanced Materials, 16*(9–10), 1015–1019. http://joam.inoe.ro/index.php?option=magazine&op=view&idu=3544&catid=86.

Sharma, S., Kumar, R., Singh, K. S., Kumar, V., & Kumar, A. (2014b). Single channel tunable wavelength demultiplexer using nonlinear one dimensional defect mode photonic crystal. *Optik, 125*(17), 4895–4897. https://doi.org/10.1016/j.ijleo.2014.04.034.

Sharma, S., Kumar, R., Singh, K. S., Kumar, A., & Kumar, V. (2015). Omnidirectional reflector using linearly graded refractive index profile of 1D binary and ternary photonic crystal. *Optik, 126*(11 – 12), 1146–1149. https://doi.org/10.1016/j.ijleo.2015.03.029.

Sharma, S., Kumar, A., Singh, K. S., & Tyagi, H. (2021). 2D photonic crystal based biosensor for the detection of chikungunya virus. *Optik, 237*, 166575.

Sharma, P., & Sharan, P. (2015a). An analysis and design of photonic crystal-based biochip for detection of glycosuria. *IEEE Sensors Journal, 15*(10), 5569–5575. https://doi.org/10.1109/JSEN.2015.2441651.

Sharma, P., & Sharan, P. (2015b). Design of photonic crystal-based biosensor for detection of glucose concentration in urine. *IEEE Sensors Journal, 15*(2), 1035–1042. https://doi.org/10.1109/JSEN.2014.2359799.

Sharma, S., & Tewari, R. (2020). Design of an ODR Mirror by using temperature dependence ZNS based 1D photonic crystals. *Solid State Technology, 63*, 4138–4142.

Sharma, S., & Vimal. (2019). Enhanced ODR reflection by using Sio2 exponentially graded materials for 1D photonic crystals. *International Journal of Advanced Science and Technology, 28*(16), 604–610. http://sersc.org/journals/index.php/IJAST/article/download/1883/1343.

Shelby, R. A., Smith, D. R., & Schultz, S. (2001). Experimental verification of a negative index of refraction. *Science, 292*(5514), 77–79. https://doi.org/10.1126/science.1058847.

Srivastava, S. K., & Ojha, S. P. (2009). Broadband optical reflector based on Si/SiO2 one-dimensional graded photonic crystal structure. *Journal of Modern Optics, 56*(1), 33–40. https://doi.org/10.1080/09500340802428330.

Suthar, B., & Bhargava, A. (2021). Biosensor application of one-dimensional photonic crystal for malaria diagnosis. *Plasmonics, 16*(1), 59–63. https://doi.org/10.1007/s11468-020-01259-8.

Suthar, B., Kumar, V., Singh, K. S., & Bhargava, A. (2012). Tuning of photonic band gaps in one dimensional chalcogenide based photonic crystal. *Optics Communications, 285*(6), 1505–1509. https://doi.org/10.1016/j.optcom.2011.10.047.

Vimal, Sharma, S., Sharma, A. K., & Tiwari, R. (2022). Tunable filter at second transmission window containing 1D ternary superconductor/dielectric photonic crystals. In *Vol. 239. Smart innovation, systems and technologies* (pp. 189–195). Springer Science and Business Media Deutschland GmbH. https://doi.org/10.1007/978-981-16-2857-3_20.

Wild, S., Roglic, G., Green, A., Sicree, R., & King, H. (2004). Global prevalence of diabetes: Estimates for the year 2000 and projections for 2030. *Diabetes Care, 27*(5), 1047–1053. https://doi.org/10.2337/diacare.27.5.1047.

Wu, X. K., Liu, S. B., Zhang, H. F., Li, C. Z., & Bian, B. R. (2011). Omnidirectional photonic bandgap of one-dimensional ternary plasma photonic crystals. *Journal of Optics, 13*.

Wu, C. J., Yang, T. J., Li, C. C., & Wu, P. Y. (2012). Investigation of effective plasma frequencies in one-dimensional plasma photonic crystals. *Progress In Electromagnetics Research, 126*, 521–538. https://doi.org/10.2528/PIER12030505.

Yamamoto, N., Noda, S., & Chutinan, A. (1998). Development of one period of a three- dimensional photonic crystal in the 5–10 nm wavelength region by wafer fusion and laser beam diffraction pattern observation techniques. *Japanese Journal of Applied Physics, 37*.

CHAPTER

20

Classification of pneumonic infections through chest radiography using textural features analysis and the pattern recognition system

Rajat Mehrotra[a,b], *M.A. Ansari*[c], *and Rajeev Agrawal*[d]

[a]Department of Electrical & Electronics Engineering, GL Bajaj Institute of Technology & Management, Greater Noida, UP, India [b]Department of Electrical Engineering, School of Engineering, Gautam Buddha University, Greater Noida, India [c]Department of Electrical Engineering, School of Engineering, Gautam Buddha University, Greater Noida, UP, India [d]Department of Electronics and Communications, Llyod Institute of Engineering and Technology, Greater Noida, UP, India

Introduction

The novel coronavirus disease (COVID-19) that started in China in December 2019 had swelled throughout the whole world at this point. It represents the "Coronavirus" (CoV) family and was termed as "Severe Acute Respiratory Syndrome Coronavirus-2" (SARS-CoV-2) prior to being named as COVID-19 by WHO. This epidemic was confirmed as Public Health Emergency on January 30, 2020 (Coronavirus disease (COVID-19), 2020) and at last, on 11 March in the same year, the World Health Organization declared this coronavirus disease as pandemic. Thereafter, the number of everyday cases started incrementing exponentially and arrived at 176,724,995 cases and around 3,819,671 deaths worldwide till June 14, 2021. The infection has inundated the majority of the powerful countries, among which the United States, India, and Brazil have been sternly affected with 34,321,158, 29,510,410, 17,413,996 total cases, and 615,053, 374,305, and 487,476 deaths, respectively, till now (Mahase, 2020). Fig. 20.1 depicts the worldwide COVID-19 cases as of date.

When infected with COVID-19, a patient may create different manifestations and indications of disease which incorporate cough, fever, and respiratory sickness. In extreme conditions, this

Computational Intelligence in Healthcare Applications
https://doi.org/10.1016/B978-0-323-99031-8.00002-8

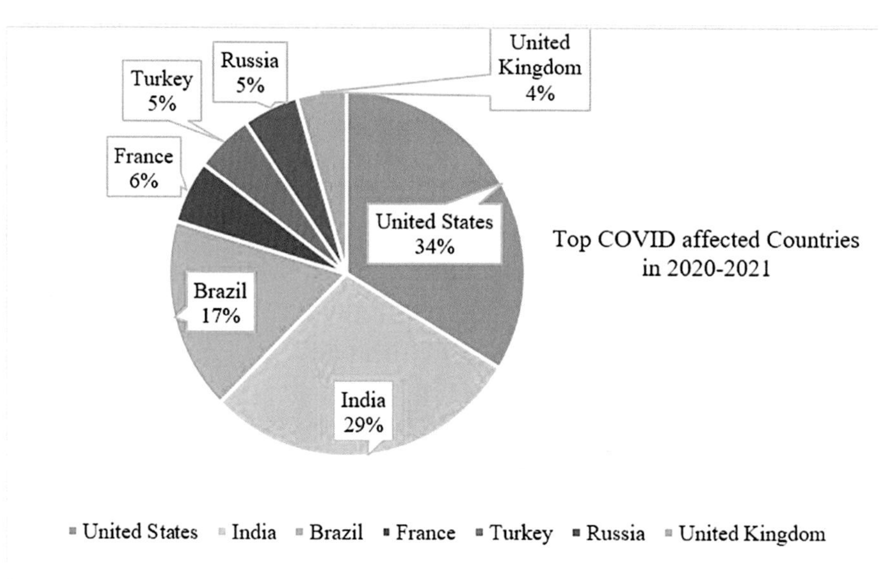

FIG. 20.1 Worldwide COVID-19 cases. *From Coronavirus Graphs: Worldwide Cases and Deaths—Worldometer. (n.d.). Retrieved June 13. (2021).*

disease may develop into pneumonia, trouble breathing, and failure of multiple organs which may lead to a casualty (Coronavirus, 2021). Because of the exponential growth of coronavirus cases, the health care services of various developed nations have reached the verge of a breakdown. They faced the deficiency of oxygen support systems as well as testing units. Numerous nations have announced absolute lockdown and requested their people to stay inside and carefully maintain a strategic distance from social affairs; however, despite such measures, the brutal effect of this virus has claimed many lives and the numbers are still increasing.

Effectual segregation of the infected person is a pivotal stride in battling COVID-19, with the end goal that truly infected patients can be given the proper and timely treatment. At present, RT-PCR is the primary testing technique utilized for recognizing COVID-19-infected persons (Corman et al., 2020; Wang, Xu, et al., 2020). The samples from the patient's respiratory system are used for this test, and the patient gets the test result in 24–48 h. A substitute for the PCR technique is a chest X-ray (CXR). Different published research in reputed journals of radiology (Bernheim et al., 2020; Xie et al., 2020) shows that CXR images may be helpful in recognizing COVID-19. Specialists in this field found that the COVID-infected lungs show some visual imprints like hazy dark spots, ground-glass opacities based on which COVID-19 patients can be separated from nonnormal ones (Fang et al., 2020). The scientists accept that a CXR-based framework can be a viable tool in the identification and evaluation of COVID-19. Intelligent computational methods for classification and recognition of ailments have been the primary area of research for the improvement of computerized diagnosis systems (CDSs) (Baker et al., 2003; Cheng et al., 2016; Doi, 2007; Horváth et al., 2009); this includes both working prototypes and software which offer data to help the clinical expert in conclusive diagnosis by inspecting computerized digital images acquired from various imaging modalities that include X-rays, magnetic resonance image, ultrasounds, and computer tomography (CT) (Van Ginneken, Ter Haar Romeny, & Viergever, 2001), which are used to discover potential anomalies that conceivably demonstrate an ailment. CDS used for the analysis of clinical images has long been the main concern for researchers. There have been signs of progress in diverse quarters, which include breast injury (Baker et al., 2003) or microcalcification identification (Rubio, Montiel, & Sepúlveda, 2017), lung ailments (Horváth et al., 2009), and even in the diagnosis of heart-related diseases (González, Melin, Valdez, & Prado-Arechiga, 2018; González, Valdez, Melin, & Prado-Arechiga, 2014, 2015). A CXR-based identification framework can have numerous points of interest over customary techniques. It tends to be quick, examine different cases at the same time, have noteworthy accessibility, and all the more significantly, such a framework can be helpful in situations with no or inadequate kits for testing. Also, the major significance of radiology is that it is accessible in each clinic, no matter big or small, in this way making it more helpful and effectively accessible. The CXR images of normal, pneumonia, and COVID-infected lung are shown in Figs. 20.2–20.4.

Keeping this in mind, we present an effective technique for the classification of COVID-19-infected lungs through CXR using textural feature analysis and an artificial neural network (ANN). It gave promising results when compared to other state-of-the-art techniques and may be used to assist radiologists in authenticating their diagnostic results and in starting timely treatment of this deadly disease. The rest of the paper is arranged as follows: "State of the art" section discusses the various state-of-the-art techniques, "Proposed methodology" section will present the proposed methodology, "Results of network evaluation" section will discuss the experimental results, along with "Conclusion" section, which will give the conclusion of the paper.

State of the art

At present, specialists from all over the globe, from diverse research areas, are working round the clock to battle this virulent disease. Numerous researches have been done and published exhibiting the techniques and methodology for the identification of coronavirus

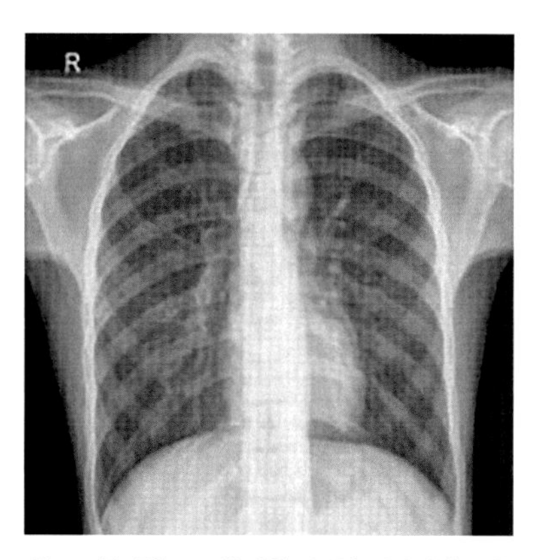

FIG. 20.2 Normal CXR image. *From Fang, Y., Zhang, H., Xie, J., Lin, M., Ying, L., Pang, P., et al. (2020). Sensitivity of chest CT for COVID-19: Comparison to RT-PCR.* Radiology, *296(2), E115–E117. https://doi.org/10.1148/radiol.2020200432.*

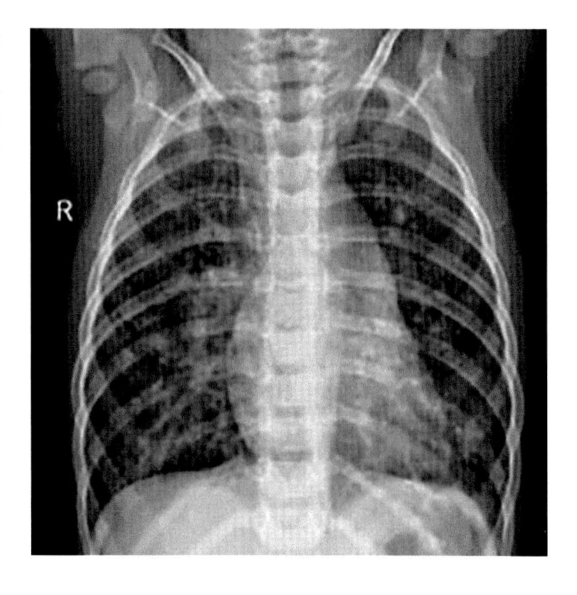

FIG. 20.3 Pneumonia infected CXR image. *From Kermany, D. S., Goldbaum, M., Cai, W., Valentim, C. C. S., Liang, H., Baxter, S. L., et al. (2018). Identifying medical diagnoses and treatable diseases by image-based deep learning.* Cell, *1122–1131.e9. https://doi.org/10.1016/j.cell.2018.02.010.*

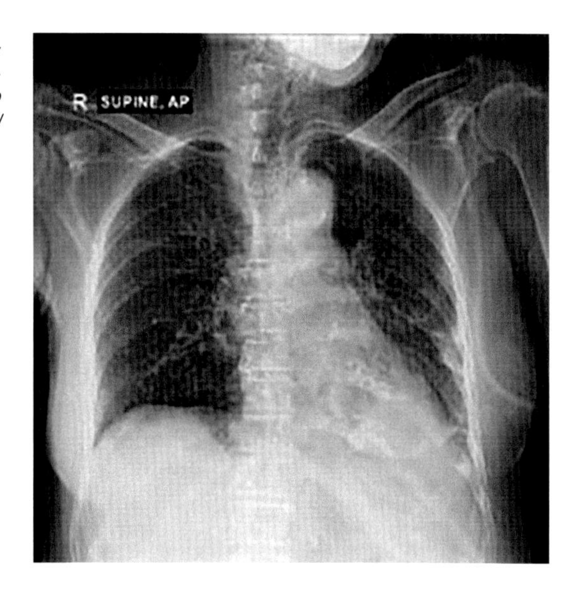

FIG. 20.4 COVID infected CXR image. *From Fang, Y., Zhang, H., Xie, J., Lin, M., Ying, L., Pang, P., et al. (2020). Sensitivity of chest CT for COVID-19: Comparison to RT-PCR.* Radiology, *296(2), E115–E117. https://doi.org/10.1148/radiol.2020200432.*

through images of chest radiography. Some related major state-of-the-art work done in this field are presented here.

In Wang et al. (2017), the authors used a novel dataset termed ChestXRay8, and thereafter, ChestXRay14 is introduced for the identification of infections in a frontal CXR image. This dataset holds 112,120 frontal CXR images partitioned into 14 classes of the nonrelative type that belongs to 13 distinct ailments and a normal class that contains images with a normal

CXR. These datasets are not balanced as they have a diverse quantity of test samples in each class that belongs to a specific ailment type. For the best results, they used ResNet-50 convolutional neural network (CNN) and reported precision of 63.3% for the pneumonia class, utilizing each of the 14 classes of the dataset. Various methodologies have utilized these datasets to identify one or more of these illnesses, to work with the 14 classes simultaneously. In Rajpurkar et al. (2017), the author presented a model called CheXNet, utilizing 121 layers to identify each of the 14 diseases with an accuracy of 76.8% for the pneumonia class. Likewise, his proposed framework gives a heat map localization of disease as predicted by CNN. Antin, Kravitz, and Martayan (2017) used the k-mean to identify the pneumonia class obtained through the ChestXRay14 after using the antialiasing filter on it. Khobragade, Tiwari, Patil, and Narke (2017) proposed a technique of utilizing thresholding and discontinuity-based preprocessing methods like Canny Filter succeeded by a geometrically based extraction of features using FFNN as the classifier. They revealed that the classification outcomes of their experiment claimed were of 92% accuracy. The author (Pattar, 2015) also utilized FFNN on a database of 116 images to classify the images into pneumonia or lung cancer with an accuracy of 94.8%. Ebrahimian, Rijal, Noor, Yunus, and Mahyuddin (2014) put forward an algorithm to take care of issues of separating two comparable diseases, like lobar pneumonia and pulmonary tuberculosis utilizing the congruency of phase method. This method was put into action to get the best capability for differentiating between healthy, lobar pneumonia, and pulmonary tuberculosis. A total of eight textural features were then explored as universal measures for the identification of diseases. Noor et al. (2014) worked to segregate three classes of lung infections, to be specific lobar pneumonia and pulmonary tuberculosis, and cancer in the lungs utilizing CXR.

An updated principal component analysis is used to measure wavelet-based textures yielding feature vectors for the statistical discrimination technique. Karargyris, Antani, and Thoma (2011) built up a methodology for distinguishing ribs in chest radiography. This methodology calculates wavelet-based features alongside various anatomic structure orientations for categorization. In Hina, Khalid, and Akbar (2017), authors zeroed in on the methods of segmentation, extraction of features, and identification of images of X-rays as healthy and infected dependent on the features trained classifier. Barrientos et al. (2017b) introduced a technique for automated pneumonia detection by examination of patterns present in pneumonic lungs ultrasound images; the author obtained certain particular features from the characteristic vectors, and their classification is done with regular neural networks. Barrientos et al. (2017a) proposed a strategy to perceive and remove the segment of the skin as noise in the image of lung ultrasound as part of the skin makes the interpretation of pneumonia-infected lung difficult in ultrasound. Cisneros-Velarde et al. (2016) put forward a technique for the identification of pneumonia utilizing ultrasound videos. The author investigated the videos in small fragments to develop statistics of the complete video for identifying pneumonia. Zenteno, Castaneda, and Lavarello (2016) proposed a technique based on fundamental frequency downshift measurement for the identification of pneumonia from a dataset containing 6 months to 5 years of subject test samples. Liu and Mehrotra (2016) proposed a counteraction technique against pneumonia by measuring how frequently and up to what extent the person is active by utilizing Microsoft Kinect V2. Hospital-acquired pneumonia manages patient activity during their stay in the hospital. The depicted model perceived the activity by utilizing the extraction of features from the Kinect model. The results of this model give data to medical services staff to better understand the hospital-acquired pneumonia arrangement and to alter the treatment time.

Proposed methodology

The motivation behind this research is to improve the accuracy in detecting and differentiating COVID positive cases from others through CXR images, obtained from a publicly available repository, using textural analysis and ANNs. Extraction of various textural features from CXR images is done and classification is performed using an ANN. The steps utilized in this experiment are shown with the help of a block diagram in Fig. 20.5. It starts with the acquiring of the CXR dataset which is organized into three classes: class 1 contains COVID-19-infected images, class 2 consist of normal CXR images, and class 3 accommodates pneumonia-infected CXR images. These images then go through the following stages: CXR image preprocessing, feature extraction using DWT and GLCM, feature reduction through PCA, division of data, and classification using neural network into three classes.

CXR dataset for experiment

The CXR dataset for the purpose of assessment of the proposed research framework is taken from two different repositories (Fang et al., 2020; Kermany et al., 2018). This dataset integrates 981 CXR images in total, out of which 234 images are tagged as normal, 390 are tagged as pneumonia images, and 357 as COVID-infected images. The image dimensions of all the images are set to 225×225 in JPG/JPEG format to aid the proposed framework.

Preprocessing of CXR images

Prior to dealing with the CXR images for the proposed framework, it is preprocessed. First, the raw images are downsized to $225 \times 225 \times 1$ pixels to reduce dimensionality counts and aid the proposed framework to exhibit better results in considerably less time. For augmenting the quality of the CXR images and for the elimination of noise and extraneous information, the median filter is used. In order to extract features from the infected area, it is pivotal to find the region of interest (ROI) within the image (Apostolopoulos & Mpesiana, 2020; Sethy, Behera, Ratha, & Biswas, 2020; Wang, Lin, & Wong, 2020). For fragmenting an ROI, the potential spots of disease should be known in the image. For example, in the case of pneumonia, a shady white spot appears within the lung region. In our proposed method, we have found the ROI by focusing only on the area of the lungs and removing the rest from the images (Hemdan, Shouman, & Karar, 2020; Song et al., 2020). We have also done contrast adjustment of the CXR images to see the area inside the ROI more evidently as shown in Figs. 20.6–20.8, which presents a comparison between the extracted ROI and the contrast adjusted CXR image together with their normal correlation.

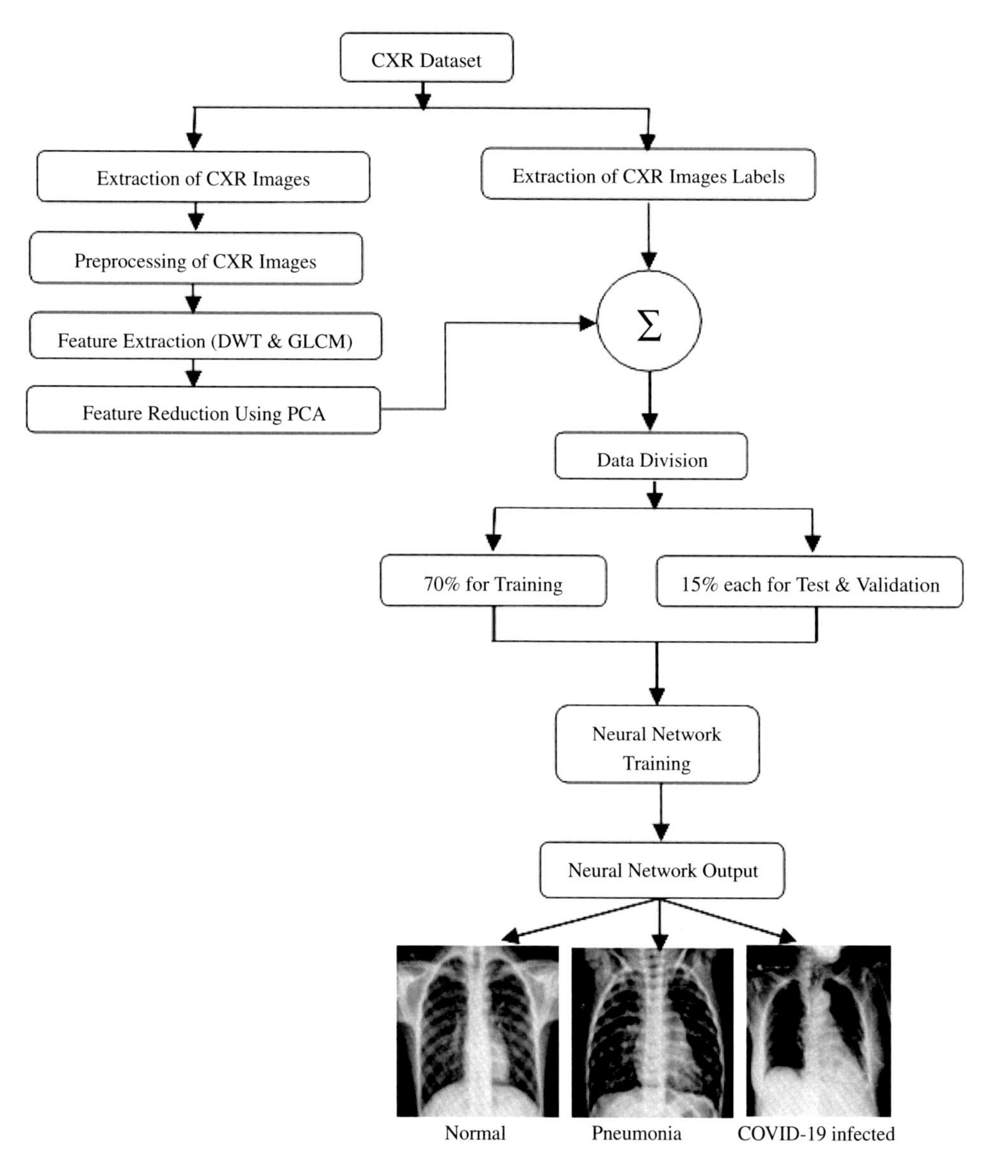

FIG. 20.5 Proposed classification framework of feature analysis and classification through ANN. *No Permission Required.*

Extraction of DWT and GLCM-based features

Extraction of DWT-based features

The wavelet transform technique is used to break down a picture into its subbands. Subband pictures may comprise a low- and high-frequency picture. Extraction of numerous textural features can be done from each of these subband pictures. Subsequently,

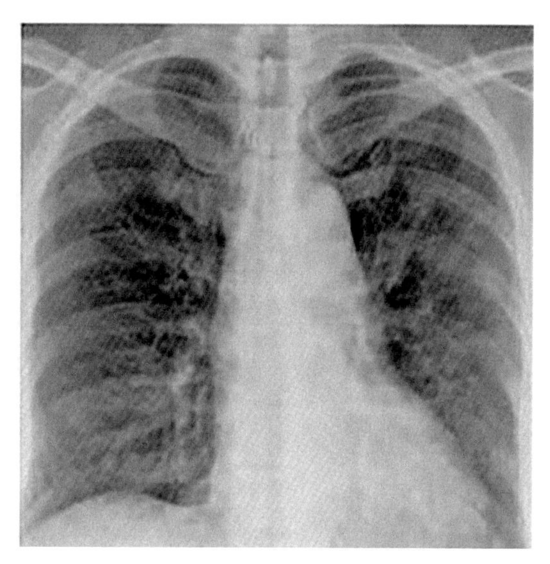

FIG. 20.6 Extracted ROI. *No Permission Required.*

classification and identification of pictures can be done using these extracted features. The fundamental idea of the wavelet is clarified in the disintegrating one-dimensional signal. It disintegrates the signal $S(x)$ into its constituent subbands, utilizing translation as well as a scaling phenomenon termed as the mother wavelet, which is utilized for creating different window functions. The expression for mother wavelet is given by (Minarno, Munarko, Kurniawardhani, Bimantoro, & Suciati, 2014)

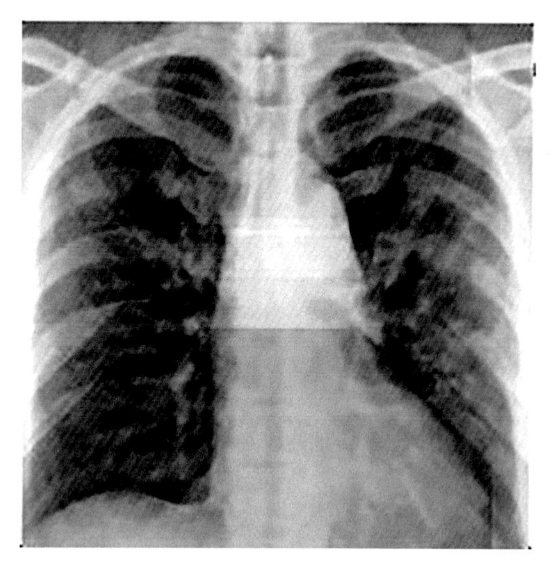

FIG. 20.7 Contrast-enhanced ROI. *No Permission Required.*

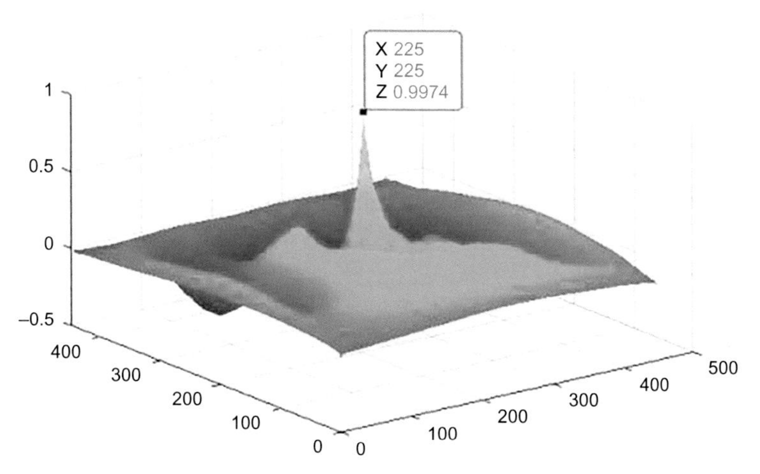

FIG. 20.8 Correlation between extracted and contrast-enhanced ROI. *No Permission Required.*

$$\varphi_{r,t}(x) = \frac{1}{\sqrt{|r|}}\varphi\left(\frac{x-r}{t}\right) \tag{20.1}$$

where t and r denote the translation and scaling factors, respectively. The previously mentioned Eq. (20.1) is applied to the signal $S(x)$ as

$$W_{\varphi}S(r,t) = \{S, \varphi_{r,t}(x)\} = \int S(x)\cdot\varphi^{*}_{r,t}(x)dx \tag{20.2}$$

Normally, images represent two-dimensional signals containing rows and columns. Hence, for decomposing such signals, we require a two-dimensional discrete wavelet transform (2-DWT). 2-DWT is calculated through two-dimensional filters succeeded by two-dimensional downsampling. The 2-DWT is calculated by applying a one-dimensional discrete wavelet transform to the rows and columns of the images individually. The filters are initially convoluted independently in rows and columns to obtain a two-dimensional transformation. The filters for two-dimensional wavelets are acquired by the product of one-dimensional wavelet filters as

$$WF_{LL} = l_f^T * l_f \tag{20.3}$$

$$WF_{LH} = l_f^T * h_f \tag{20.4}$$

$$WF_{HL} = h_f^T * l_f \tag{20.5}$$

$$WF_{HH} = h_f^T * h_f \tag{20.6}$$

where l_f and h_f represent one-dimensional low- and high-pass filters. Every image received from the process of transformation is downsampled, to squeeze the band size to half, in all directions by a factor of 2. At all levels of disintegration, the 2-DWT gives four image subbands, specifically HH, HL, LH, and LL. Each of these image subbands carries data at a particular scale and direction. LL relates to the image approximation. LH presents the image details in the horizontal direction, containing details of high-vertical and low-horizontal frequencies. HL presents the image details in the vertical direction containing details of low-vertical and high-horizontal frequencies. HH presents the image details in the diagonal direction containing details of high-vertical and high-horizontal frequencies. To achieve the next degree of decomposition, 2-DWT is used to progressively disintegrate images of the LL subband. The discrete wavelet transform contains different banks of filters, like Coiflets, Haar, Biorthogonal, Daubechies, and so on. This bank of wavelet filters creates varied frequency ranges for distinctive wavelet functions.

Extraction of GLCM-based features

The gray-level cooccurrence matrix is an $R \times R$ matrix, where R denotes the gray-level values in the original image. This matrix displays the probability of the gray-level intensity of ith and jth pixels r distance apart and situated at an angle δ. Hence, their probability can be expressed as $P(i, j, r, \delta)$. The distance "r" can take any value between 1 and 8 and at any angle between 0 and 360 degrees as illustrated in Fig. 20.9.

GLCM can be determined, for the proposed system, after doing some fundamental analysis of it; [0,2] is found to be a decent setup to utilize, that is, equivalent to 0 degrees with 2 pixels comparison. Therefore in the proposed system, the gray-level cooccurrence matrix is developed by evaluating 2 pixels from the pixel of significance in the horizontal direction. The GLCM setup utilized for the proposed work is shown in Fig. 20.10. We then acquire the image features through the texture of the image that corresponds to the pixel behavior inside each image grayscale dependent on the spatial comparison of pixels with each other (Mehrotra, Ansari, Agrawal, & Anand, 2020).

The image statistical features are calculated in order to illustrate the texture using a gray level co-occurrence matrix. Some important features are:

(1) Angular second moment or energy: it calculates the uniformity in the texture of an image and gives a high value for a perfectly homogenous image and is given by

$$E = \sum_{i,j=0}^{R-1} P^2(i, j, r, \delta) \tag{20.7}$$

(2) Entropy: it is a procedure of measuring the uncertainty of an image and is opposite to energy and is expressed as

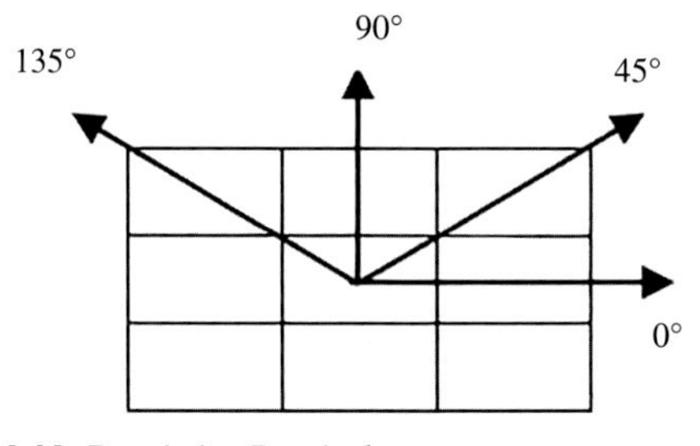

FIG. 20.9 Offset array in GLCM. *No Permission Required.*

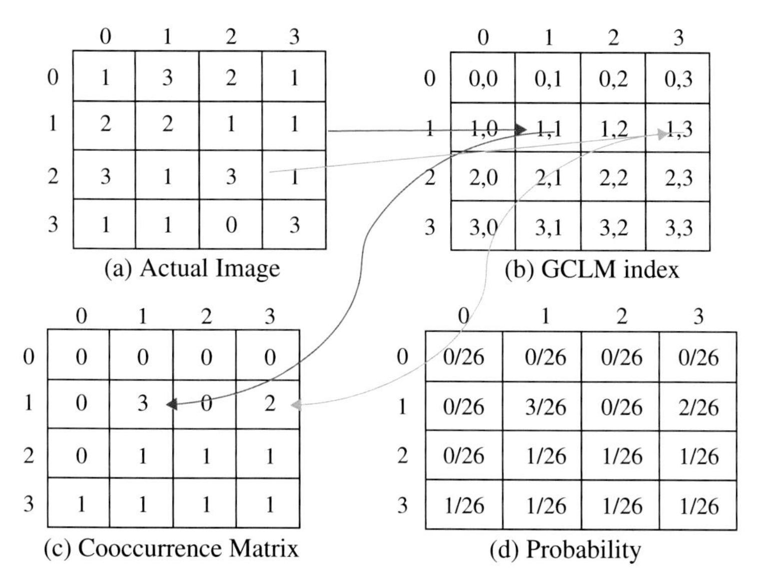

FIG. 20.10 Construction of GLCM. *No Permission Required.*

$$H=\sum_{i,j=0}^{R-1}P(i,j,r,\delta)\cdot\log P(i,j,r,\delta) \tag{20.8}$$

(3) Contrast: it calculates the gray-level changes in an image that points out the difference in the image's gray level. Smooth images show low values in opposition to a coarse image.

$$Con=\sum_{i,j=0}^{R-1}(i-j)^2\cdot P(i,j,r,\delta) \tag{20.9}$$

(4) Inverse difference moment: it displays a higher value for images with lower contrast.

$$IDM=\sum_{i,j=0}^{R-1}\frac{P(i,j,r,\delta)}{i+|i+j|^2} \tag{20.10}$$

(5) Correlation: it is the procedure of measuring how pixels are linearly dependent on each other and is expressed as

$$Cor=\sum_{i,j=0}^{R-1}\frac{(i-\mu_s)(j-\mu_t)P(i,j,r,\delta)}{\sigma_s\sigma_t} \tag{20.11}$$

$$\mu_s=\sum_{i,j=0}^{R-1}i\cdot P(i,j,r,\delta);\ \mu_t=\sum_{i,j=0}^{R-1}j\cdot P(i,j,r,\delta); \tag{20.12}$$

where

$$\sigma_s=\sum_{i,j=0}^{R-1}(i-\mu_s)^2\cdot P(i,j,r,\delta); \tag{20.13}$$

$$\sigma_t=\sum_{i,j=0}^{R-1}(i-\mu_t)^2\cdot P(i,j,r,\delta) \tag{20.14}$$

(6) Mean: it is derived by each pixel value summation of the image over the net quantity of the image pixels.

$$M = \frac{1}{R}\sum_{i,j=0}^{R-1} P(i,j,r,\delta) \tag{20.15}$$

(7) Standard deviation: it serves as a way of calculating nonhomogeneity and provides the probable distribution of an experimental population.

$$Std = \sqrt{\frac{1}{R}\sum_{i,j=0}^{R-1} P(i,j,r,\delta) - M} \tag{20.16}$$

(8) Skewness: it is a way of measuring symmetry. It is used for giving judgments regarding the surface of an image as it describes what number of the pixels are "lopsided."

$$Skw = \frac{1}{R}\frac{\sum_{i,j=0}^{R-1} (P(i,j,r,\delta) - M)^3}{Std^2} \tag{20.17}$$

(9) Kurtosis: it measures whether or not the image pixels are sharp or smooth as compared to the normal distribution.

$$Kur = \frac{1}{R}\frac{\sum_{i,j=0}^{R-1} (P(i,j,r,\delta) - M)^4}{Std^3} \tag{20.18}$$

(10) Homogeneity: it calculates the element's proximity distribution in the gray level co-occurrence matrix to its diagonal.

$$Homo = \sum_{i,j=0}^{R-1} \frac{P(i,j,r,\delta)}{1+|i-j|} \tag{20.19}$$

Tables 20.1–20.3 present the textural features derived from preprocessed CXR images for each of the used classes.

TABLE 20.1 Features derived from class 1 COVID-19 infected CXR images.

Contrast	Correlation	Energy	Homo	Mean	Std	Entropy	Kurtosis	Skewness	IDM
0.2171	0.088	0.7672	0.9357	0.0038	0.0857	3.5983	6.3089	0.5313	0.9593
0.2135	0.0704	0.7678	0.9352	0.0047	0.0856	3.5845	6.3568	0.4404	−0.0325
0.225	0.0743	0.7647	0.9343	0.0035	0.0857	3.5237	6.7931	0.5583	−1.0341
0.2259	0.09	0.7628	0.933	0.0025	0.0857	3.4885	6.9042	0.624	0.7313
0.2086	0.1099	0.7534	0.9324	0.0051	0.0856	3.5555	5.8569	0.4389	0.9616
0.2011	0.0923	0.7615	0.9346	0.0034	0.0857	3.4841	6.6653	0.4313	0.8142
0.1974	0.1098	0.7646	0.9357	0.0028	0.0857	3.5333	6.1466	0.5023	0.517
0.1889	0.1163	0.7624	0.9346	0.0015	0.0857	3.5007	5.9774	0.3919	2.2574
0.2052	0.1103	0.7537	0.9335	0.0044	0.0856	3.5307	5.8967	0.4683	−1.162
0.1965	0.1002	0.7521	0.9329	0.0028	0.0857	3.5537	5.4199	0.317	1.6105

TABLE 20.2 Features derived from class 2 Normal CXR images.

Contrast	Correlation	Energy	Homo	Mean	Std	Entropy	Kurtosis	Skewness	IDM
0.207	0.1286	0.7744	0.9379	0.0031	0.0857	3.5415	6.1594	0.4783	0.6777
0.2068	0.0971	0.7648	0.9352	0.0028	0.0857	3.4948	6.3224	0.4369	−0.7087
0.1976	0.0974	0.7727	0.9382	0.003	0.0857	3.5384	5.7489	0.3805	0.4969
0.1932	0.1315	0.7738	0.9385	0.0023	0.0857	3.5147	6.1363	0.3572	−0.6483
0.2128	0.0753	0.7617	0.9341	0.0034	0.0857	3.5615	6.1523	0.5404	0.7265
0.2158	0.079	0.7663	0.9359	0.0035	0.0857	3.5229	5.8753	0.501	0.7606
0.2031	0.061	0.7603	0.9348	0.0035	0.0857	3.5535	5.7177	0.3012	−0.3626
0.224	0.0823	0.7646	0.9345	0.0035	0.0857	3.5077	6.6259	0.6658	0.6811
0.2238	0.1098	0.7604	0.9339	0.0038	0.0857	3.4056	7.4036	0.5427	0.297
0.208	0.1245	0.7556	0.9338	0.0032	0.0857	3.4526	6.5439	0.4818	0.5396

The CXR dataset is then assorted before dividing it into training, testing, and validation so that the system works on jumbled data and avoids concentrating on a smaller part of the complete dataset. The CXR dataset is then parted into a training, testing, and validation set, with 70% of the data for training the network and 15% each for testing and validation separately.

Artificial neural network architecture

For identifying the class of disease (class1, class2, or class3) from the CXR dataset, it is required to train the neural network using the features acquired from the CXR images so as to make the network understand the patterns of the three image classes using the value of each feature, the neural network architecture utilized is presented in Fig. 20.11. We have derived 10 pivotal features with 1036 feature values from a total of 981 CXR images, which includes

TABLE 20.3 Features derived from class 3 pneumonia CXR images.

Contrast	Correlation	Energy	Homo	Mean	Std	Entropy	Kurtosis	Skewness	IDM
0.2054	0.1135	0.7565	0.934	0.0042	0.0857	3.5395	6.2057	0.4057	1.0357
0.2165	0.0893	0.7653	0.9346	0.0048	0.0856	3.5744	6.5493	0.4854	1.2797
0.204	0.1094	0.7625	0.9343	0.0042	0.0857	3.4894	6.2718	0.5901	0.8517
0.2011	0.1135	0.7697	0.9365	0.0035	0.0857	3.5104	6.5745	0.5053	0.4629
0.2066	0.1249	0.7707	0.9372	0.0028	0.0857	3.2965	6.9785	0.5514	0.4329
0.2332	0.0747	0.7782	0.938	0.0036	0.0857	3.4231	8.3217	0.7233	0.1769
0.2185	0.0887	0.7839	0.939	0.0029	0.0857	3.4757	7.5701	0.5189	0.5458
0.2247	0.1048	0.7752	0.9377	0.0035	0.0857	3.4267	7.3652	0.6194	0.9683
0.2013	0.1357	0.7558	0.9345	0.0035	0.0857	3.5737	6.2241	0.4344	0.0737
0.2169	0.1311	0.7724	0.9375	0.0031	0.0857	3.4051	8.0466	0.6889	2.3533

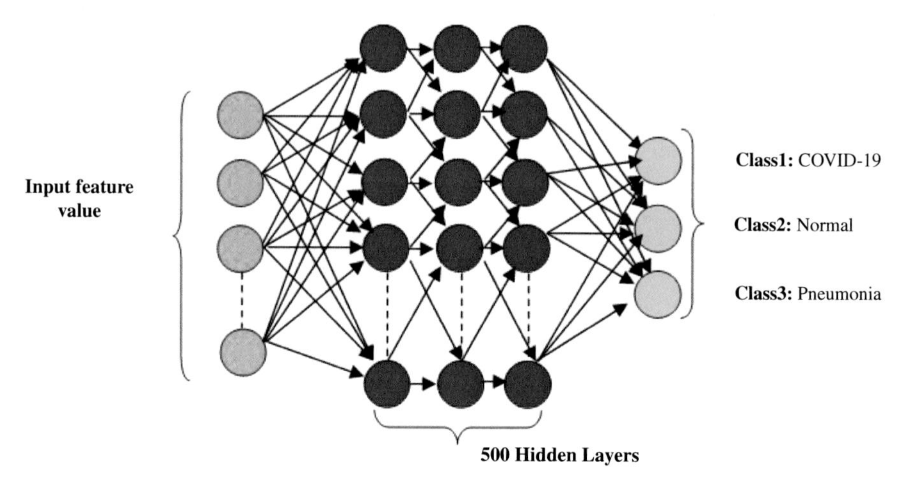

FIG. 20.11 The architecture of the neural network for the proposed work. *No Permission Required.*

class 1: COVID infected, class 2: normal, and class 3: pneumonia-infected CXR images. These feature values are then divided into the ratio of 70:15:15. In total, 70% is used for training while 15% is used for testing and validation.

Results of network evaluation

In this paper, a novel framework for the detection and classification of COVID-infected lungs from normal and pneumonia-infected lungs using CXR is presented. We have done the extraction of various textural features from CXR images and performed classification using an ANN. Table 20.4 presents the architectural parameters for the proposed neural

TABLE 20.4 Architectural parameters for proposed neural network.

Neural network parameters	Values
No. of inputs	1036 feature values
No. of hidden layers	500
Epoch	1000
Output	3
Learning rate	0.01
Momentum constant	0.9
Minimum performance gradient	1.00E-05
Training Algo	Gradient descent with momentum and adaptive learning rate backpropagation
Performance evaluation	Cross entropy
Calculation	Parallel with MEX

network model used in the presented work. The network is trained using 1036 image feature values with 500 hidden layers to identify and classify the image into three classes. The training algorithm used for the training of the network is gradient descent with momentum and adaptive learning rate backpropagation with cross-entropy used to evaluate the performance of the proposed network. The proposed framework gave promising results with an overall accuracy of 97.6% when compared to other state-of-the-art techniques. This framework may be used to assist radiologists in authenticating their diagnostic results and in starting timely treatment of this deadly disease. The training results of the proposed architecture are shown later. Fig. 20.12 shows the training state of the proposed network. Fig. 20.13 presents the best validation performance of 0.12354 at epoch 125.

Figs. 20.14 and 20.15 exhibit the matrix of confusion as well as the receiver operating characteristics for training, testing, validation as well as overall accuracy of the proposed network. From Fig. 20.10, it is evident that the proposed framework represents topnotch accuracy of 97.7% and 100% in testing and validation phases in identifying COVID-19 cases with a superior overall accuracy of 97.6%. The mathematical expressions for recall, specificity, precision, Cohen's kappa coefficient (κ), error rate, F1, and accuracy are presented in Table 20.5. Here TP is the forecasted positive cases that are truly positive; TN is the anticipated negative case, which is in reality negative; and FN is the prophesied negative cases that are actually positive. Such cases are termed type 2 errors. FP is the number of cases that are predicted to be positive but are actually negative. Such cases are termed type 1 errors. Table 20.6 contains the analysis of the overall confusion matrix that shows that the system is sufficiently robust in identifying

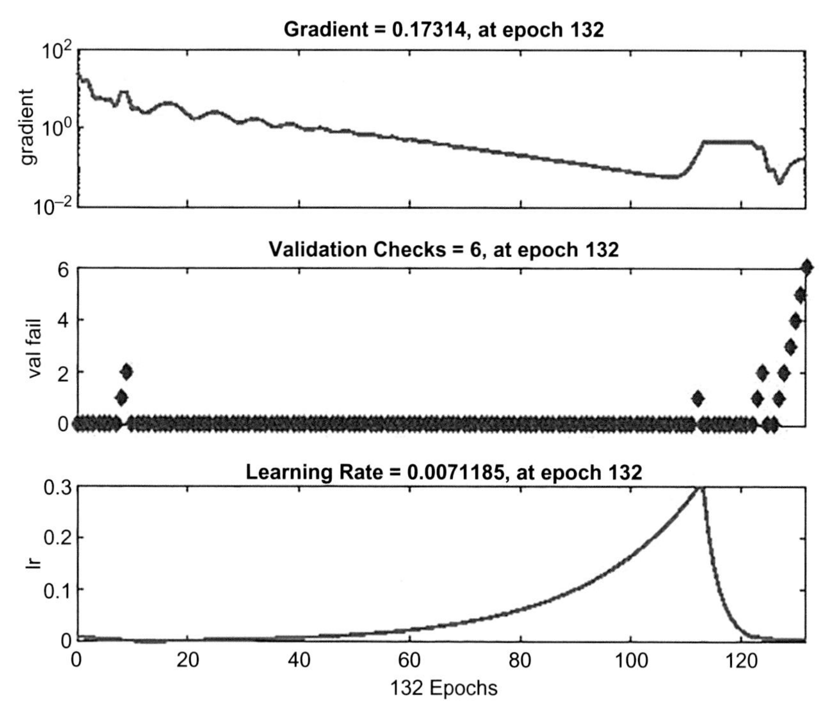

FIG. 20.12 Training state of the proposed network. *No Permission Required.*

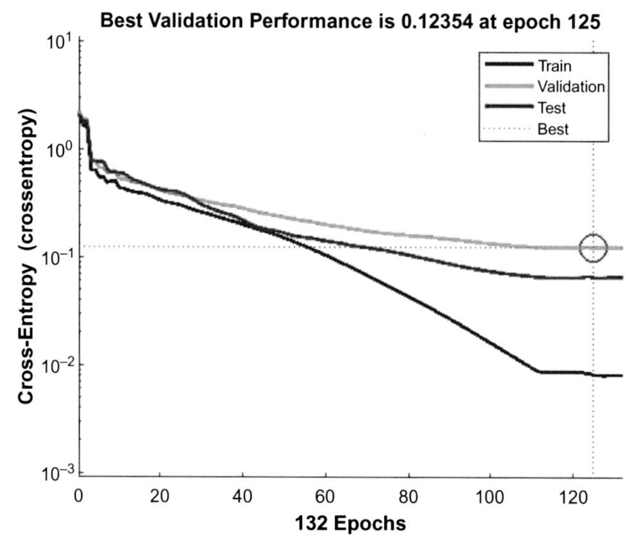

FIG. 20.13 Validation performance of the proposed network. *No Permission Required.*

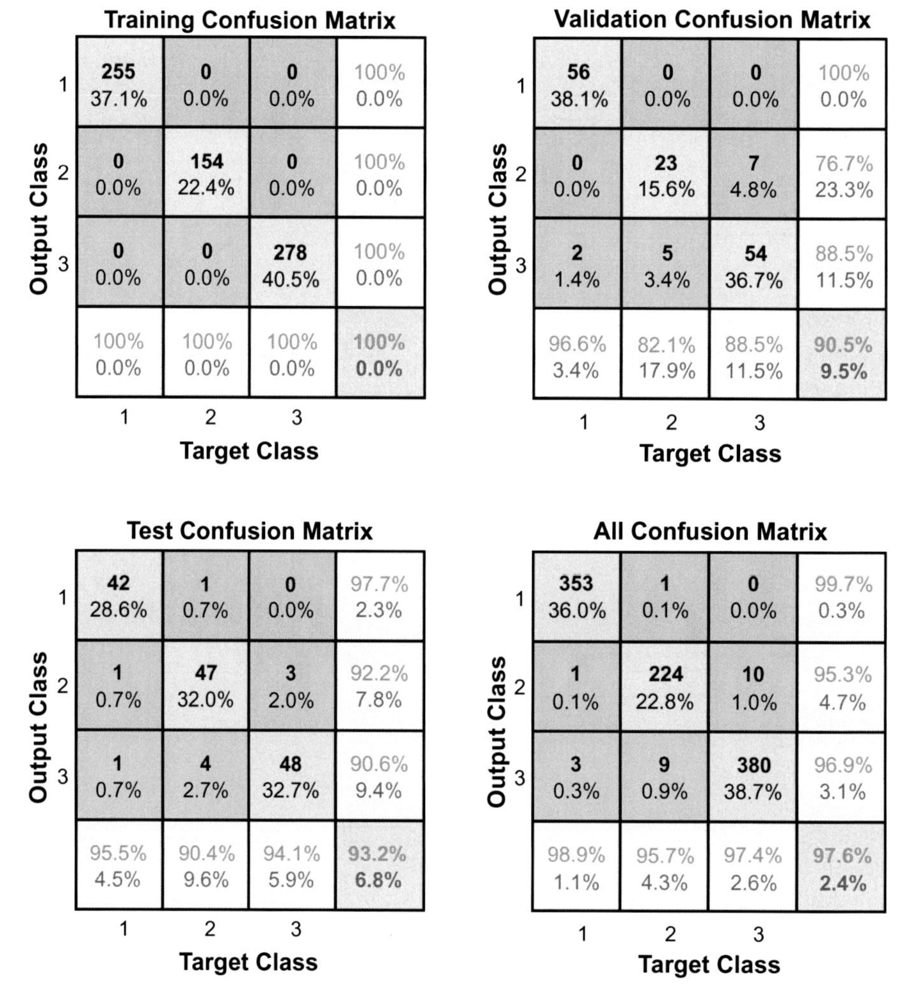

FIG. 20.14 Matrix of confusion of the proposed network. *No Permission Required.*

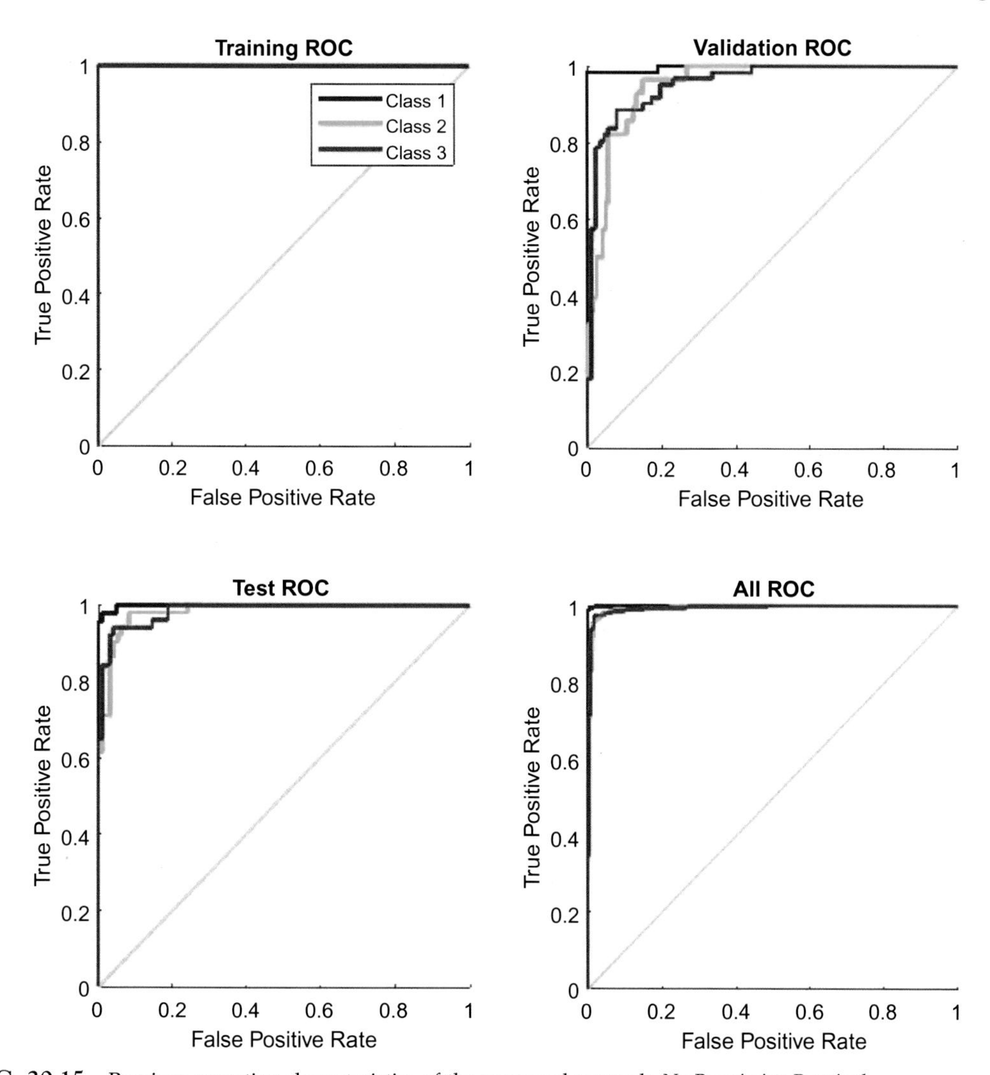

FIG. 20.15 Receiver operating characteristics of the proposed network. *No Permission Required.*

and differentiating COVID-19 cases from pneumonia and normal cases with a kappa value of 0.9890, F1 value of 0.9976, and accuracy of 0.9949. Fig. 20.16 shows the error histogram of the proposed network.

The work of the proposed system in emergencies is depicted in Fig. 20.17. As the presented system has kappa and F1 values of 0.9890 and 0.9976, respectively, for COVID-19; hence, it can segregate COVID-19-infected images with the utmost accuracy and can be used in identifying more serious patients than others (Yoon et al., 2020). All the more explicitly, we can reserve our limited life-saving medical equipment by barring ordinary pneumonia and others at the beginning phase and use this equipment for seriously affected COVID-19 patients.

TABLE 20.5 Parameters for classification of image.

Parameters	Expressions
Specificity	$\frac{TN}{TN+FP}$
Accuracy	$\frac{TP+TN}{TP+TN+FP+FN}$
Sensitivity	$\frac{TP}{TP+FN}$
Precision	$\frac{TP}{TP+FP}$
Cohen's kappa coeff. (κ)	$\frac{\text{Accuracy}-\text{Random Accuracy}}{1-\text{Random Accuracy}}$
Error rate	$\frac{FP+FN}{TP+TN+FP+FN}$
F1 score	$\frac{2\times(\text{Precision}\times\text{Specificity})}{\text{Precision}+\text{Specificity}}$

TABLE 20.6 Confusion matrix analysis.

Type of infection	TP	TN	FP	FN	Recall	Specificity	Precision	Cohen's kappa coeff. (κ)	Error rate	F1 score	Accuracy
COVID-19	353	623	1	4	0.9887	0.9983	0.9971	0.9890	0.0050	0.9976	0.9949
Normal	224	736	11	10	0.9572	0.9852	0.9531	0.9411	0.0214	0.9688	0.9785
Pneumonia	380	579	12	10	0.9743	0.9796	0.9743	0.9530	0.0224	0.9769	0.9775
Overall Accuracy					**97.6%**						

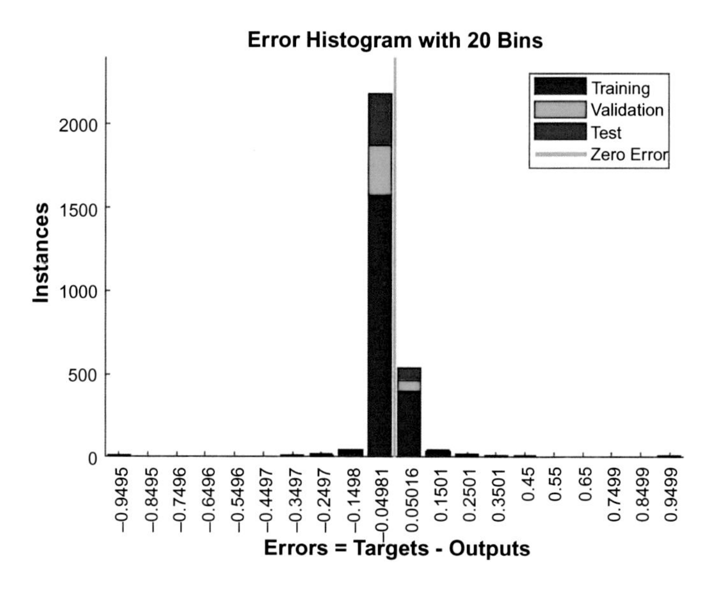

FIG. 20.16 Error histogram of the proposed network. *No Permission Required.*

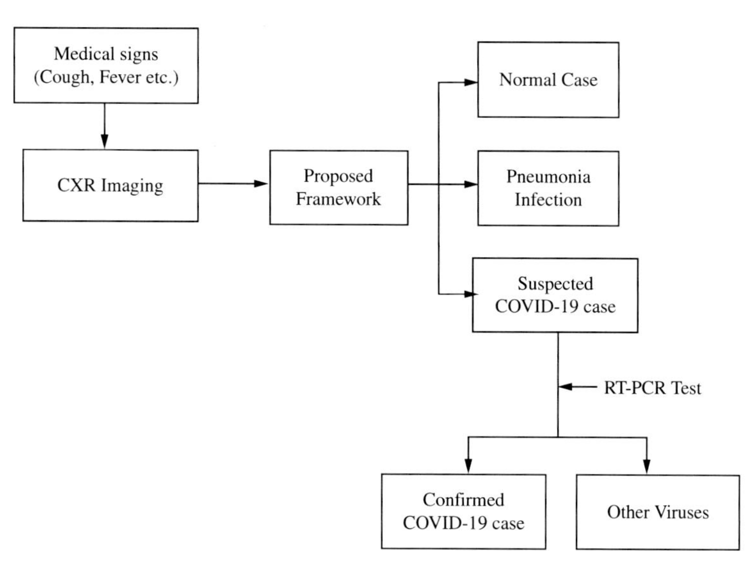

FIG. 20.17 Working of the proposed system in medical situations. *No Permission Required.*

TABLE 20.7 Potential state of the art techniques.

Author	Image type	Techniques used	Prediction accuracy
Wang, Lin, and Wong (2020)	CXR	ResNet-50	63.3%
Rajpurkar et al. (2017)	CXR	CNN	76.8%
Khobragade et al. (2017)	CXR	FFNN	92%
Barrientos et al.	Ultrasound Images	ANN	Sensitivity and specificity (91.52% and 100%, respectively)
Pavithra and Pattar	CXR	FFNN	94.8%.
Apostolopoulos and Mpesiana (2020)	CXR	VGG-19	93.48%
Wang and Wong	CXR	COIVD-Net	92.4%
Sethy and Behera	CXR	ResNet-50 + SVM	95.38%
Hemdan et al. (2020)	CXR	COVIDX-Net	90.0%
Song et al. (2020)	Chest CT	DRE-Net	86%
Proposed Method	CXR	ANN	**97.6%**

Conclusion

Textural-based feature analysis symbolizes a superior method of identifying diseases using medical images. It fills in as an initial step in judging the superior features that can aid in the identification of COVID-19-infected CXR images from pneumonia and normal images. The proposed framework has given promising results with an overall accuracy of 97.6%

and can prove useful to medical practitioners dealing with such diseases. To improve the outcomes of the network additionally, some supplementary techniques may be used to obtain a few more geometrical features in order to have more potential for classification and provide ANN with a better class discrimination criterion. A Genetic Algorithm can be further utilized for optimizing the classifier in order to train the network with better accuracy. Table 20.7 presents a comparison with the state-of-the-art techniques that shows the potential of the proposed network.

References

Antin, B., Kravitz, J., & Martayan, E. (2017). *Detecting pneumonia in chest X-rays with supervised learning*.

Apostolopoulos, I. D., & Mpesiana, T. A. (2020). Covid-19: Automatic detection from X-ray images utilizing transfer learning with convolutional neural networks. *Physical and Engineering Sciences in Medicine*, *43*(2), 635–640. https://doi.org/10.1007/s13246-020-00865-4.

Baker, J. A., Rosen, E. L., Lo, J. Y., Gimenez, E. I., Walsh, R., & Soo, M. S. (2003). Computer-aided detection (CAD) in screening mammography: Sensitivity of commercial CAD systems for detecting architectural distortion. *American Journal of Roentgenology*, *181*(4), 1083–1088. https://doi.org/10.2214/ajr.181.4.1811083.

Barrientos, F., Roman-Gonzalez, A., Barrientos, R., Solis, L., Alva, A., Correa, M., et al. (2017a). Filtering of the skin portion on lung ultrasound digital images to facilitate automatic diagnostics of pneumonia. In *2016 IEEE 36th Central American and Panama convention, CONCAPAN 2016*Institute of Electrical and Electronics Engineers Inc. https://doi.org/10.1109/CONCAPAN.2016.7942376.

Barrientos, R., Roman-Gonzalez, A., Barrientos, F., Solis, L., Correa, M., Pajuelo, M., et al. (2017b). Automatic detection of pneumonia analyzing ultrasound digital images. In *2016 IEEE 36th Central American and Panama convention, CONCAPAN 2016*Institute of Electrical and Electronics Engineers Inc. https://doi.org/10.1109/CONCAPAN.2016.7942375.

Bernheim, A., Mei, X., Huang, M., Yang, Y., Fayad, Z. A., Zhang, N., et al. (2020). Chest CT findings in coronavirus disease 2019 (COVID-19): Relationship to duration of infection. *Radiology*, *295*(3), 685–691. https://doi.org/10.1148/radiol.2020200463.

Cheng, J. Z., Ni, D., Chou, Y. H., Qin, J., Tiu, C. M., Chang, Y. C., et al. (2016). Computer-aided diagnosis with deep learning architecture: Applications to breast lesions in US images and pulmonary nodules in CT scans. *Scientific Reports*, *6*. https://doi.org/10.1038/srep24454.

Cisneros-Velarde, P., Correa, M., Mayta, H., Anticona, C., Pajuelo, M., Oberhelman, R., et al. (2016). Automatic pneumonia detection based on ultrasound video analysis. In *Proceedings of the annual international conference of the IEEE Engineering in Medicine and Biology Society, EMBS* (pp. 4117–4120). Institute of Electrical and Electronics Engineers Inc. https://doi.org/10.1109/EMBC.2016.7591632. Vols. 2016-.

Corman, V. M., Landt, O., Kaiser, M., Molenkamp, R., Meijer, A., Chu, D. K. W., et al. (2020). Detection of 2019 novel coronavirus (2019-nCoV) by real-time RT-PCR. *Euro Surveillance*, *25*(3). https://doi.org/10.2807/1560-7917.ES.2020.25.3.2000045.

Coronavirus disease (COVID-19). (2020). https://www.who.int/emergencies/diseases/novel-coronavirus-2019/question-and-answers-hub/q-a-detail/coronavirus-disease-covid-19.

Coronavirus. (2021). https://www.who.int/health-topics/coronavirus/coronavirus#tab=tab_1.

Doi, K. (2007). Computer-aided diagnosis in medical imaging: Historical review, current status and future potential. *Computerized Medical Imaging and Graphics*, *31*(4–5), 198–211. https://doi.org/10.1016/j.compmedimag.2007.02.002.

Ebrahimian, H., Rijal, O. M., Noor, N. M., Yunus, A., & Mahyuddin, A. A. (2014). Phase congruency parameter estimation and discrimination ability in detecting lung disease chest radiograph. In *IECBES 2014, conference proceedings—2014 IEEE conference on biomedical engineering and sciences: "Miri, where engineering in medicine and biology and humanity meet"* (pp. 729–734). Institute of Electrical and Electronics Engineers Inc. https://doi.org/10.1109/IECBES.2014.7047604.

Fang, Y., Zhang, H., Xie, J., Lin, M., Ying, L., Pang, P., et al. (2020). Sensitivity of chest CT for COVID-19: Comparison to RT-PCR. *Radiology*, *296*(2), E115–E117. https://doi.org/10.1148/radiol.2020200432.

González, B., Melin, P., Valdez, F., & Prado-Arechiga, G. (2018). Ensemble neural network optimization using a gravitational search algorithm with interval type-1 and type-2 fuzzy parameter adaptation in pattern recognition applications. In *Vol. 749. Studies in Computational Intelligence* (pp. 17–27). Springer Verlag. https://doi.org/10.1007/978-3-319-71008-2_2.

González, B., Valdez, F., Melin, P., & Prado-Arechiga, G. (2014). Echocardiogram image recognition using neural networks. *Studies in Computational Intelligence, 547*, 427–435. https://doi.org/10.1007/978-3-319-05170-3_29.

González, B., Valdez, F., Melin, P., & Prado-Arechiga, G. (2015). A gravitational search algorithm for optimization of modular neural networks in pattern recognition. *Studies in Computational Intelligence, 574*, 127–137. https://doi.org/10.1007/978-3-319-10960-2_8.

Hemdan, E. E. D., Shouman, M. A., & Karar, M. E. (2020). COVIDX-net: A framework of deep learning classifiers to diagnose COVID-19 in X-ray images. *arXiv*. https://arxiv.org.

Hina, K., Khalid, S., & Akbar, M. U. (2017). A review on automatic tuberculosis screening using chest radiographs. In *2016 6th international conference on innovative computing technology, INTECH 2016* (pp. 285–289). Institute of Electrical and Electronics Engineers Inc. https://doi.org/10.1109/INTECH.2016.7845039.

Horváth, G., Orbán, G., Horváth, Á., Simkó, G., Pataki, B., Máday, P., et al. (2009). *A CAD system for screening X-ray chest radiography. Vol. 25, Issue 5* (pp. 210–213). Springer Science and Business Media LLC. https://doi.org/10.1007/978-3-642-03904-1_59.

Karargyris, A., Antani, S., & Thoma, G. (2011). Segmenting anatomy in chest x-rays for tuberculosis screening. In *Proceedings of the annual international conference of the IEEE Engineering in Medicine and Biology Society, EMBS* (pp. 7779–7782). https://doi.org/10.1109/IEMBS.2011.6091917.

Kermany, D. S., Goldbaum, M., Cai, W., Valentim, C. C. S., Liang, H., Baxter, S. L., et al. (2018). Identifying medical diagnoses and treatable diseases by image-based deep learning. *Cell*, 1122–1131.e9. https://doi.org/10.1016/j.cell.2018.02.010.

Khobragade, S., Tiwari, A., Patil, C. Y., & Narke, V. (2017). Automatic detection of major lung diseases using chest radiographs and classification by feed-forward artificial neural network. In *1st IEEE international conference on power electronics, intelligent control and energy systems, ICPEICES 2016*Institute of Electrical and Electronics Engineers Inc. https://doi.org/10.1109/ICPEICES.2016.7853683.

Liu, L., & Mehrotra, S. (2016). Detecting out-of-bed activities to prevent pneumonia for hospitalized patient using Microsoft Kinect V2. In *Proceedings—2016 IEEE 1st international conference on connected health: Applications, systems and engineering technologies, CHASE 2016* (pp. 364–365). Institute of Electrical and Electronics Engineers Inc. https://doi.org/10.1109/CHASE.2016.74.

Mahase, E. (2020). Coronavirus covid-19 has killed more people than SARS and MERS combined, despite lower case fatality rate. *BMJ (Clinical Research Ed.), 368*, m641. https://doi.org/10.1136/bmj.m641.

Mehrotra, R., Ansari, M. A., Agrawal, R., & Anand, R. S. (2020). A transfer learning approach for AI-based classification of brain tumors. *Machine Learning with Applications*, *2*, 100003. https://doi.org/10.1016/j.mlwa.2020.100003.

Minarno, A. E., Munarko, Y., Kurniawardhani, A., Bimantoro, F., & Suciati, N. (2014). Texture feature extraction using co-occurrence matrices of sub-band image for batik image classification. In *2014 2nd international conference on information and communication technology, ICoICT 2014* (pp. 249–254). Institute of Electrical and Electronics Engineers Inc. https://doi.org/10.1109/ICoICT.2014.6914074.

Noor, N. M., Rijal, O. M., Yunus, A., Mahayiddin, A. A., Peng, G. C., Ling, O. E., et al. (2014). Pair-wise discrimination of some lung diseases using chest radiography. In *IEEE TENSYMP 2014–2014 IEEE region 10 symposium* (pp. 151–156). Institute of Electrical and Electronics Engineers Inc. https://doi.org/10.1109/tenconspring.2014.6863015.

Pattar, S. (2015). Detection and classification of lung disease—Pneumonia and lung cancer in chest radiology using artificial neural network. *International Journal of Scientific and Research Publications, 5*(1), 2250–3153.

Rajpurkar, P., Irvin, J., Zhu, K., Yang, B., Mehta, H., Duan, T., et al. (2017). CheXNet: Radiologist-level pneumonia detection on chest X-rays with deep learning. *ArXiv*.

Rubio, Y., Montiel, O., & Sepúlveda, R. (2017). Microcalcification detection in mammograms based on fuzzy logic and cellular automata. In *Vol. 667. Studies in computational intelligence* (pp. 583–602). Springer Verlag. https://doi.org/10.1007/978-3-319-47054-2_38.

Sethy, P. K., Behera, S. K., Ratha, P. K., & Biswas, P. (2020). *Detection of coronavirus disease (COVID-19) based on deep features*. http://pdfs.semanticscholar.org.

Song, Y., Zheng, S., Li, L., Zhang, X., Zhang, X., Huang, Z., et al. (2020). Deep learning enables accurate diagnosis of novel coronavirus (COVID-19) with CT images. *IEEE/ACM Transactions on Computational Biology and Bioinformatics*. https://doi.org/10.1101/2020.02.23.20026930.

Van Ginneken, B., Ter Haar Romeny, B. M., & Viergever, M. A. (2001). Computer-aided diagnosis in chest radiography: A survey. *IEEE Transactions on Medical Imaging, 20*(12), 1228–1241. https://doi.org/10.1109/42.974918.

Wang, X., Peng, Y., Lu, L., Lu, Z., Bagheri, M., & Summers, R. M. (2017). ChestX-ray8: Hospital-scale chest X-ray database and benchmarks on weakly-supervised classification and localization of common thorax diseases. In *Proceedings—30th IEEE conference on computer vision and pattern recognition, CVPR 2017* (pp. 3462–3471). Institute of Electrical and Electronics Engineers Inc. https://doi.org/10.1109/CVPR.2017.369. Vols. 2017-.

Wang, W., Xu, Y., Gao, R., Lu, R., Han, K., Wu, G., et al. (2020). Detection of SARS-CoV-2 in different types of clinical specimens. *JAMA—Journal of the American Medical Association, 323*(18), 1843–1844. https://doi.org/10.1001/jama.2020.3786.

Wang, L., Lin, Z. Q., & Wong, A. (2020). COVID-net: A tailored deep convolutional neural network design for detection of COVID-19 cases from chest X-ray images. *Scientific Reports, 10*, 19549. https://doi.org/10.1038/s41598-020-76550-z. https://doi.org/arXiv:2003.09871v4.

Xie, X., Zhong, Z., Zhao, W., Zheng, C., Wang, F., & Liu, J. (2020). Chest CT for typical coronavirus disease 2019 (COVID-19) pneumonia: Relationship to negative RT-PCR testing. *Radiology, 296*(2), E41–E45. https://doi.org/10.1148/radiol.2020200343.

Yoon, S. H., Lee, K. H., Kim, J. Y., Lee, Y. K., Ko, H., Kim, K. H., et al. (2020). Chest radiographic and ct findings of the 2019 novel coronavirus disease (Covid-19): Analysis of nine patients treated in Korea. *Korean Journal of Radiology, 21*(4), 498–504. https://doi.org/10.3348/kjr.2020.0132.

Zenteno, O., Castaneda, B., & Lavarello, R. (2016). Spectral-based pneumonia detection tool using ultrasound data from pediatric populations. In *Proceedings of the annual international conference of the IEEE Engineering in Medicine and Biology Society, EMBS* (pp. 4129–4132). Institute of Electrical and Electronics Engineers Inc. https://doi.org/10.1109/EMBC.2016.7591635. Vols. 2016-.

CHAPTER

21

Convolutional bi-directional long-short-term-memory based model to forecast COVID-19 in Algeria

Sourabh Shastri[a], *Kuljeet Singh*[a], *Astha Sharma*[b], *Mohamed Lounis*[c], *Sachin Kumar*[a], *and Vibhakar Mansotra*[a]

[a]Department of Computer Science & IT, University of Jammu, Jammu, Jammu & Kashmir, India [b]Department of Electronics and Communication Engineering, G.L. Bajaj Institute of Technology and Management, Greater Noida, Uttar Pradesh, India [c]Department of Agro-veterinary Science, Faculty of Natural and Life Sciences, University of Ziane Achour, Djelfa, Algeria

Introduction

In 9 months, the new coronavirus disease (COVID-19) has affected about 155 million persons around the world and more than 3.2 million among them died (Johns Hopkins University of Medicine, Coronavirus Resource Center, 2021). This respiratory viral disease is caused by a new type of coronavirus called SARS-Cov2 due to its similarity with SARS-Cov (Liu et al., 2020).

Since the appearance of the first cases in the city of Wuhan in China, multiple types of research have been tried and are ongoing to understand the epidemiological features of this "new" disease. Comprehension of epidemiological characteristics is of great importance in forming the best measures of prevention of the fast spread of this disease in the absence of efficient treatments or vaccines. In this way, prediction of the epidemic evolution and describing the epidemic cure are very helpful (Peng, Yang, Zhang, Zhuge, & Liu, 2020).

Multiple models were employed to predict the future of the COVID-19 epidemic curve. These models include the conventional mathematical models, such as susceptible infected recovered (SIR) (Cooper, Mondal, & Antonopoulos, 2020) and susceptible exposed infected recovered (SEIR) models (Carcione, Santos, Bagaini, & Ba, 2020), and their derivates and

Computational Intelligence in Healthcare Applications
https://doi.org/10.1016/B978-0-323-99031-8.00003-X

machine learning (ML) or artificial intelligence models that include autoregressive integrated moving average (ARIMA) models (Khan, Saeed, & Ali, 2020), simple recurrent neural network (RNN) (Zeroual, Harrou, Dairi, & Sun, 2020), long-short-term memory (LSTM) (Shahid, Zameer, & Muneeb, 2020), linear regression (Rustam et al., 2020), least absolute shrinkage and selection operator (LASSO) regression, exponential smoothing (ES) (Shastri, Singh, Kumar, Kour, & Mansotra, 2021), support vector machine (SVM), variational autoencoder (VAE), and gated recurrent unit (GRU) models (Rustam et al., 2020).

LSTM, which was applied for the first time in 1997 by Hochreiter and Schmidhuber (1997), is a recurrent neural network (RNN) that assures the network to continue and save the long-term dependencies at a present time from multiple previous time steps, which perform well both in univariate and multivariate time-series forecasting (Sengupta, 2020).

Widely used in various tasks, such as natural language processing (NLP) image generation and video analysis (Zheng et al., 2020), LSTMs have been successfully and widely used in multiple fields like highway trajectory, stock price, and air pollution forecasting (Wang, Zheng, Ai, Liu, & Zhu, 2020). It has also seen an explosion used in the medical field (Sedik et al., 2020).

In the current COVID-19 context, LSTMs are widely used to predict the epidemic curve and multiple variants are proposed including the simple LSTM model (Shahid et al., 2020; Zeroual et al., 2020), vanilla LSTM (Barman, 2006) models, stacked LSTM (Shastri, Singh, Kumar, Kour, & Mansotra, 2020), convolutional neural network LSTM (CNN-LSTM) (Sedik et al., 2020), bidirectional recurrent neural network LSTM (BRNN-LSTM or bidirectional LSTM (bi-LSTM)), K-Means-LSTM model (Vadyala, Betgeri, Sherer, & Amritphale, 2020), shallow LSTM model (Pal, Sekh, Kar, & Prasad, 2020), hybrid LSTM models (Zandavi, Rashidi, & Vafaee, 2020; Zheng et al., 2020), variational LSTM-autoencoder model, and Weibull-based long-short-term-memory (W-LSTM) model (Ibrahim et al., 2021).

Related works

Several studies have been done using different LSTM models since the onset of the COVID-19 epidemic. These studies varied from those using LSTM for short-term or long-term prediction to those using different LSTM variants or comparing the performance of LSTM with other models. Some of these works are cited later.

In the study of Chimmula and Zhang (2020), the authors tried to estimate the end of the COVID-19 epidemic in Canada using the LSTM model. Their predictions showed that the disease would disappear in June 2020.

Ben, Abdelkarim, Hussein, and Abdelmomen (2021) showed that including lockdown information in a bi-LSTM model improves significantly its accuracy in forecasting daily cases of coronavirus disease in Qatar.

Ibrahim et al. (2021) introduced a novel variational LSTM-autoencoder model to forecast coronavirus disease evolution in different countries. The model which includes factors related to urban and demographic characteristics like density, urbanization, and fertility, in addition to the historical data of the virus spreading, has shown good performance in short- and long-term forecasting.

Hu, Ge, Li, and Xiong (2020) employed a modified stacked autoencoder LSTM model using previous data of COVID-19 in China to estimate different parameters like the final number, the duration, and the end of the epidemic.

Pal et al. (2020) proposed a shallow LSTM model with a Bayesian optimization to predict the COVID-19 evolution. They concluded that this model is well adapted for long-term prediction in the case of COVID-19. This model, which has shown that the weather does not have a significant role, had a better prediction than other deep learning models.

Bedi, Dhiman, Gole, Gupta, and Jindal (2021) compared the performance of LSTM and a modified SEIRD model for forecasting 30 days in India. The authors reported that the two models showed the same prediction in terms of computed case fatality and recovery rate in all the country, but some differences were observed at the states level. In parallel, Liu et al. (2020) demonstrated that the modified SEIRD, the LSTM, and the geographically weighted regression (GWR) models have remarkable prediction capabilities and comparable performance.

Chatterjee, Gerdes, and Martinez (2020) used different LSTM models to forecast COVID-19 in the top 17 countries from January 1 to April 22 (The vanilla, stacked, bi-LSTM, and multilayer LSTM models). The authors concluded that any of the tested models is accurate at 100%. Also, the vanilla, stacked, and bi-LSTM models have shown the best performance. These three models were also used by Hridoy et al. (2020) in association with the logistic curve method to forecast the dynamics and the endpoint of COVID-19 in Bangladesh.

Azarafza, Azarafza, and Tanha (2020) used LSTM to forecast COVID-19 in Iran and compared its results to other deep learning models, namely, RNN, SARIMA, HWES (Holt winter's exponential smoothing), and MA (moving averages) models. The authors reported that LSTM has the best performance and gave fewer error values than the other cited methods.

Kafieh et al. (2021) applied different ML models like random forest (RF), multilayer perceptron (MLP), and different LSTM variants like LSTM with extended features (LSTM-E), LSTM with regular features (LSTM-R), and multivariate LSTM (M-LSTM) for a prediction of the number of COVID-19-infected cases in Iran. They reported that the M-LSTM was more accurate than the other models.

Tuli, Tuli, Verma, and Tuli (2020) used W-LSTM to predict the propagation and the severity of COVID-19 in 50 countries. The model has shown good satisfaction in the 50 studied countries and performed better than other LSTMs and ARIMA models.

Yudistira (2020) used an LSTM model to forecast the COVID-19 curve in four countries (Argentina, Indonesia, Saudi Arabia, and Sweden). He remarked that LSTM has a good prediction for a short time period better than the RNN. The author suggests that LSTM could be a good tool in long-term forecasting future epidemic outbreaks.

In the work of Kırbaş, Sözen, Tuncer, and Kazancıoğlu (2020), the authors found that the LSTM model is more accurate than ARIMA and nonlinear autoregression neural network (NARNN) in forecasting COVID-19 in eight European countries (Belgium, Denmark, Finland, France, Germany, Switzerland, Turkey, and the United Kingdom).

Sengupta (2020) applied a stacked LSTM model using time series generated by three other models: SIR, ARIMA, and ML regression models for forecasting daily cases for 4 months in India. The model has shown that the number of new cases per day will increase till the first week of November 2020, when it can reach 90,000 cases per day before it starts to decline.

In the paper of Shahid et al. (2020), the authors used different ML models that were used to assess ill, deceased, and recovered persons in the 10 most touched countries comprising ARIMA, SVR, LSTM, bi-LSTM, and GRU models. They classified these models according to their performance, where biLSTM has the highest and ARIMA showed the lowest performance.

Simple RNN, LSTM, bi-LSTM, GRUs, and VAE have been applied in the study of Zeroual et al. (2020) for a short period prediction of COVID-19 cases in six countries including three European countries (France, Italy, Spain), China, Australia, and the United States. They stated that VAE has a better performance than the other models.

In the same way, Shastri et al. (2020) compared the performance of three LSTM variants (stacked LSTM, bi-LSTM, and convolutional LSTM (Conv-LSTM)) to forecast the COVID-19 number of cases in India and the United States for a period of 30 days. They showed that the Conv-LSTM model has better skills than the others. Inversely, Arora, Kumar, and Panigrahi (2020) found that bi-LSTM has a lower mean range error than convolutional and stacked LSTM models and is more suitable for COVID-19 prediction in India.

Other works using hybrid LSTM models were also conducted. Du et al. (2020) proposed an ISI hybrid model embedded with natural language processing (NLP) and LSTM in China which has shown high performance in short-term prediction. Also, Zandavi (Zandavi et al., 2020) developed a hybrid model combining an LSTM with dynamic behavioral models, which can predict the propagation of COVID-19 with high accuracy.

In the current work, we have used two deep learning-based models including bi-LSTM and Conv-LSTM to predict COVID-19-infected and deceased cases in Algeria for a future period of 1 month based on data of reported cases, deaths, and recovery by the Algerian Ministry of Health. To the best of our knowledge, no such work was done in Algeria on the COVID-19 pandemic before.

Data sources

In this work, datasets of COVID-19 infected, cured, and deceased cases were used. Datasets were collected from the Algerian Ministry of Health and hospital reform official website for the period from February 25, 2020 to September 25, 2020 (Algerian health and hospital reform minister: Carte épidémiologique, 2020). The confirmed cases were calculated according to the RT-PCR positive tests. This number was estimated at 50,754 cases on September 25, 2020. On the same date, the number of the cured cases was 35,654 and the death number reached 1707 cases.

Methods

Traditional RNN can save historical information a few times and are able to perform only if current information is available to predict the present state. In order to find the next state, RNN can process the hidden layer activation function of the back state only, and it also cannot overcome the problem of the vanishing gradient. RNN cannot model the long-term dependencies, which lead to the inception of LSTM cells. LSTM replaces the traditional hidden layer concept of RNN with memory cells, which can model the long-term dependencies and also overcome the problem of vanishing gradient problems.

Bidirectional LSTM

In this work, we used two LSTMs models, bi-LSTM and Conv-LSTM. Bi-LSTM can process the information of forward and backward hidden layers simultaneously in both directions, as shown in Fig. 21.1. The backward layer outputs cannot be used as an input for forward layers and vice versa (Graves & Schmidhuber, 2005). The mathematical formulation of these models is cited later.

Mathematically, layer L of bi-LSTM can be expressed at time t as shown in Eqs. (21.1)–(21.7) (Shastri et al., 2020).

$$\overleftarrow{f}_t^L = \sigma\left(W_{\overleftarrow{f}h}^L h_{t+1}^L + W_{\overleftarrow{f}h}^L h_t^{L-1} + b_{\overleftarrow{f}}^L\right) \tag{21.1}$$

$$\overleftarrow{i}_t^L = \sigma\left(W_{\overleftarrow{i}h}^L h_{t+1}^L + W_{\overleftarrow{i}x}^L h_t^{L+1} + b_{\overleftarrow{i}}^L\right) \tag{21.2}$$

$$\overleftarrow{\tilde{c}}_t^L = \tanh\left(W_{\overleftarrow{\tilde{c}}} h^L h_{t+1}^L + W_{\overleftarrow{\tilde{c}}} x^L h_t^{L-1} + b_{\overleftarrow{\tilde{c}}}^L\right) \tag{21.3}$$

$$\overleftarrow{c}_t^L = \overleftarrow{f}_t^L \cdot \overleftarrow{c}_{t+1}^L + \overleftarrow{i}_t^L \cdot \overleftarrow{\tilde{c}}_t^L \tag{21.4}$$

$$\overleftarrow{o}_t^L = \sigma\left(W_{\overleftarrow{o}h}^L h_{t+1}^L + W_{\overleftarrow{o}x}^L h_t^{L-1} + b_{\overleftarrow{o}}^L\right) \tag{21.5}$$

$$\overleftarrow{h}_t^L = \overleftarrow{o}_t^L \cdot \tanh\left(\overleftarrow{c}_t^L\right) \tag{21.6}$$

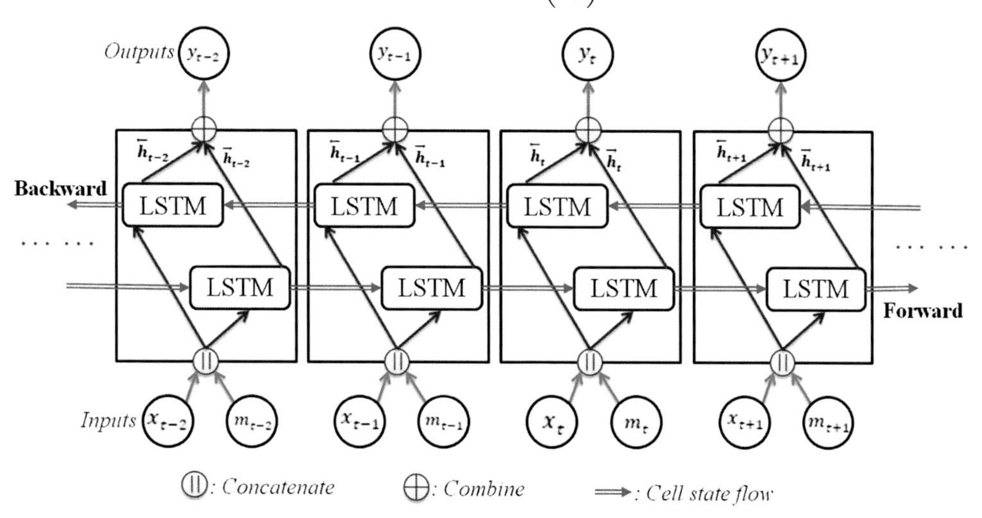

FIG. 21.1 Architecture of bidirectional LSTM. *No Permission Required.*

The output of bi-LSTM is a summative result of forward and backward layer outputs such as $\overrightarrow{h}_t$ and $\overleftarrow{h}_t$, respectively.

$$\text{Therefore } y_t = W_{\overrightarrow{h}y}\overrightarrow{h}_t + W_{\overleftarrow{h}y}\overleftarrow{h}_t + b_y \tag{21.7}$$

Bi-LSTM uses backward pass, forward pass, and upgrading the weights over time to train the model. For time, $1 \leq t \leq T$ it can run the forward pass to find forecasted values. For forward states, it performs the forward pass for time $t=1$ to T, and for backward states, it performs the backward pass for time $t=T$ to 1.

Convolutional LSTM

Conv-LSTM predicts the spatiotemporal data along with the features such as inputs $x_1, x_2, \ldots,$ and x_t, output as $c_1, c_2, \ldots,$ and c_t, gates i_t, o_t, f_t, and hidden states as $h_1, h_2, \ldots,$ and h_t. Using the Conv-LSTM operator (*), transitions are possible between input to state and state to state, as shown in Fig. 21.2. Conv-LSTM can derive the future state which is determined by its previous intermediate states and inputs (Shastri, Singh, Deswal, Kumar, & Mansotra, 2021).

A mathematical formulation of Conv-LSTM is shown in Eqs. (21.8)–(21.12), where "*" is a convolutional operator and "·" represents the Hadamard product (Arora et al., 2020).

$$i_t = \sigma(W_{xi} * x_t + W_{hi} * h_{t-1} + W_{ci} \cdot c_{t-1} + b_i) \tag{21.8}$$

$$f_t = \sigma\left(W_{xf} * x_t + W_{hf} * h_{t-1} + W_{cf} \cdot c_{t-1} + b_f\right) \tag{21.9}$$

$$c_t = f_t \cdot c_{t-1} + i_t \cdot \tanh(W_{xc} * x_t + W_{hc} * h_{t-1} + b_c) \tag{21.10}$$

$$o_t = \sigma(W_{xo} * x_t + W_{ho} * h_{t-1} + W_{co} \cdot c_t + b_o) \tag{21.11}$$

$$h_t = o_t \cdot \tanh(c_t) \tag{21.12}$$

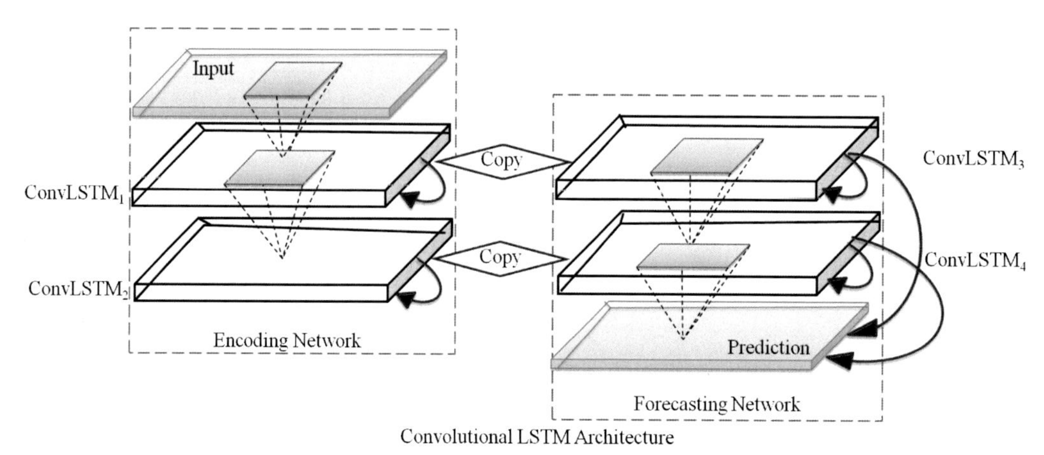

FIG. 21.2 Encoding-forecasting architecture of convolutional LSTM. *No Permission Required.*

Experiment

The experiment has been carried out at Google Collaboratory using Python 3.0 with open source libraries like Tensorflow, Pandas, Numpy, and Keras. The experimental setup is based on a working environment having Intel(R) Core (TM) i5-7400 CPU @ 3.00 GHz with 8 GB RAM under 64-bit Windows 10 Pro Operating system. Bi-LSTM and Conv-LSTM are used to carry out experimental analysis on time series prediction of COVID-19 cases. We forecast the confirmed, death, and recovered COVID-19 cases using proposed models. We split our dataset into 80% and 20% for training and testing purposes, respectively.

In bi-LSTM, we shape our data in 3D input as samples, time steps, and features and feed it to our model. MinMaxScaler is being used to normalize the data before applying the model to it. A hidden layer of 200 neurons along with the ReLu activation function is being used. Bi-LSTM is trained for 500 epochs with validation split as 0.2 and verbose as 1. Mean square error is used for loss function along with the Adam optimizer. The final results of the model are the summative output of the backward and forward layers.

In Conv-LSTM, a convolutional operator (*) is being used that replaces the internal matrix multiplication of simple traditional LSTM models. We normalize our datasets using MinMaxScaler and shape our data to make it fit for model acceptance. Input data is passed through the split_sequence() function to make the input in correct form as samples, time steps, features, rows, and columns. A single 2D Conv-LSTM layer is used with 64 filters and (1, 2) as its kernel size, which indicates the number of rows and columns as 1 and 2, respectively. The model is trained for 500 epochs with validation split as 0.2 and verbose equals 1. Results are flattened and connected through dense layer; also, ReLu activation is used to avoid vanishing gradient problems. To evaluate the model summary mean square error with Adam optimizer is used.

Results

The principal objective of the present work is to forecast and predict the COVID-19 evolution in Algeria based on daily infected, cured, and deceased cases reported by the Ministry of health and hospital reform using bi-LSTM and Conv-LSTM. The concerned datasets were collected from the onset of the epidemic (February 25 to September 25, 2020) for a period of 214 days.

Prediction of COVID-19 confirmed, cured, and deceased cases

The graphical forecasted findings of confirmed COVID-19 cases, cured, and deceased are shown in Figs. 21.3 and 21.4.

Results show that the numbers of ill, recovered, and dead persons continue increasing in the next month. The Conv-LSTM model shows a significant upward trend than the bi-LSTM. The estimated number of sick persons with the Conv-LSTM will attain approximately 80 k cases, while the number of cured and deceased will be nearly 60 k and more than 2 k cases respectively.

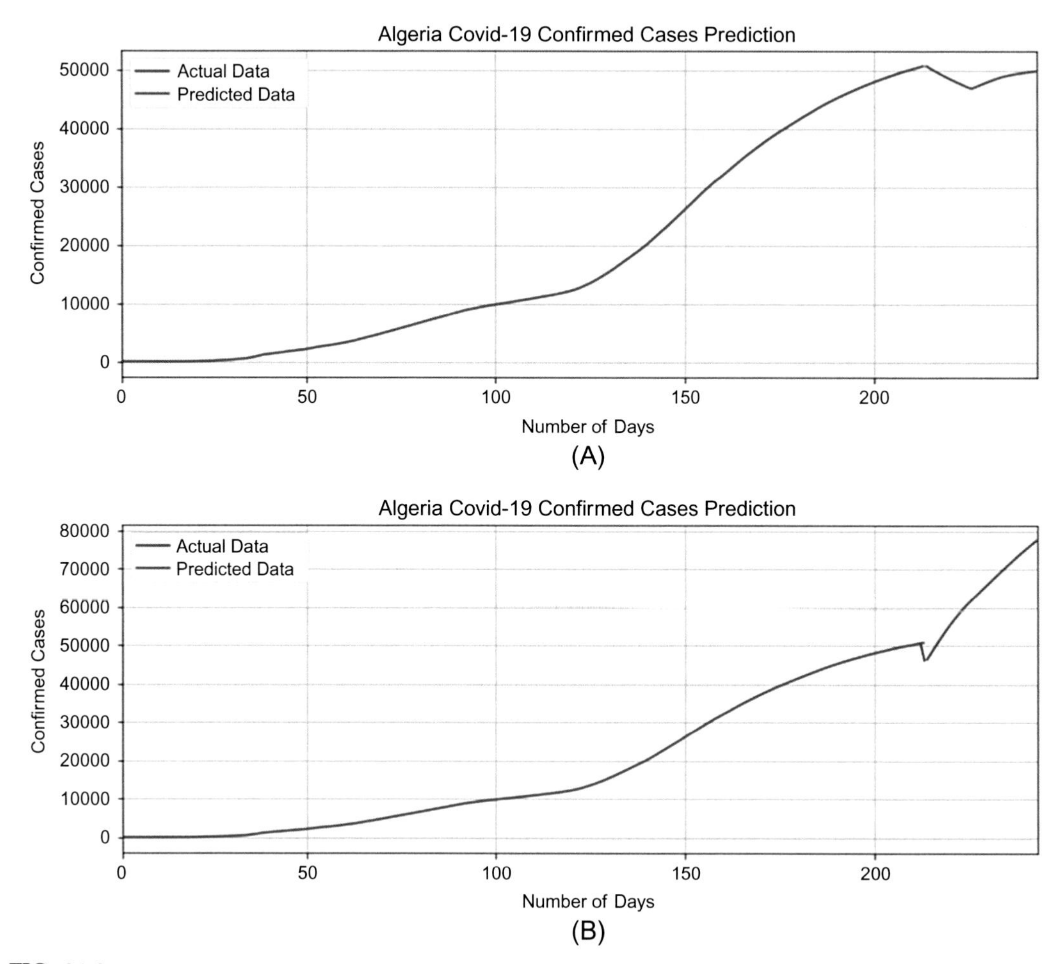

FIG. 21.3 Forecasting of cumulative COVID-19 cases for Algeria. *No Permission Required.*

Fig. 21.3A shows actual (blue line) forecasted (red line) results for COVID-19 confirmed cases of Algeria using bi-LSTM. Fig. 21.3B shows actual and predicted cumulative cases using Conv-LSTM showing an upward trend for 1 month later.

Fig. 21.5C shows actual (blue line) forecasted (red line) results for COVID-19 death cases of Algeria using bi-LSTM, which shows a downward trend in the predicted deaths. Fig. 21.5D shows actual and predicted deaths using Conv-LSTM, where we can observe an upward trend in the death number for a future month.

Fig. 21.4E shows actual (blue line) forecasted (red line) results for COVID-19 recovered cases of Algeria using bi-LSTM with a downward trend in the future recoveries of COVID-19 cases. Fig. 21.4F describes the actual and predicted trends using Conv-LSTM and shows that the number of recoveries will increase for 1 month in the near future.

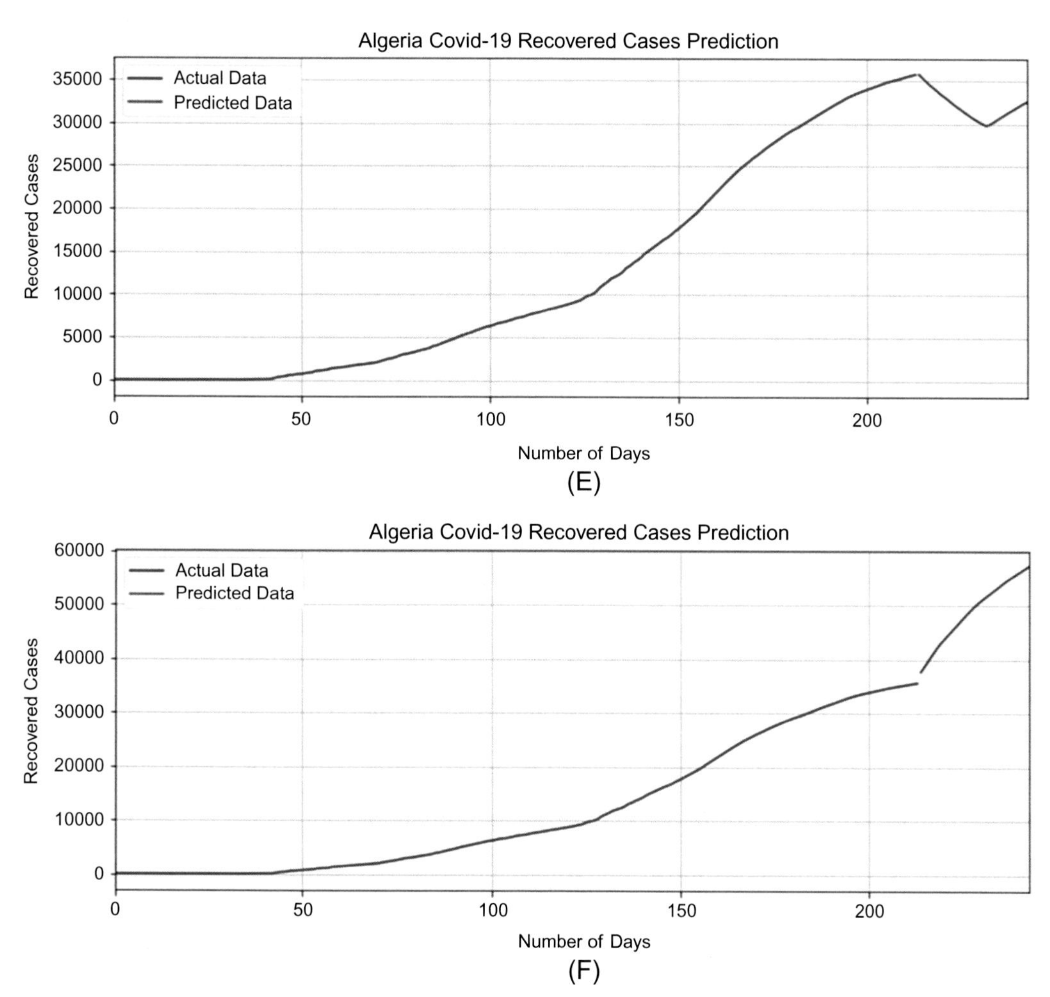

FIG. 21.4 Forecasting of COVID-19-recovered cases for Algeria. *No Permission Required.*

To evaluate and compare the performance of these two models, different measures were used, like the mean absolute percentage error (MAPE) (Table 21.1) and accuracy (Table 21.2).

Results showed that the Conv-LSTM showed better accuracy for confirmed cases (95.55% vs 90%), recoveries (96.66% vs 93.33%), and deaths (97.77% vs 96.25%) than the bi-LSTM model. These results are confirmed in a previous study by Shastri et al. (2020) showing that the Conv-LSTM has better accuracy than the bi-LSTM and the stacked LSTM for confirmed cases and deaths in India and the United States.

MAPE was also compared between the two models (Table 21.2). Results showed that the Conv-LSTM provides better results with very fewer errors than the bi-LSTM model. These results were also confirmed for cumulative cases and deaths in India and the United States (Shastri et al., 2020).

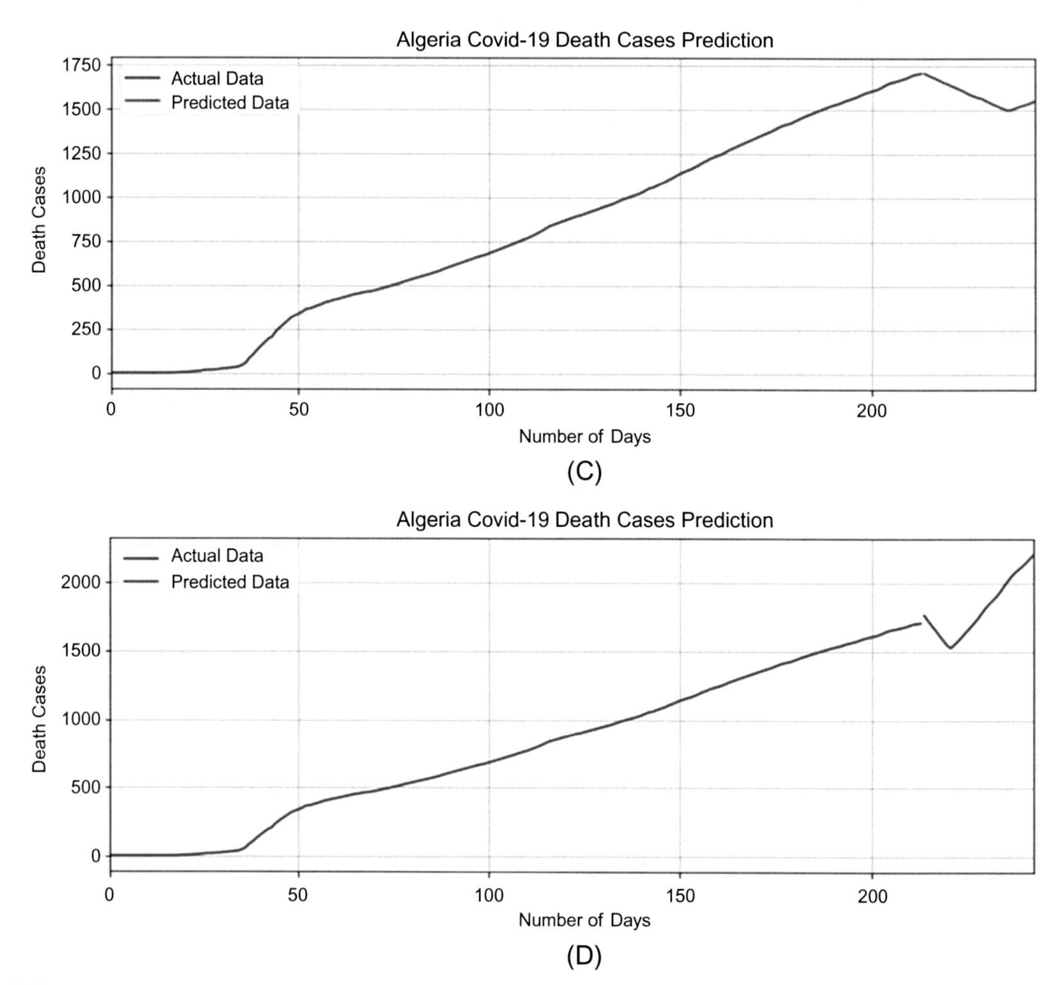

FIG. 21.5 Forecasting of COVID-19-deceased cases for Algeria. *No Permission Required.*

TABLE 21.1 Metrics comparison of bi-LSTM and Conv-LSTM on Algeria COVID-19.

Algeria COVID-19 cases	Models							
	Bidirectional LSTM				Convolutional LSTM			
	Accuracy	Precision	Recall	F-measure	Accuracy	Precision	Recall	F-measure
Confirmed Cases	90.00	100.00	50.00	0.66	95.55	100.00	81.81	0.90
Death Cases	96.25	100.00	72.72	0.84	97.77	100.00	85.71	0.92
Recovered Cases	93.33	100.00	77.77	0.87	96.66	100.00	90.0	0.94

TABLE 21.2 MAPE comparison of bi-LSTM and Conv-LSTM for Algeria COVID-19 cases.

	Models	
Algeria COVID-19 cases	**Bidirectional LSTM MAPE**	**Convolutional LSTM MAPE**
Confirmed Cases	10.00	4.44
Death Cases	3.75	2.22
Recovered Cases	6.66	3.33

Conclusion

In the current paper, two deep learning models, namely, bi-LSTM and Conv-LSTM, were used to forecast confirmed cases, recovered, and dead persons in Algeria. We demonstrated that the Conv-LSTM has better accuracy and fewer MAPEs than the bi-LSTM. Prediction of the Conv-LSTM showed that the COVID-19 will continue its spreading in Algeria with an increase of infected and deceased cases. The number of cured patients seems also to increase in the next month. This is the first work in Algeria comparing the performance of LSTM models in Algeria. Results of the Conv-LSTM models may be taken into consideration by the political and sanitary authorities and will be helpful for future preventive strategies against the COVID-19 pandemic.

Acknowledgment

This research work is dedicated to COVID-19 frontline workers.

Funding information

This research did not receive any specific grant from funding agencies in the public, commercial, or not-for-profit sectors.

Conflicts of interest

The authors declare that they have no conflict of interest.

Ethical approval

This article does not contain any studies with human participants or animals performed by any of the authors.

Informed consent

Informed consent was not required as no humans or animals were involved.

References

Algerian health and hospital reform minister: Carte épidémiologique. (2020).

Arora, P., Kumar, H., & Panigrahi, B. K. (2020). Prediction and analysis of COVID-19 positive cases using deep learning models: A descriptive case study of India. *Chaos, Solitons and Fractals, 139*. https://doi.org/10.1016/j.chaos.2020.110017.

Azarafza, M., Azarafza, M., & Tanha, J. (2020). COVID-19 infection forecasting based on deep learning in Iran. *medRxiv*. https://doi.org/10.1101/2020.05.16.20104182.

Barman, A. (2006). Times series analysis and forecasting of COVID-10 cases using LSTM and ARIMA models. *Arxiv*.

Bedi, P., Dhiman, S., Gole, P., Gupta, N., & Jindal, V. (2021). Prediction of COVID-19 trend in India and its four worst-affected states using modified SEIRD and LSTM models. *SN Computer Science, 2*(3). https://doi.org/10.1007/s42979-021-00598-5.

Ben, A., Abdelkarim, E., Hussein, A. A., & Abdelmomen, M. (2021). Predicting COVID-19 cases using bidirectional LSTM on multivariate time series. *Environmental Science and Pollution Research International, 28*, 56043–56052. https://doi.org/10.1007/s11356-021-14286-7.

Carcione, J. M., Santos, J. E., Bagaini, C., & Ba, J. (2020). A simulation of a COVID-19 epidemic based on a deterministic SEIR model. *Frontiers in Public Health, 8*. https://doi.org/10.3389/fpubh.2020.00230.

Chatterjee, A., Gerdes, M. W., & Martinez, S. G. (2020). Statistical explorations and univariate timeseries analysis on COVID-19 datasets to understand the trend of disease spreading and death. *Sensors, 20*(11), 3089. https://doi.org/10.3390/s20113089.

Chimmula, V. K. R., & Zhang, L. (2020). Time series forecasting of COVID-19 transmission in Canada using LSTM networks. *Chaos, Solitons and Fractals, 135*. https://doi.org/10.1016/j.chaos.2020.109864.

Cooper, I., Mondal, A., & Antonopoulos, C. G. (2020). A SIR model assumption for the spread of COVID-19 in different communities. *Chaos, Solitons and Fractals, 139*. https://doi.org/10.1016/j.chaos.2020.110057.

Du, S., Wang, J., Zhang, H., Cui, W., Kang, Z., Yang, T., et al. (2020). Predicting COVID-19 using hybrid AI model. *SSRN Electronic Journal*. https://doi.org/10.2139/ssrn.3555202.

Graves, A., & Schmidhuber, J. (2005). Framewise phoneme classification with bidirectional LSTM and other neural network architectures. *Neural Networks, 18*(5–6), 602–610. https://doi.org/10.1016/j.neunet.2005.06.042.

Hochreiter, S., & Schmidhuber, J. (1997). Long short-term memory. *Neural Computation, 9*(8), 1735–1780. https://doi.org/10.1162/neco.1997.9.8.1735.

Hridoy, A. E. E., Naim, M., Emon, N. U., Tipo, I. H., Alam, S., Mamun, A. A., et al. (2020). Forecasting COVID-19 dynamics and endpoint in Bangladesh: A data-driven approach. *medRxiv*. https://doi.org/10.1101/2020.06.26.20140905.

Hu, Z., Ge, Q., Li, S., & Xiong, M. (2020). Artificial intelligence forecasting of Covid-19 in China. *International Journal of Educational Excellence, 6*(1), 71–94. https://doi.org/10.18562/ijee.054.

Ibrahim, M. R., Haworth, J., Lipani, A., Aslam, N., Cheng, T., Christie, N., et al. (2021). Variational-LSTM autoencoder to forecast the spread of coronavirus across the globe. *PLoS One, 16*(1), e0246120. https://doi.org/10.1371/journal.pone.0246120.

Johns Hopkins University of Medicine, Coronavirus Resource Center. (2021). https://coronavirus.jhu.edu/map.html.

Kafieh, R., Arian, R., Saeedizadeh, N., Amini, Z., Serej, N. D., Minaee, S., et al. (2021). COVID-19 in Iran: Forecasting pandemic using deep learning. *Computational and Mathematical Methods in Medicine, 2021*, 1–16. https://doi.org/10.1155/2021/6927985.

Khan, F., Saeed, A., & Ali, S. (2020). Modelling and forecasting of new cases, deaths and recover cases of COVID-19 by using Vector Autoregressive model in Pakistan. *Chaos, Solitons and Fractals, 140*. https://doi.org/10.1016/j.chaos.2020.110189.

Kırbaş, İ., Sözen, A., Tuncer, A. D., & Kazancıoğlu, F.Ş. (2020). Comparative analysis and forecasting of COVID-19 cases in various European countries with ARIMA, NARNN and LSTM approaches. *Chaos, Solitons and Fractals, 138*. https://doi.org/10.1016/j.chaos.2020.110015.

Liu, F., Wang, J., Liu, J., Li, Y., Liu, D., Tong, J., et al. (2020). Predicting and analyzing the COVID-19 epidemic in China: Based on SEIRD, LSTM and GWR models. *PLoS One, 15*(8), e0238280. https://doi.org/10.1371/journal.pone.0238280.

Pal, R., Sekh, A. A., Kar, S., & Prasad, D. K. (2020). Neural network based country wise risk prediction of COVID-19. *Applied Sciences*, 6448. https://doi.org/10.3390/app10186448.

Peng, L., Yang, W., Zhang, D., Zhuge, C., & Liu, H. (2020). Epidemic analysis of COVID-19 in China by dynamical modeling. *medXiv*. preprint.

Rustam, F., Reshi, A. A., Mehmood, A., Ullah, S., On, B.-W., Aslam, W., et al. (2020). COVID-19 future forecasting using supervised machine learning models. *IEEE Access*, *8*, 101489–101499. https://doi.org/10.1109/access.2020.2997311.

Sedik, A., Iliyasu, A. M., Abd El-Rahiem, B., Abdel Samea, M. E., Abdel-Raheem, A., Hammad, M., et al. (2020). Deploying machine and deep learning models for efficient data-augmented detection of COVID-19 infections. *Viruses*, *12*(7), 769. https://doi.org/10.3390/v12070769.

Sengupta, S. (2020). *Forecasting the peak of COVID-19 daily cases in India using time series analysis and multivariate LSTM*. EasyChair Preprint.

Shahid, F., Zameer, A., & Muneeb, M. (2020). Predictions for COVID-19 with deep learning models of LSTM, GRU and Bi-LSTM. *Chaos, Solitons and Fractals*, *140*. https://doi.org/10.1016/j.chaos.2020.110212.

Shastri, S., Singh, K., Kumar, S., Kour, P., & Mansotra, V. (2020). Time series forecasting of Covid-19 using deep learning models: India-USA comparative case study. *Chaos, Solitons and Fractals*, *140*. https://doi.org/10.1016/j.chaos.2020.110227.

Shastri, S., Singh, K., Deswal, M., Kumar, S., & Mansotra, V. (2021). CoBiD-net: A tailored deep learning ensemble model for time series forecasting of covid-19. *Spatial Information Research*. https://doi.org/10.1007/s41324-021-00408-3.

Shastri, S., Singh, K., Kumar, S., Kour, P., & Mansotra, V. (2021). Deep-LSTM ensemble framework to forecast Covid-19: An insight to the global pandemic. *International Journal of Information Technology*, *13*(4), 1291–1301. https://doi.org/10.1007/s41870-020-00571-0.

Tuli, S., Tuli, S., Verma, R., & Tuli, R. (2020). Modelling for prediction of the spread and severity of COVID-19 and its association with socioeconomic factors and virus types. *medRxiv*. https://doi.org/10.1101/2020.06.18.20134874.

Vadyala, S. R., Betgeri, S. N., Sherer, E. A., & Amritphale, A. (2020). Prediction of the number of COVID-19 confirmed cases based on K-means-LSTM. *arXiv*. https://arxiv.org.

Wang, P., Zheng, X., Ai, G., Liu, D., & Zhu, B. (2020). Time series prediction for the epidemic trends of COVID-19 using the improved LSTM deep learning method: Case studies in Russia, Peru and Iran. *Chaos, Solitons and Fractals*, *140*. https://doi.org/10.1016/j.chaos.2020.110214.

Yudistira, N. (2020). *COVID-19 growth prediction using multivariate long short term memory*. ArXiv, abs/2005.04809.

Zandavi, M., Rashidi, T. H., & Vafaee, F. (2020). *Forecasting the spread of COVID-19 under control scenarios using LSTM and dynamic behavioral models* (pp. 1–14). ArXiv. abs/2005.12270.

Zeroual, A., Harrou, F., Dairi, A., & Sun, Y. (2020). Deep learning methods for forecasting COVID-19 time-series data: A comparative study. *Chaos, Solitons & Fractals*, *140*, 110121. https://doi.org/10.1016/j.chaos.2020.110121.

Zheng, N., Du, S., Wang, J., Zhang, H., Cui, W., Kang, Z., et al. (2020). Predicting COVID-19 in China using hybrid AI model. *IEEE Transactions on Cybernetics*, *50*(7), 2891–2904. https://doi.org/10.1109/tcyb.2020.2990162.

Index

Note: Page numbers followed by *f* indicate figures and *t* indicate tables.

B

C

Printed in the United States
by Baker & Taylor Publisher Services